The Duffing Equation

The Duffing Equation

Nonlinear Oscillators and their Behaviour

Edited by

Ivana Kovacic

University of Novi Sad
Faculty of Technical Sciences
Serbia

Michael J. Brennan

University of Southampton
Institute of Sound and Vibration Research
United Kingdom

WILEY

A John Wiley and Sons, Ltd., Publication

This edition first published 2011

Registered office
John Wiley & Sons Ltd, The Atrium, Southern Gate, Chichester, West Sussex, PO19 8SQ, United Kingdom

For details of our global editorial offices, for customer services and for information about how to apply for permission to reuse the copyright material in this book please see our website at www.wiley.com.

Library of Congress Cataloging-in-Publication Data

The Duffing equation : nonlinear oscillators and their phenomena / edited by
Ivana Kovacic, Michael J. Brennan.
p. cm.
Includes bibliographical references and index.
ISBN 978-0-470-71549-9 (cloth)
1. Duffing equations. 2. Nonlinear oscillators–Mathematical models. I.
Kovacic, Ivana, 1972- II. Brennan, Michael J. (Michael John), 1956-
QA372.D83 2011
515′.35–dc22

2010034587

A catalogue record for this book is available from the British Library.

Print ISBN: 9780470715499
E-PDF ISBN: 9780470977866
O-book ISBN: 9780470977859
E-Pub ISBN: 9780470977835

Cover : Photo of Duffing reprinted from F.P.J. Rimrott, Georg Duffing (1861–1944), Technische Mechanik, 14(1), 77–82, 1994. Copyright 1994, reprinted with permission from Technische Mechanik.

Set in 10/12 pt Times, by Thomson Digital, Noida, India.
Printed in Malaysia by Ho Printing (M) Sdn Bhd

Contents

Contributors

Balakumar Balachandran received his BTech in Naval Architecture from the Indian Institute of Technology, Madras, India, MS in Aerospace Engineering from Virginia Tech, and PhD in Engineering Mechanics from Virginia Tech. Currently, he is a Professor of Mechanical Engineering at the University of Maryland. He serves on the Editorial Board of the Journal of Vibration and Control, is a Deputy Editor of the AIAA Journal, and is an Associate Editor of the ASME Journal of Computational and Nonlinear Dynamics. He is a Fellow of ASME and AIAA. He has served as the Chair of the ASME Applied Mechanics Division Technical Committee on Dynamics and Control of Structures and Systems, and he currently serves as the Chair of the ASME Design Engineering Division Technical Committee on Multibody Systems and Nonlinear Dynamics. His research interests include nonlinear phenomena, dynamics and vibrations, and control.

Michael J. Brennan graduated from the Open University while he was serving in the Royal Navy. He received an MSc in Sound and Vibration Studies and a PhD in the active control of vibration, both from the University of Southampton, United Kingdom. He is a retired Professor of Engineering Dynamics at the Institute of Sound and Vibration Research (ISVR), the University of Southampton, UK, and is currently a Visiting Professor at UNESP, Ilha Solteira in Brazil. He is a past President of the European Association of Structural Dynamics, Associate Editor of the Transactions of the ASME Journal of Vibration and Acoustics and Guest Professor at Harbin Engineering University in China. He has a wide range of research interests, encompassing active and passive control of vibration, acoustics, vibroacoustics and rotor dynamics.

Livija Cveticanin graduated from the Faculty of Mechanical Engineering, University of Novi Sad, Serbia. She obtained her MSc in Mechanics from the Faculty of Natural Sciences, University of Belgrade, Serbia and PhD in the Technical Sciences at the Faculty of Technical Sciences (FTN), University of Novi Sad, Serbia. She is currently a Full Professor in the Department of Mechanics at the FTN and the Head of the Graphical Engineering and Design Department. She is a former Vice-Dean of the FTN and former President of the Yugoslav Society of Mechanics. Her research interest is directed towards nonlinear vibrations, rotor dynamics and dynamics of systems and mechanisms with time varying parameters.

Tamás Kalmár-Nagy received his MSc in Engineering Mathematics from the Technical University of Budapest and his PhD degree in Theoretical and Applied Mechanics from Cornell University. He is now an Assistant Professor in the Department of Aerospace Engineering at Texas A&M University. He serves on the Editorial Board of the Mathematical Problems in Engineering and Fluctuation and Noise Letters. He is a member of the ASME Design Engineering Division Technical Committee on Multibody Systems and Nonlinear Dynamics, as well as the Technical Committee on Vibration and Sound. His research interests are in delay-differential equations, perturbation methods, nonlinear vibrations, dynamics and control of uncertain and stochastic systems.

Ivana Kovacic graduated in Mechanical Engineering from the Faculty of Technical Sciences (FTN), University of Novi Sad, Serbia. She obtained her MSc and PhD in the Theory of Nonlinear Vibrations at the FTN. She is currently an Associate Professor in the Department of Mechanics at the FTN. She is also a Visiting Professor in the Institute of Sound and Vibration Research (ISVR) at the University of Southampton, UK, Assistant Editor of the Journal of Sound and Vibration and Book Reviews Editor for the Journal of Mechanical Engineering Science-Part C of the Proceedings of the Institution of Mechanical Engineers, UK. Her research involves the use of quantitative and qualitative methods to study differential equations arising from nonlinear dynamics problems mainly in mechanical engineering.

Stefano Lenci graduated in Civil Engineering from the University of Ancona, Italy. He obtained his PhD in Structural Engineering at the University of Florence. He had a two-year post-doc position at the University of Paris VI. He became an Assistant Professor at the Sapienza University of Rome and later Associate Professor at the Polytechnic University of Marche, Ancona, Italy, where he now serves as a Full Professor. He is a member of the Academy of Science of the Marche Region, Italy, and of the Editorial Board of the International Journal of Nonlinear Mechanics. He is Head of the PhD program in Structural Engineering, and member of ASME, Aimeta, SICC and Euromech. His research interests cover all fields of nonlinear dynamics and chaos of mechanical systems and structures. He also works on continuum mechanics, in particular in problems involving interfaces.

Asok Kumar Mallik received his bachelor and master degrees in Mechanical Engineering from the University of Calcutta and PhD from IIT Kanpur. He is currently an Honorary Distinguished Professor at the Bengal Engineering and Science University, Shibpur and an INSA Senior Scientist at S.N. Bose National Centre for Basic Sciences at Kolkata, India. He was a Professor of Mechanical Engineering and the first occupant of S. Sampath Institute Chair at the Indian Institute of Technology Kanpur. He was a commonwealth scholar at the Institute of Sound and Vibration Research at Southampton, UK and an Alexander von Humboldt Fellow at TH Aachen and TU Darmstadt, Germany. He received the Distinguished Teacher Award of IIT Kanpur. He is an elected Fellow of The Indian National Academy of Engineering and all the Science Academies in India. He has authored/coauthored 6 books and more than 80 research papers. Areas of his research include vibration engineering, nonlinear dynamics and kinematics. He also writes popular articles on mathematics and physics.

Giuseppe Rega is a Professor of Solid and Structural Mechanics at the Sapienza University of Rome, Italy, Chairman of the PhD Program in Structural Engineering, Director of the Doctoral School in Civil Engineering and Architecture, and past President of Italian Association of Theoretical and Applied Mechanics (AIMETA). He is a Member of the EUROMECH Nonlinear Oscillations Conference Committee and of the IUTAM General Assembly, Chairman of Euromech Colloquia and IUTAM Symposia, including ENOC 2011 in Rome, Associate

Editor, Guest Editor or Editorial Board Member of several international journals. He was honored with an International Conference and a Special Issue for his sixtieth birthday. His research interests are in cable dynamics, nonlinear vibrations in applied mechanics and structural dynamics, bifurcation and chaos, control of oscillations and chaos, reduced-order modeling, dynamic integrity, wave propagation and smart materials.

Keith Worden started his academic life as a theoretical physicist with a BSc from the University of York. This was followed by a PhD in Mechanical Engineering from Heriot-Watt University and he has been an engineer (of sorts) since. He is currently a Professor of Mechanical Engineering at the University of Sheffield. His main research interests are in nonlinear systems and structural health monitoring. He has been struggling with the Duffing equation for the best part of 25 years now with very little to show for it, but he does not mind because it has largely been fun.

Heather Worden has degrees in linguistics and speech therapy and is a practising speech therapist. A German specialist, she has helped to translate a number of scientific documents.

Hiroshi Yabuno graduated from Keio University, Japan. He received an MSc and a PhD in the Nonlinear Dynamics in Mechanical systems, both from Keio University. He is currently a Professor of Mechanical Engineering at this university. He is also a member of the Editorial Board of the Journal of Vibration and Control, an Editor of the Journal of System Design and Dynamics, and a member of the Working Party II of IUTAM. He is a past Professor of the University of Tsukuba and a past Visiting Professor of the University of Rome La Sapienza. His research interests include analysis, control, and utilization of nonlinear dynamics of mechanical systems; especially stabilization control of vehicle systems, motion control of underactuated manipulators, and bifurcation control of advanced atomic force microscopy.

Preface

The nonlinear equation describing an oscillator with a cubic nonlinearity is called the Duffing equation. Georg Duffing, a German engineer, wrote a comprehensive book about this in 1918. Since then there has been a tremendous amount of work done on this equation, including the development of solution methods (both analytical and numerical), and the use of these methods to investigate the dynamic behaviour of physical systems that are described by the various forms of the Duffing equation. Because of its apparent and enigmatic simplicity, and because so much is now known about the Duffing equation, it is used by many researchers as an approximate model of many physical systems, or as a convenient mathematical model to investigate new solution methods. This equation exhibits an enormous range of well-known behaviour in nonlinear dynamical systems and is used by many educators and researchers to illustrate such behaviour. Since the 1970s, it has became really popular with researchers into chaos, as it is possibly one of the simplest equations that describes chaotic behaviour of a system.

The idea to write this book came to us a couple of years after we had started working together on nonlinear problems in 2006. Although we are both mechanical engineers, we have a very different viewpoint on what is important when it comes to working on engineering research topics; Ivana very much specialises in theoretical mechanics and Michael is firmly in the practical engineering camp. The one thing that we did agree on, however, was that there was a real need to synthesise the huge amount of research conducted over the past 90 years or so on the Duffing equation, both for the academic and the engineering community. As working academics, this task was thought to be too large for us to undertake alone, so we decided to put together an edited book, drawing on the expertise of specialists working in nonlinear dynamics from around the world. The result is this book; in which each of the contributors was given a specific brief to write about one particular form of the Duffing equation. It should be noted that all of these forms were not in Duffing's original book, but the contemporary view seems to be that any differential equation that contains a cubic nonlinearity seems to be known as the Duffing equation; we have accepted this popular view.

A particularly interesting part of this project was tracking the development of the subject of nonlinear dynamics with specific regard to the Duffing equation. This was not an easy task for us, as Duffing's original book was written in German and so the early papers citing his book were also in German, and were not necessarily cited by the more popular papers written in English. It was of particular interest to the Editors to find out (a) how Duffing's work became well known to researchers and engineers in the world of the English language, and (b) when the equation he is now famous for, took his name. This is revealed in Chapter 1 of this book. The remainder of the book is a collection of chapters written by experts in the field of nonlinear dynamics. It contains a comprehensive treatment of the various forms of the Duffing equation, relates these equations to real oscillatory problems, and demonstrates the rich dynamics that can be exhibited by systems described by this equation. Thus, for the first time, all this information has been assembled in one book. An overview of each chapter is given at the end of Chapter 1. Because there are eleven contributors there is some inevitable overlap between some of the chapters. We have agonised over this, but on balance we have decided that this has some advantages, as each chapter can be read as a standalone piece of work. However, to help the reader, relevant links to the other chapters have been inserted.

We hope that this book will have broad appeal to a wide range of readers, from experienced researchers who would like to have this book in their reference collection, to young/new researchers in the field of nonlinear dynamics/vibrations who wish to learn some basic methods, and to engineers, who would like to see the effect that nonlinearities will have on the dynamic behaviour of their systems.

Finally, we would like to thank all our contributors for their efforts and support over the past two years.

Ivana Kovačić and Michael J. Brennan
Southampton, July 2010

1

Background: On Georg Duffing and the Duffing equation

Ivana Kovacic[1] and Michael J. Brennan[2]
[1]*University of Novi Sad, Faculty of Technical Sciences, Serbia*
[2]*University of Southampton, Institute of Sound and Vibration Research, United Kingdom*

1.1 Introduction

It is possibly the dream of many researchers to have an equation named after them. One person who achieved this was Georg Duffing, and this book is devoted to various aspects of his equation. This equation is enigmatic. In its original form, it essentially has only one extra nonlinear stiffness term compared to the linear second-order differential equation, which is the bedrock of vibrations theory, and this opens the door to a whole new world of interesting phenomena. Much of this was not known at the time of Georg Duffing, and is described in this book. The story behind the equation is also very interesting, because Georg Duffing was not an academic; he was an engineer, who carried out academic work in his spare time, as will be described later. In the present day when academics are being constantly reminded about the impact of their research work, and are constantly being judged by their output, in terms of publications, it is also interesting to look at the academic output from Georg Duffing and the impact of his work. Rarely is a paper or textbook written on nonlinear dynamics today without some reference to the Duffing equation, such is the impact of his work, yet he wrote less than ten publications in his life.

The Duffing Equation: Nonlinear Oscillators and their Behaviour, First Edition. Edited by Ivana Kovacic and Michael J. Brennan.

The aim of this book is twofold. The first is to give a historical background to Duffing's work, and to track the evolution of his work to the present day. This is done in this chapter. The second aim is to provide a thorough treatment of the different forms of his equation through the various chapters written by the contributing authors. This will involve qualitative and quantitative analysis coupled with descriptions of the many physical phenomena that are described by the various forms of his equation.

Nowadays, the term 'Duffing equation' is used for any equation that describes an oscillator that has a cubic stiffness term, regardless of the type of damping or excitation. This, however, was not the case in Duffing's original work, in which he restricted his attention to the *free* and *forced harmonic vibration* of an oscillator in which the stiffness force had quadratic and cubic terms, and the damping considered was of the linear viscous type. In this book the contemporary view is taken and many forms of the Duffing equation are studied, with the notable exceptions of a randomly or parametrically excited oscillator.

1.2 Historical perspective

In any historical perspective, the authors undoubtedly provide their own interpretation of events, and this is also the case here. The history of nonlinear dynamics is vast and has many different threads to it, from the highly mathematical to the physical. It is not the intention of the authors to give a detailed history here – for this, the reader is referred to a review paper written by Holmes that covers the period 1885–1975 [1] and a slightly more recent paper by Shaw and Balachandran [2]. The authors restrict their attention to the historical perspective with respect to Duffing's work.

The concept of nonlinear vibrations was known long before Duffing wrote his book on oscillations [3], in which his famous equation is given. However, Duffing was the one to tackle the problem of a nonlinear oscillator in a systematic way starting with the linear oscillator, and examining the effects of quadratic and cubic stiffness nonlinearities. He emphasised the differences between the linear and the nonlinear oscillators for both free and forced vibration, also considering the effects of damping. Prior to Duffing, there had been some work on the mathematical analysis of nonlinear oscillators, for example by Hermann von Helmholtz [4] and Baron Rayleigh [5]. Two contemporaries of Duffing, Henri Poincaré (1854–1912) and Aleksandr Lyapunov (1857–1918), who were both giants in the history of nonlinear dynamics, did not appear to influence Duffing's work – at least they were not cited in his book.

In the story of nonlinear dynamics, as well as in Duffing's book, the pendulum plays a dominant role, and so it is appropriate to start the story with Galileo.

Galileo Galilei: 1564–1642. Galileo studied the pendulum and noticed that the natural frequency of oscillation was roughly independent of the amplitude of oscillation, i.e., they are isochronous. For it to be used in a time-keeping instrument, it needed to be forced because the oscillations diminished with time due to damping. He invented a mechanism to do this called an escapement [6]. This work was quickly followed by that of Huygens, who realised that the pendulum was inherently nonlinear.

Christiaan Huygens: 1629–1695. Huygens patented the pendulum clock in 1657. The early clocks had wide pendulum swings of up to 100°. Huygens discovered that wide swings made the pendulum inaccurate because he observed that the natural period was dependent upon the amplitude of motion, i.e., it was a nonlinear system. Subsequently the clocks were modified with a new escapement so that the pendulum swing was reduced to about 4–6°. Huygens also discovered that if the pendulum had a length that varied during the oscillation, according to an isochronous curve, then the frequency of oscillation became independent of the amplitude (effectively he linearised a nonlinear system) [7].

In many vibrating systems, it is the interaction between stiffness and mass that causes the 'interesting' dynamic behaviour. The first person to introduce the concept of stiffness theoretically was Hooke.

Robert Hooke: 1635–1703. Hooke is famous for his law [8], which gives the linear relationship between the applied force and resulting displacement of a linear spring. At the same time that Hooke was formulating the constitutive law for a spring, Newton was formulating his laws of motion, the most important of which for dynamical systems, is his second law.

Isaac Newton: 1643–1727. Newton, of course, is famous for his three laws of motion [9]. According to Truesdell [10], at the time of Newton and Hooke, simple harmonic motion (SHM) was not understood in the context of elastic bodies. However, Galileo was well aware of SHM in his study of the pendulum. Although vibration is often studied using rigid-body, lumped parameter systems (especially the study of nonlinear vibrations), a key area of practical interest is the vibrations of elastic bodies, such as beams, plates and shells. The first person to extend Hooke's law to such a system (a beam) was Liebnitz.

Gottfried Wilhelm Leibniz: 1646–1716. Leibniz is attributed with applying Hooke's law to a system containing moments; i.e., the bending moment is proportional to the second moment of area of a beam. This is thought to be the first application of calculus to a continuous system [10].

Although Hooke and Newton introduced some very important fundamental building blocks for mechanics, a general framework for the study of mechanics was lacking. The first person to provide some rudimentary tools for analysis was James Bernoulli.

James Bernoulli: 1654–1716. James Bernoulli developed the following approaches to solving problems in mechanics: balance of forces resolved in two fixed orthogonal directions; balance of forces normal and tangential to the line; virtual work; balance of moments. Truesdell [10] also attributes the first nonlinear law of elasticity to James Bernoulli. Around the same time, James Bernoulli's bother, John, was studying the vibration of a catenary, and then the vibration of a weighted string. During this study he formulated the equation for the natural frequency of a system.

John Bernoulli: 1667–1748. John Bernoulli studied the case of a string in tension loaded with weights. In this work he determined that the natural frequency of a system is equal to the square root of its stiffness divided by its mass, $\omega_\mathrm{n} = \sqrt{k/m}$, [11]. This is believed to be the first publication to state this relationship.

Some seventy years or so after Newton and Hooke formulated their laws for stiffness and mass, Euler connected them together in the form of a harmonically excited differential equation. This equation is the one that is taught to all students of vibration as a mathematical description of an undamped forced single-degree-of-freedom system.

Leonhard Euler: 1707–1783. Euler was the first person to write down the equation of motion of a harmonically forced, undamped linear oscillator, $m\ddot{x} + kx = F \sin \omega t$. He formally introduced the nondimensional driving frequency $\Omega = \omega/\omega_n$ and noted that the response becomes infinite when $\Omega = 1$. Hence, he was the first person to explain the phenomenon of resonance [12].

More than 100 years later, Helmholtz was the first person to add a nonlinear stiffness term to Euler's equation of motion.

Hermann Von Helmholtz: 1821–1894. Helmholtz was the first person to include nonlinearity into the equation of motion for a harmonically forced undamped single degree-of-freedom oscillator. He postulated that the eardrum behaved as an asymmetric oscillator, such that the restoring force was $f = k_1x + k_2x^2$, which gave rise to additional harmonics in the response for a tonal input [4]. In the context of nonlinear dynamics, the equation $m\ddot{x} + k_1x + k_2x^2 = F \sin \omega t$ is now commonly known as the Helmholtz equation.

Around the same time that Helmholtz published his work, Rayleigh published his classic book on acoustics and vibration – The Theory of Sound [5]. This book had two volumes and covered an enormous amount of fundamental material in acoustics and vibration. In one small part of the first volume he considered a nonlinear oscillator.

John William Strutt, Third Baron Rayleigh: 1842–1919. Rayleigh considered the *free vibration* of a nonlinear single-degree-of-freedom oscillator. He studied the same system as Helmholtz, in which the force–deflection characteristic was quadratic, and he also investigated a system in which the force–deflection characteristic was symmetrical, given by $f = k_1x + k_3x^3$ [5]. In the latter case the equation of motion for this was given as $m\ddot{x} + k_1x + k_2x^3 = 0$. This is very close to Duffing's equation, but does not have a forcing term, and Rayleigh only provided a small amount of analysis, showing that nonlinear systems will vibrate at a fundamental frequency and harmonics of this frequency depending on the amplitude of vibration and the type of nonlinear stiffness force.

Also, around the time that Helmholtz and Rayleigh published their books [4,5], concerning vibrations and acoustics, Routh published his book on the dynamics of rigid bodies [13]. Among other things, he considered the *free vibration* of a system with a linear-plus cubic-stiffness force. For an undamped system he showed that the frequency of oscillation is affected by the amplitude.

Apart from the great pioneers mentioned above, who, motivated by acoustics, laid down the foundations for vibration theory, two other authors deserve a mention, because they directly inspired Duffing in his work. They are Von. O. Martienssen [14] and J. Biermanns [15]. In both of these papers an electrical system was studied in which included an inductor. For high current levels, the relationship between the current, i and the flux, ϕ is nonlinear. Biermanns showed that the nonlinear relationship between the current and the flux could be written as a power series, and if this is truncated at the third power as shown above, then the resulting equation for current is very similar to that given by Rayleigh, i.e., it can be modelled as $i = A_1\phi + A_3\phi^3$. This results in a 'hardening' characteristic, i.e., the current and the flux have the nonlinear relationship in the same way that force and displacement have in the mechanical system when the nonlinear term is positive. Martienssen observed this behaviour experimentally and reported the existence of the *jump-down phenomenon* as frequency was increased and the *jump-up phenomenon* as frequency was decreased. He also modelled the system and showed that between the jump-up and jump-down frequencies, three steady-state conditions could occur.

1.3 A brief biography of Georg Duffing

In 1994, F.P.J. Rimrott published a brief biography in Technische Mechanik [16] and part of this is translated in this chapter. The photograph of Georg Duffing is taken from this article and is shown in Figure 1.1.

Georg Wilhelm Christian Caspar Duffing was born on 11 April 1861 in Waldshut in Baden, Germany. He was the oldest of six children of the merchant Christian

Figure 1.1 George Duffing . Reprinted from [16], Copyright 1977, with permission from Technische Mechanik.

Duffing and his wife Julie, whose maiden name was Spies. A year after he was born the family moved to Mannheim, where the grandfather Spies, a carpenter, had a large woodyard on the shore of the river Neckar.

Georg Duffing had a gift for mathematics and a natural musical talent. He studied the violin and performed in public as a child, and played in a band in his youth.

From 1878 to 1883, Duffing embarked on his formal higher education. He spent one year at mathematical school, one year at engineering school and three years at the Mechanical Engineering school at the Polytechnic, which is the University Fridericiana in Karlsruhe, today [17]. Although he had a heart problem, which subsequently prevented him from doing military service, Duffing was among the best runners in Baden.

After his graduation, Duffing went to Köln to work for *Deutzer Motorenwerken*, where he developed steam engines, which were produced in 1905.

At age 46, he married Elizabeth Lofde from Berlin. They had four children.

In 1910 Duffing was invited to Westinghouse in the USA. He stayed there for several months and came home with enough money to work as a self-employed inventor and scientist.

The Duffing family moved to Berlin in 1913, mainly because he wanted to listen to the lectures of Max Planck on quantum theory. This was typical behaviour – he always wanted to gain more knowledge.

When the First World War broke out and money lost its value, Duffing was working on vibrations, brakes, gears and engines, by trial and error. On Sundays he would go to the laboratory of the Royal Technical Faculty with his oldest daughter, where Professor Eugen Meyer allowed him to conduct experiments. He patented his inventions; however, it did not improve his financial situation. During that time, he was studying vibrations described by particular differential equations. In 1917 he completed his 134-page monograph numbered 41/42, with the title "Forced oscillations with variable natural frequency and their technical significance" [3]. It was published in 1918 by Vieweg & Sohn and cost five Deutsch Marks. This is the work for which he is famous.

In 1921, when the Duffing family encountered financial difficulties, he received an invitation of work from Ölgesellschaft Stern & Sonneborn A.G., which became the famous *Shell* company. He was offered the position as head of a laboratory where he invented a viscosimeter for lubricants.

The family moved to Hamburg, where Duffing suffered from severe flu and thrombosis of his leg, the consequences of which remained with him for the rest of his life.

A tragical part of his life came in 1927 when the ship 'Cap Arcona', had technical problems during a voyage. Stern & Sonneborn, had provided the oil that was made in accordance with Duffing's instructions. During the voyage, an engine failed. Duffing checked the oil and found out that many types of oil had been mixed, probably to save money. There was a trial, where Duffing presented facts clearly and honestly. He was resolute as he had been throughout his life. However, because he had testified against Stern & Sonneborn, he lost his job.

The Duffing family moved back to Berlin in 1931. Although he was 70 years old he carried on his research and inventing activities. During the Second World War, he had particular difficulties during the bombing attacks, as he could not easily take shelter in the cellar because of the problems with his leg. They subsequently moved to a small peaceful town called Schwedt on the river Oder.

Georg Duffing died there on 5 April 1944 aged 83 years. He is buried in the Jerusalem Graveyard in Halleschen Tor in Berlin.

1.4 The work of Georg Duffing

Written records of Georg Duffing's work comprise his patents and publications. His patents were registered both in the USA and Germany. The very first patent seems to have been registered as a 'Speed regulator for explosion engines' in the USA in 1905, and has the number 799459 [18], the illustration of which is shown in Figure 1.2. In the first

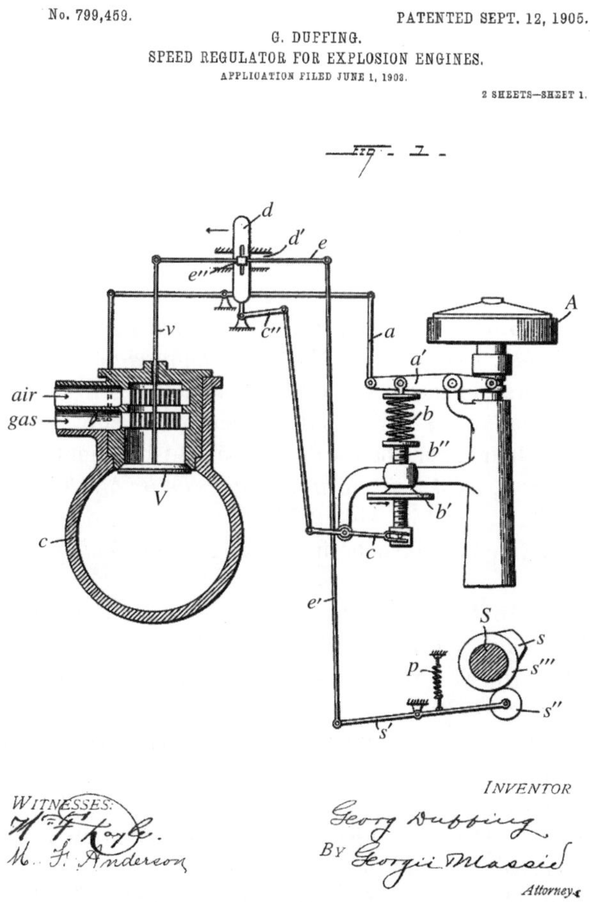

Figure 1.2 Illustration of Duffing's patent 'Speed regulator for explosion engines' [18].

paragraph of the written part of his application [18], Duffing wrote: "Be it known that I, Georg Duffing, engineer, a subject of the German Emperor, residing at 93 Deutzerstrasse, Mülheim-on-the-Rhine, Germany, have invented certain new and useful Improvements in Speed-Regulators for Explosion-Engine; and I do hereby declare the following to be a full, clear, and exact description of the invention, such as will enable others skilled in the art to which it apertains to make and use the same." During the following decades Duffing invented many 'new and useful improvements', fewer of which were registered in the USA than in Germany. Some of the USA patents can be seen, for example in [19–21], while Rimrott [16] gave an extensive list of German patents.

In terms of publictions, it is hard to qualify Georg Duffing's work as prolific, as he was the author of only nine publications. On the other hand, the fact that he was not an academic and that he was active only for about 25 years in the 20th century, make this number respectable. His publications includes books, book chapters and journal articles. They are listed in chronological order in Table 1.1.

Table 1.1 List of Duffing's publications.

No	Publications
1.	G. Duffing, *Beitrag zur Bestimmung der Formveränderung gekröpfter Kurbelwellen*. Verlag von Julius, Berlin, 1906.
2.	G. Duffing, *Erzwungene schwingungen bei veränderlicher eigenfrequenz und ihre technische bedeutung*, Series: Sammlung Vieweg, No 41/42. Vieweg & Sohn, Braunschweig, 1918.
3.	L. Gümbel; G. Duffing, *Der heutige Stand der Schmierungsfrage. Zur numerischen Integration gewöhnlicher Differentialgleichungen I. und II*, Series: Forschungsarbeiten auf dem Gebiete des Ingenieurwesens, No 224. Verlag des Vereines deutschen Ingenieure, Berlin 1920.
4.	G. Duffing, Beitrag zur Theorie der Flüssigkeitsbewegung zwischen Zapfen und Lager. *ZAMM Zeitschrift für Angewandte Mathematik und Mechanik*, 4, 296 Fig. 314, 1924.
5.	G. Duffing, Reibungsversuche am Gleitlager. *VDI - Zeitschrift*, 72, 495–499, 1928.
6.	G. Duffing, *Elastizität und Reibung beim Riementrieb*, Series: Sonderheft des Verbandes der Ledertreibriemen-Fabrikanten Deutschlands E. V., No 12, Berlin W 35, Kurfürstenstr. 148: Ledertreibriemen u. Techn. Lederartikel, 1930.
7.	G. Duffing, Die Schmiermittelreibung bei Gleitflächen von endlicher Breite; in *Handbuch der Physikalischen und Technischen Mechanik* Edited by F, Auerbach, W. Hort. Barth, Leipzig 1931.
8.	G. Duffing, Elastizität und Reibung beim Riementrieb. *Forschung im Ingenieurwesen*, 2, 99 Fig. 104, 1931.
9.	G. Duffing, Messung der Zähigkeit durch gleichförmige koachsiale Bewegung einer Kugel in einem Kreiszylinder, *ZAMM – Zeitschrift für Angewandte Mathematik und Mechanik*, 13, 366 Fig. 373, 1933.

The motivation for this book and the publication for which Duffing is recognised, is the monograph listed as number 2, mentioned in the previous section and listed as [3] in the referencess of this chapter. The next section is devoted to the monograph and contains the description of its content. It should be noted that, although it is not the only book that Duffing wrote, the phrase 'Duffing's book' will be used in relation to this particular publication only.

1.5 Contents of Duffing's book

The title page of Duffing's famous book is shown in Figure 1.3. It can be seen that the book was written in German, which was Duffing's native tongue. To help the reader understand what was written in the book some key pages have been translated by Keith and Heather Worden, and these are shown in the Appendix of this book. A brief summary of the contents of Duffing's book is given below.

1.5.1 Description of Duffing's book

George Duffing was not an academic, but an engineer, as was clearly written on the title page of his book 'Forced oscillations with variable natural frequency and their technical significance'. His motivation for the research reported in the book stemmed from his personal practical experience and observations of engineering systems. However, he was hoping that "the work would raise some interest in mathematical circles, because it requires some additional tools/knowledge and more time than one technician has." Duffing repeated this wish several times through the book, wanting to "be timely" and admitting that it was the reason for him to deliver the results despite the fact that "they had not been completed".

The book comprises seven chapters and five Appendices. It contains results on the response of both linear and nonlinear oscillatory systems obtained analytically, graphically, numerically and experimentally. The majority of the Appendices cover some necessary mathematical background work, which Duffing included to help the nonmathematician understand the content without having to search the literature.

In Chapter I a linear single-degree-of-freedom system excited by an arbitrary time-varying external force is considered using the convolution integral. An undamped system is analysed first, and this is followed by a damped system for periodic excitation only. As a special case, the response of the system under harmonic excitation is determined for resonance and offresonance conditions. This chapter serves as a reference, and describes simple systems for which results for nonlinear systems given in the subsequent chapters can be compared.

Chapter II is the most comprehensive – it is where free and forced undamped oscillations of the systems with a nonlinear restoring force are treated, and where the first significant results for these types of problems are given. Some of these systems are subsequently named after Duffing. The restoring force is assumed to

Erzwungene Schwingungen bei veränderlicher Eigenfrequenz und ihre technische Bedeutung

Von

Georg Duffing

Ingenieur

Mit 23 Abbildungen

Braunschweig

Druck und Verlag von Friedr. Vieweg & Sohn

1918

Figure 1.3 The title page of Duffing's book.

contain small quadratic and/or cubic nonlinearity. Duffing first considered free vibrations of a system with such a restoring force and obtained the first integral of motion corresponding to the energy conservation law. He then expressed the motion using Weierstrass elliptic functions. Separately, the case with softening cubic nonlinearity, corresponding to a symmetrical *potential well*, and the case with quadratic nonlinearity, corresponding to an asymmetrical *potential well*, are treated in this way. Further, Duffing studied forced vibrations for a system with cubic softening nonlinearity. Using previously obtained results, he applied the method of variation of constants to derive a fifth-order polynomial expression in one specifically defined parameter. He showed graphically that depending on the value of the forcing frequency compared to the natural frequency, the number of its roots can vary from one to three. The multivaluedness of the steady-state response is also confirmed by developing an iterative method – the method of successive approximation, which was subsequently called Duffing's method [22]. This technique is first validated on the linear system by demonstrating that its solution is equivalent to the sum of the complementary function and particular integral of the equation of motion. So that he could apply it to the forced vibrations of a softening system, he showed that the first approximation could be assumed to be harmonic at the frequency of excitation. As a result, he derived the frequency–amplitude equation, which is a cubic function of the amplitude. Graphical interpretation of the result shows that the multivaluedness of the response is dependent on how the excitation frequency compares with the natural frequency of the linearised system. Although Duffing was aware that to predict the response, one must examine the history of the response, i.e., of the hysteretic behaviour of nonlinear systems, surprisingly there is not a single frequency-response diagram in his book. To illustrate his findings, he provided the example of a forced pendulum, approximating the equation of motion to a system with a softening cubic restoring force. Duffing also analysed forced vibration for a system with quadratic nonlinearity by applying the method of successive approximation, assuming the first approximation to be the sum of a bias (DC) term and a harmonic term. After the derivation of the amplitude–frequency equation, it is solved graphically, and it is demonstrated that there can be multiple values of the amplitude for each frequency, where the number and the values of solutions are dependent upon whether or not the natural frequency is greater or less than the excitation frequency. In addition, the response of a forced system with negative quadratic and cubic nonlinearity is determined by means of Duffing's iterative method. The results are illustrated by investigating a pendulum that is excited by a constant plus harmonic force, whose equation of motion is transformed appropriately. The chapter is concluded with a summary of the main findings, and includes a table in which the differences between the responses of linear and nonlinear externally excited systems are highlighted.

Chapter III is devoted entirely to the experimental illustration and analysis of a system whose general equation of motion covers all the cases of *forced vibrations* considered in the previous chapter. The rig consisted of a pendulum which could be adjusted so that it corresponded to either a symmetrical or asymmetrical system. Duffing compared his theoretical results with those obtained experimentally and found satisfactory numerical agreement.

Chapter IV contains only one section, which is concerned with the influence of damping on the response of a softening cubic system with harmonic forced excitation. Again, the possibility of a multivalued response is shown graphically. Comparing the case of weak damping with the corresponding diagram for the undamped case,

Duffing remarked that there is no qualitative difference below and above the natural frequency.

Stability analysis of the periodic motion of the harmonically excited oscillator with cubic nonlinearity is investigated in Chapter V. With this aim, a linearised variational equation of the perturbed solution is considered, but with regard to the pendulum.

In Chapter VI some real systems are considered that are of interest from a practical point of view and whose governing equations correspond to those considered earlier: first, an electrical circuit analysed in Martienssen's paper [14], which is related to a free oscillating cubic system; then, a synchronous generator whose equation of motion corresponds to the asymmetric pendulum equation; and finally, a three-phase generator whose equation of motion corresponds to the symmetric pendulum equation and for which the multivaluedness of the response is shown analytically and graphically.

Chapter VII, entitled 'Generalisations', is concerned with the application of Duffing's method to the study of the systems excited by a sum of several harmonic forces. The cases of a quadratic and cubic restoring force are dealt with separately. Duffing also pointed out the necessity to study nonlinear systems with many degrees of freedom, due to their technical significance.

In Appendix 1 some details about the Weierstrass elliptic functions are given, while the integration of elliptic differential equations is commented on in Appendix 2. Appendix 3 contains the algorithm on how to transform a certain differential form to the Weierstrass normal form. Free vibrations of a pendulum are studied in Appendix 4 by means of elliptic functions. In Appendix 5 the Ritz method is applied to the forced vibrations with either cubic or quadratic nonlinearity with the aim of obtaining the amplitude-frequency expression.

1.5.2 Reviews of Duffing's book

The appearance of Duffing's book was announced and its contribution recognised soon after it had been published. Two reviews appeared in scientific journals, both written by Professor Georg Hamel from Berlin.

The first review was in the Annual Bulletin of Mathematics ('Jahbrbuch der Mathematik' 1916–1918) [23]. According to Professor Hamel, the main aim of the book was to explain several significant phenomena that appear during oscillatory motion of an externally excited asymmetric pendulum. The reviewer highlighted the difference between the number and stability properties of the steady-state solutions of its approximate equation, in which the restoring force contains quadratic and cubic nonlinearity, and the linearised equation.

Another review was submitted in 1920 to the ZAMM-Journal of Applied Mathematics and Mechanics ('Zeitschrift fur Angewandte Mathematik und Mechanik') and published in its very first issue in 1921 [24]. At the beginning of the review, the main characteristics of the resonance response of a harmonically excited linear oscillator are listed. Then, stating that "the equations that describe numerous vibration problems are more complex", the example of a forced pendulum was given as an illustration. It was emphasised that its solution was obtained for free vibrations by using elliptic functions, but in case of forced vibrations this appeared to be unattainable. In order to overcome this problem, Duffing approximated the equation of the pendulum to an equation with softening cubic nonlinearity, assumed the solution of motion in the form of the first harmonic and applied three methods (the method of the variation of parameters, the method of successive approximation and the Ritz method), showing the possibility of a multivalued response. It is also noted that Duffing succeeded in confirming some results experimentally as well as discussing the equation of motion with both quadratic and cubic nonlinearity, and damped vibrations.

It is worth mentioning that the reviewer recognised and supported Duffing's wish and intention, writing [24]: "Strange vibration phenomena in relatively simple cases are enlightened in this study, as a reward for an engineer and as an inspiration for a mathematician to gain deeper insight."

1.6 Research inspired by Duffing's work

1.6.1 1918–1952

Following Duffing's book, it took some time for his work to become known. This could have been due to the fact that it was published in German. In what follows, a potted history of the research work that followed on from Duffing's book is given. It will be seen that it took about ten years for the book to be cited in a publication written in English, and this was in Timoshenko's book. It is possible that Timoshenko got to know of Duffing's work when he was at Westinghouse in the United States. He went there in 1922, more than a decade after Duffing's visit which was in 1910.

Possibly the earliest paper that cites Duffing's work was written by Hamel [25]. Hamel also wrote reviews of Duffing's book [23,24]. In this paper, Hamel studied the pendulum, but did not approximate the restoring force as a linear plus a cubic term as in Duffing's book. Rather, using the variational approach, he minimised the action integral, deriving the amplitude–frequency equation, obtaining a more accurate result.

Rüdenberg considered both mechanical and electrical systems with nonlinear restoring-force characteristics [26], continuing the work of Martienssen [14], Biermanns [15] and Duffing [3]. He considered both *free* and *forced oscillations*. For free oscillations, he considered undamped systems and for forced vibrations he considered both undamped and damped systems using a combination of analytical and geometrical approaches similar to that taken by Duffing. He assumed a harmonic response, but considered a generalised restoring force instead of than one of polynomial form, which permits graphical rather than closed-form solutions.

In 1924, Appleton studied the softening nonlinear behaviour of a galvanometer used in the laboratory in Cambridge University [27]. He observed that the output from the galvanometer could have two different values for certain current inputs. He modelled the system as Duffing had done for the pendulum and produced frequency-response curves that were similar to those observed in the laboratory. He also considered the stability of his solutions. Remarkably, Appleton did not refer to any literature, except to note that a paper by Waibal in *Annal der Physik* had observed hysteresis behaviour in a galvanometer.

Lachman wrote a paper concerning the solution of the exact equation describing the forced vibration of a pendulum in 1928 [28]. He used Duffing's name in the title of the paper, demonstrating that he was directly inspired by this equation appearing in Duffing's book.

Timoshenko's classic textbook was published in 1928 [29]. In it he considered simple mechanical systems with geometric nonlinearity and cites Duffing's book. This appears to be the first time that it was cited in a publication written in English, and is possibly the beginning of international acknowledgement of the importance of Duffing's pioneering work.

Five years later, Den Hartog, who was also employed by Westinghouse (1924–1932), developed a graphical method for solving the forced vibration of a system with a nonlinear spring, and compares his results directly with the method developed by Martienssen, the work that inspired Duffing. Duffing's book is cited in this paper [30].

In 1938, Rauscher developed an iterative method to determine the response of a forced nonlinear oscillator with a general nonlinear restoring force characteristic using the amplitude of free vibration of the oscillator as an initial guess [31]. He cited Duffing's book as being the long-established text on the subject.

Von Kármán published a paper in the Bulletin of the American Mathematical Society in 1940 [32], based on the fifteenth Josiah Willard Gibbs lecture that he gave in 1939. In this paper he described several nonlinear engineering problems, one of which involved *subharmonic resonance* due to nonlinear restoring forces. Duffing's book was listed in the bibliography.

In the late 1930s and 1940s a group of applied mathematicians worked on nonlinear problems in New York University. These were led by Richard Courant, who left Germany in the mid-1930s, where he had been an assistant to Hilbert at Göttingen. His group included Kurt Friedrichs, his former student who left Germany in 1937 to join him, and James Stoker, who subsequently wrote the seminal book on nonlinear vibrations [22]. In 1942 in a series of lectures given at Brown University [33] based on a course given by Friedrichs and Stoker at New York University, the equation of an oscillator in which the restoring force consists of a linear and a cubic term was described as Duffing's equation. This was 24 years after Duffing's book, and the authors believe this was the first occasion in which the equation was named in such a way.

In 1949, Levenson published a paper in the Journal of Applied Physics based on his doctoral work at New York University in which the Duffing equation and his name appears in the title [34]. In this paper, Levenson considered the harmonic and subharmonic response of the system. It appears that by 1949, some 31 years after Duffing's book, the equation describing the oscillator with a linear and cubic nonlinear restoring force had become known as the Duffing equation.

Following Stoker's book in 1950 [22], the Duffing equation sat alongside van der Pol's, equation as one of the classic equations in nonlinear vibrations and was being cited in a wide range of literature from physics, for example [35], to mathematics, for example [36]. This last paper was a review paper published in 1952; it appears to signal the end of the work in this area, apart from one paper on the transient behaviour of a ferroresonance circuit in 1956, for about a decade.

1.6.2 1962 to the present day

Since the 1960s, many journal papers have been published related to the Duffing equation. A survey has been carried out via SCOPUS to track the journal papers that used the word 'Duffing' in the title, abstract or keywords. The number of such papers published per year is shown in Figure 1.4. It can be seen that until the 1970s, only a few papers appeared per year, concerned mainly with finding an approximation for the displacement of the oscillator. Then, this number dramatically increased, which was because people started to recognise the Duffing equation as a model for different systems. Also, digital computers started to be used to solve analytically nonlinear

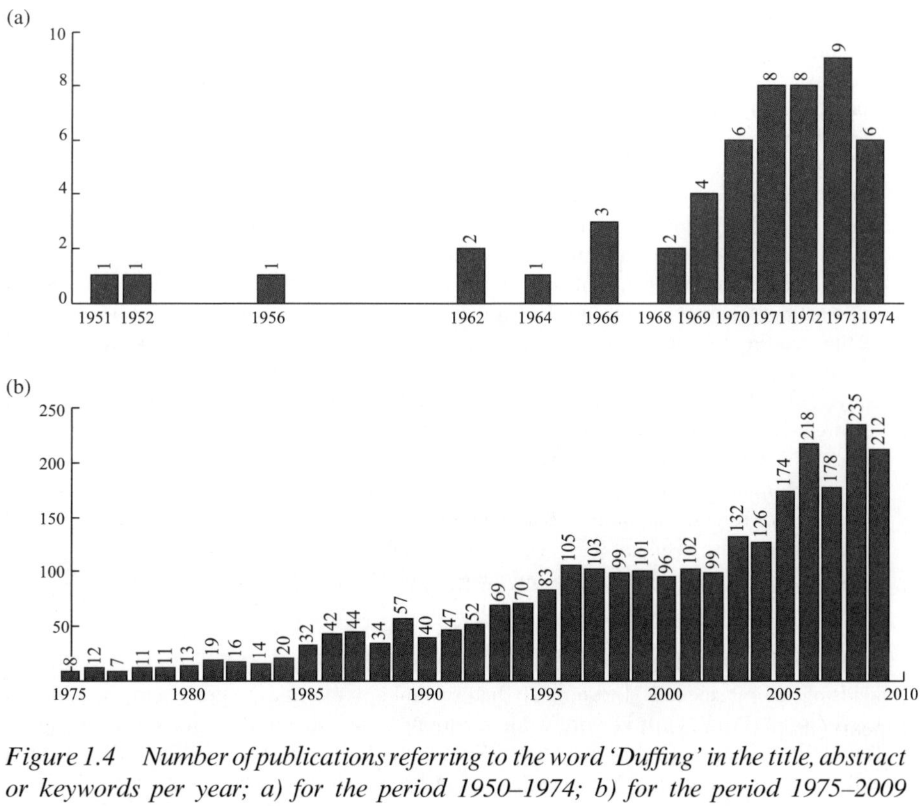

Figure 1.4 Number of publications referring to the word 'Duffing' in the title, abstract or keywords per year; a) for the period 1950–1974; b) for the period 1975–2009 (Source: Elsevier Scopus™, accessed 9 August 2008 and updated 30 March 2010).

ordinary differential equations. This increasing trend continued even further, when in 1976 Holmes and Rand published their paper on bifurcations of Duffing's equation and the application of catastrophe theory [37].

In the 1980s Ueda published his work on chaos, initially named 'randomly transitional phenomena' [38] and 'random phenomena' [39]. Reference [39] was a translated version of an earlier paper published in Japanese in the *Transactions of the Institute of Electrical Engineers of Japan*, Vol. A98, March 1978. The postscript in reference [39] sheds some interesting light on the discovery of chaotic behaviour in the purely cubic Duffing oscillator. Because this is such an important milestone in the study of the Duffing equation, it is copied in full below.

POSTSCRIPT by YOSHISUKE UEDA

I deem it a great honour to be given the opportunity to translate my article into English and I would like to express my thanks to the members of the editorial board. In the following I am writing down some comments and fond memories of days past when I was preparing the manuscript with tremendous difficulty.

It was on November 1961 when I met with chaotic motions in an analogue computer simulating a forced self-oscillatory system. Since then my interest has been held by the phenomenon, and I have been fascinated by the problem "what are steady states in non-linear systems?" After nearly ten years I understood "randomly transitional phenomenon", I published my findings in the *Transactions of the Institute of Electronics and Communication Engineers of Japan*, Vol. 56, April 1973 [10]. My paper then received a number of unfavorable criticisms from some of my colleagues: such as, "Your results are of no importance because you have not examined the effects of simulation and/or calculation errors at all", "Your paper is of little importance because it is merely an experimental result", "Your result is no more than a periodic oscillation. Don't form a selfish concept of steady states", and so forth. Professor Hiromu Momota of the Institute of Plasma Physics was the first to appreciate the worth of my work. He said "Your results give an important feature relating to stochastic phenomena" on 3 March 1974. Through his good offices I joined the Collaborating Research Program at the Institute of Plasma Physics at Nagoya University. These events gave me such unforgettable impressions that I continued the research with tenacity. At this moment I yearn for those days with great appreciation for their criticisms and encouragements.

By the middle of the 1970s, I had obtained many data of strange attractors for some systems of differential equations; but I had no idea to what journals and/or conferences I might submit these results. I was then lucky enough to meet with Professor David Ruelle who was visiting Japan in the early summer of 1978. He advised me to submit my results to the *Journal of Statistical Physics* [P1]. Further, he named the strange attractor of Fig. 3 "Japanese Attractor" and introduced it to the whole world [P2–P5]. At the same time chaotic behavior in deterministic systems began to come under the spotlight in various fields of natural sciences. I fortunately had several opportunities to present my accumulated results [P6–P11]. It is worthwhile mentioning that, due to the efforts of Professor David Ruelle and Professor Jean-Michel Kantor, the Japanese Attractor will be displayed at the National Museum of Sciences, Techniques and Industries which will open in Paris, 1986. In these circumstances this paper is a commemorative for me and I sincerely appreciate their kindness on these matters.

As the reader will notice in this translation and also in ref. [P1] I was rather nervous of using the term "strange attractor", because I had no understanding of its mathematical definition in those days. Although I do not think I fully understand the definition of it even today, I begin to use the term "strange attractor" without hesitation because it seems to agree with reality. However, it seems to me that the term "chaos", although it is short and simple, is a little bit exaggerated. In the universe one does have a lot more complicated, mysterious and incomprehensible phenomena! I should be interested in readers' views of my opinion.

REFERENCES TO POSTSCRIPT

P1. Y. Ueda, *J. statist. Phys.* **20**, 181 (1979).

P2. D. Ruelle, *La Recherche*, **11**, 132 (1980).

P3. D. Ruelle, *The Mathematical Intelligence* **2**, 126 (1980).

P4. D. Ruelle, *Mathematical Calendar*. Springer, Berlin (November 1981).

P5. D. Ruelle, *Czech. J. Phys.* **A32**, 99 (1982).

P6. Y. Ueda, New approaches to non-linear problems in dynamics, *SIAM J. Appl. Math.* 311 (1980).

P7. Y. Ueda, Annis N.Y. Acad. Sci. **357**, 422 (1980).

P8. Y. Ueda and N. Akamatsu, *IEEE Trans. Circuits and Systems,* **28**, 217 (1981).

P9. H. Ogura *et al.*, Prog. Theret. Phys. **66**, 2280 (1981).

P10. Y. Ueda, *Proc. 24thMidwest Symposium on Circuits and Systems,* p. 549. University of New Mexco (1981).

P11. Y Ueda and H. Ohta, *Chaos and Statistical Methods*, p. 161. Springer, Berlin (1984).

In 1979 Holmes's article 'A nonlinear oscillator with a strange attractor' appeared [40]. This is also a highly cited paper. The continuous growing trend of the published articles has included the development of analytical and numerical methods to find different solutions for motion and to study the phenomena associated with the equations of motion. Investigation of the transition between different regimes has also been widely researched as has identification of the systems, and different problems of synchronisation and control, etc.

What is also apparent is the diversity of disciplines in which Duffing's equation appears. As illustrated in Figure 1.5, the majority of publications in the SCOPUS survey belongs to Engineering. Around 25% of them are from Physics and Astronomy and 19% from Mathematics. Computer Science encompasses 6%, Material Science and Multidisciplinary studies 2%, while Chemical Engineering has only 1% of the publications. The rest of the disciplines, such as Earth Science, Biochemistry or Biology each have less than 1% of the total, so are given in a cumulative way.

It should be emphasised that it is not the equation of motion with a positive linear term and cubic nonlinearity that was named after Duffing, but many other homogeneous or inhomogeneous second-order ordinary differential equations were also

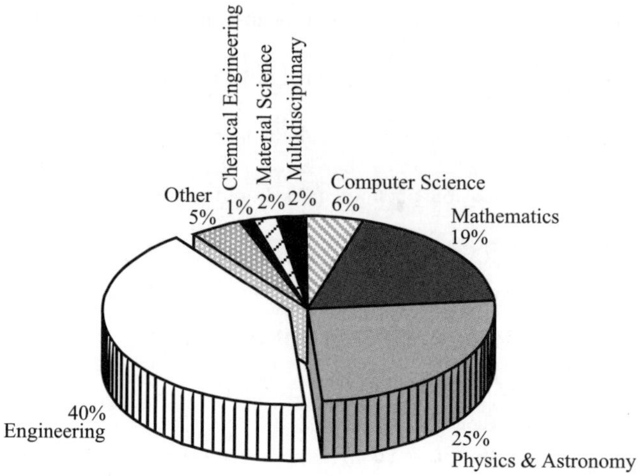

Figure 1.5 Percentage of publications referring to the word 'Duffing' in the title, abstract or keywords for the period 1950–2009 per disciplines (Source: Elsevier Scopus™, accessed 9 August 2008 and updated 30 March 2010).

called the Duffing equation(s) and formed the basis of many mathematical models of different systems. Some of the practical examples of the systems whose dynamic behaviour is described by these equations are given in Chapter 2 of this book. Subsequent chapters are concerned with different forms of the Duffing equation(s) each contains some references that can help the reader to track the most important and the most influential publications associated with each form of the equation studied.

1.7 Some other books on nonlinear dynamics

As mentioned previously, Stoker's book [22] was a key publication in the field of nonlinear vibrations in the 1950s. From the 1960s there have been many books published in this area, demonstrating the rapid development of the topic and growing interest by various communities. Several books are listed in this chapter, so that the reader can probe more deeply into the topics of their choice. The books are grouped together as follows; mathematical treatment of equations of motion [41–48], nonlinear phenomena, with a focus on chaos [49–54], and finally those which have more of an engineering bias [55–59]. The books by Hayashi [42] and Nayfeh and Mook [43] are considered to be of particular importance as they give the concepts and analytical methods for the study of nonlinear oscillators. In addition, the former provides experimental results and the latter includes an extensive bibliography.

1.8 Overview of this book

This book has been created with the aim of enabling the reader to gradually gain insight into the equations associated with Duffing's name, the oscillators that they describe, methods that are used to study their response and related phenomena. Following the historical background given in Chapter 1, Chapter 2 shows how some real dynamic systems can be modelled approximately by the Duffing equation. Chapters 3 and 4 are concerned primarily with qualitative and quantitative analysis of free vibration problems (Chapter 4 does, however, contain some quantitative analysis of forced vibrations). These two chapters set the scene for the remaining four chapters, which are concerned with different forced vibration problems. More details about each book chapter, as well as the Appendix and the Glossary, are given below.

As mentioned above, Chapter 2 gives practical examples of systems whose dynamic behaviour is described by different forms of the Duffing equation. Various physical systems are chosen to illustrate the physical phenomena that result in different forms of this equation. These equations are subsequently nondimensionalised to link with the other chapters in the book. In addition, several basic types of geometric nonlinearity are described: hardening (with positive linear stiffness and positive cubic stiffness), softening (with positive linear stiffness and negative cubic stiffness), systems with negative linear-positive nonlinear stiffness (i.e., with a double/two/twin-well potential) and, finally, a system that is purely nonlinear (no linear term). The equations that describe the systems with these types of nonlinearity are subsequently investigated in more detail in later chapters for the case of free and forced vibrations and different damping mechanisms.

Chapter 3 is concerned with free vibrations of a system with viscous damping. Qualitative analysis is conducted to demonstrate that the system undergoes *local bifurcations* when the linear stiffness and damping are changed. It is shown that negative linear stiffness and negative linear damping can produce buckling and *self-excited oscillation*, respectively. It is also shown that nonlinear stiffness characterises the postbuckling behaviour, i.e., the existence of nontrivial *fixed points* and their stability. The effect of nonlinear damping on the existence and magnitude of the *steady-state* response for the self-excited system is demonstrated. Furthermore, more global aspects of the *bifurcation* are investigated. By using a *Hamiltonian* structure, some of the qualitative characteristics of nonlinear dynamics are also studied.

Some quantitative methods for obtaining the solutions of various forms of the Duffing equation with hardening, softening, negative linear-positive nonlinear stiffness and pure cubic nonlinearity are presented in Chapter 4. Two groups of analytical methods are shown: nonperturbation and perturbation techniques. The following asymptotic methods are considered: the straightforward expansion method, the parameter-expanding method (the elliptic Lindstedt–Poincaré method), the generalised averaging method, the parameter perturbation method (elliptic Krylov–Bogolubov method), the elliptic harmonic balance method, the elliptic Galerkin method (the weighted residual method), the homotopy perturbation method and the homotopy analysis method. For all the methods discussed, the common factor is the generating solutions of the differential equations that describe the free or harmonically forced oscillations of the Duffing oscillator. These are based on *Jacobi elliptic functions*. To illustrate the use of these methods, several examples are given. To assess the accuracy of the approximate analytical solutions, they are compared with numerical solutions. It is shown that the analytical results obtained are in good agreement with the solutions from numerical integration even for the cases when the nonlinearity and/or the excitation force are not small.

In Chapter 5, forced harmonic oscillations of the Duffing oscillator with linear viscous damping are explored. For weak nonlinearities and weak damping, the perturbation method is used to obtain an analytical approximation for the primary resonance response. In order to study the stability of periodic responses of the forced Duffing oscillator, local stability analysis is carried out on the equations describing the slow timescale evolution. In addition, secondary resonance corresponding to strong (hard) excitation is also discussed. The combination of analytical and numerical investigations presented in this chapter is used to illustrate the *jump phenomenon* and the rich variety of nonlinear phenomena possible in the system with a hardening, softening and pure cubic nonlinearity.

Chapter 6 contains the study of a harmonically excited Duffing oscillator with different damping mechanisms, focusing on the effects of these damping mechanisms on the response of a system with a hardening, softening, negative linear-positive nonlinear stiffness and pure cubic nonlinearity. All velocity-dependent damping mechanisms are treated by using the concept of *equivalent viscous damping*. The *break-loose frequency* is introduced in the case of *Coulomb damping*. The stability analysis of the harmonic solution, *period-doubling bifurcation* and *Melnikov criterion* are obtained for linear and cubic damping. Some experimental and numerical results are included to investigate some typical trends in the response.

Forced harmonic vibration in a Duffing oscillator with negative linear stiffness and linear viscous damping are examined in Chapter 7. Various aspects of the dynamical behaviour of the Duffing oscillator with a *twin-well potential* are investigated by the combined use of analytical and numerical tools. Nonlinear periodic oscillations are discussed first, and the classical nonlinear *resonance* is studied in detail by the method of multiple scales. Then, transition to a complex response is investigated by using *bifurcation diagrams*, basins of attractions, and stable and unstable *invariant manifolds*, by summarising the regions of different dynamical response in a comprehensive behaviour chart. Analytical prediction of the transition to *chaos* via the *Melnikov criterion* is then presented. Finally, nonclassical issues such as control of *homoclinic bifurcation* and *chaos*, and *dynamical integrity* are discussed in detail with the aim of highlighting the most important ideas and objectives.

In the last chapter, the forced harmonic vibrations of an asymmetric nonlinear system are investigated. Two nonlinear asymmetric systems are described. The first is a pure cubic nonlinear oscillator with a constant and a harmonic force acting on it, and the second is a harmonically excited oscillator with both quadratic and cubic nonlinearity. Both of these systems have a *single-well potential*. Different analytical and numerical approaches are used to study and illustrate the rich dynamics of the systems, which include multiple *jumps* in the *hysteretic behaviour* and different routes to chaos.

This book also has an Appendix, which contains various sections of Duffing's book that have been tranlsated into English. His book has been cited many times since 1918, the year it was published, but to the editors' knowledge, it has never been translated into English. The sections have been chosen to give a flavour of the book, reflecting aspects of Duffing's work closely related to the content of this book.

This book ends with a Glossary, containing a list of some definitions and terms used. The aim of providing such a list is to enable the reader to go through the book smoothly, without any need to look elsewhere for background information, and to make this book appropriate for a wide-range of readers interested in the content. The terms in bold in the Glossary are written in italics in the main text, when they appear for the first time in each chapter.

References

[1] P. Holmes, Ninety plus thirty years of nonlinear dynamics: less is more and more is different. *International Journal of Bifurcation and Chaos*, 15, 2703–2716, 2005.

[2] S. Shaw and B. Balachandran, A review of nonlinear dynamics of mechanical systems in year 2008. *Journal of System Design and Dynamics*, 2, 3, 611–640, 2008.

[3] G. Duffing, *Erzwungene Schwingungen bei veränderlicher Eigenfrequenz und ihre technische Bedeutung*, Vieweg & Sohn, Braunschweig, 1918.

[4] H.L.F. Helmholtz, *On the Sensations of Tone as a Physiological Basis for the Theory of Music¸ App. XII, Theory of combinatorial tones*, 411–413, translated by Ellis, A.J., 3rd edn, Longmans Green and Co., 1895. Reprinted by Dover publications, New York.

[5] J.W. Strutt, Baron Rayleigh. *The Theory of Soun*d, Vol I, 77, 2nd edn, Constable and Co. Ltd., 1894. Reprinted by Dover publications, New York.

[6] http://en.wikipedia.org/wiki/Galileo_Galilei, Accessed 30 July 2009.

[7] http://en.wikipedia.org/wiki/Christiaan_Huygens, Accessed 30 July 2009.

[8] R. Hooke, *Lectures de Potentia Restitutiva of a Spring Explaining the Power of Spring Bodies*, London, John Martin, 1678. (Taken from The Rational Mechanics of Flexible or Elastic Bodies 1638–1788: Introduction to Vol. X and XI, Edited by C. Truesdell, Birkhauser, 1960.

[9] I. Newton, *Philosphiae Naturalis Principia Mathematica*, London, 1687.

[10] *The Rational Mechanics of Flexible or Elastic Bodies* 1638–1788: Introduction to Vol. X and XI, Edited by C. Truesdell, Birkhauser, 1960.

[11] J. Bernoulli, Meditations de chordis vibrantibus, cum pondusculis aequali a intervallo a se invicem dissitis ubi nimirum ex principio virium vivarum quaqeritur numerus vibrationum chordia pro una oscillatione penduli datae longitudinis *D*, *Comm. Acad. Petrop*, 3 (1728), 13–28 (1732) = Opera omnia 3, 198–210. Taken from The Rational Mechanics of Flexible or Elastic Bodies 1638–1788: Introduction to Vol. X and XI, Edited by C. Truesdell, Birkhauser, 1960).

[12] L. Euler, E126, De novo genere oscillationum, Comm. Acad. Sci. Petrop, 11 (1139), 128-149 (1750) = Opera omnia II 10, 78–97. (Taken from The Rational Mechanics of Flexible or Elastic Bodies 1638–1788: Introduction to Vol. X and XI, Edited by C. Truesdell, Birkhauser, 1960).

[13] J.E. Routh, *Dynamics of a System of Rigid Bodies*, Macmillan and Co, New York, 1892.

[14] V.O. Martienssen, Über neue, resonanzerscheinungen in wechselstromkreisen. *Physik Zeitschrift – Leipzig*, 11, 448–460, 1910.

[15] J. Biermanns, *Der schwingungskreis mit heisenhaltiger induktivität*, Archiv für Elektrotechnik, 345–353, 1915.

[16] F.P.J. Rimrott, Georg Duffing (1861–1944). *Technische Mechanik*, 14, 77–82, 1994.

[17] http://en.wikipedia.org/wiki/Karlsruhe_Institute_of_Technology, Accessed 9 January 2009.

[18] http://www.google.com/patents/about?id=sKZYAAAAEBAJ&dq=georg + duffing, Accessed 29 March 2010.

[19] http://www.google.com/patents?id=QTpOAAAAEBAJ&printsec=abstract&zoom=4#v=onepage&q=&f=false, Accessed 29 March 2010.

[20] http://www.freepatentsonline.com/1641417.pdf, Accessed 29 March 2010.

[21] http://www.freepatentsonline.com/1894369.pdf, Accessed 29 March 2010.

[22] J.J. Stoker, *Nonlinear Vibrations in Mechanical and Electrical Systems*, Interscience Publishers, New York, 1st edn, 1950.

[23] Jahrbuch Database http://www.emis.de/MATH/JFM/full.html, JFM 46.1168.01.

[24] G. Hamel, Buchbesprechungen. *Zeitschrift für Angewandte Mathematik und Mechanik*, 1, 72–73, 1921.

[25] G. Hamel, Über erzwungene schwingungen bei endlichen amplituden. *Mathematische Annalen*, 86, 1–13, 1922.

[26] R. Rüdenberg, Einige unharmonische schwingungsformen mit ġroβer amplitude. *Zeitschrift für Angewandte Mathematik und Mechanik*, 3, 454–467, 1923.

[27] E.V. Appleton, On the anomolous behaviour of a vibration galvanometer, *Philosophical Magazine*, S. 6. 47 (279), 609–619, 1924.

[28] K. Lachmann, Beitrag zum schwingungsproblem von Duffing. *Mathematische Annalen*, 99, 479–492, 1928.

[29] S. Timoshenko, *Vibration Problems in Engineering*, D. Van Nostrand Company, Inc., New York, 1928.

[30] J.P. Den Hartog, The amplitudes of non-harmonic vibrations. *Journal of the Franklin Institute*, 216, 459–473, 1933.

[31] M. Rauscher, Steady oscillations of systems with nonlinear and unsymmetrical elasticity. *Journal of Applied Mechanics*, 5, 169–177, 1938.

[32] T. von Kármán, The engineer grapples with nonlinear problems. *Bulletin of the American Mathematical Society*, 46, 615–683, 1940.

[33] K.O. Friedrichs, P. Le Corbeiller, N. Levinson, J.J. Stoker, *Lectures on Non-Linear Mechanics delivered at Brown University*, New York, 1942.

[34] M.E. Levenson, Harmonic and subharmonic response of the Duffing equation $\ddot{x} + \alpha x + \beta x^3 = F \cos \omega t \ \ (\alpha > 0)$. *Journal of Applied Physics*, 20, 1045–1051, 1949.

[35] S. Fifer, Studies in nonlinear vibration theory. *Journal of Applied Physics*, 22, 1421–1428, 1951.

[36] M.L. Cartwright, Non-linear vibrations: a chapter in mathematical history. *The Mathematical Gazzette*, 35, 82–88, 1952.

[37] P.J. Holmes, D.A. Rand, The bifurcations of Duffing's equation: An application of catastrophe theory. *Journal of Sound and Vibration*, 44, 237–253, 1976.

[38] Y. Ueda, Randomly transitional phenomena in the system governed by Duffing's equation. *Journal of Statistical Physics*, 20, 181–196, 1979.

[39] Y. Ueda, Random phenomena resulting from non-linearity in the system described by Duffing's equation. *International Journal of Non-Linear Mechanics*, 20, 481–491, 1985.

[40] P. Holmes, A nonlinear oscillator with a strange attractor philosophical. *Transactions of the Royal Society of London. Series A, Mathematical and Physical Sciences*, 292, 419–448, 1979.

[41] N. Minorsky, *Nonlinear Oscillations*, D. Van Nostrand Company, Princeton, 1962.

[42] C. Hayashi, *Nonlinear Oscillations in Physical Systems*, McGraw-Hill, New York, 1964.

[43] A.H. Nayfeh, D.T. Mook, *Nonlinear Oscillations*, Wiley, New York, 1979.

[44] P. Hagedorn, *Nonlinear Oscillations*, Clarendon Press, Oxford, 1981.

[45] J. Guckenheimer, P. Holmes, *Nonlinear Oscillations, Dynamical Systems and Bifurcation of Vector Fields*. Springer-Verlag, New York, 1986.

[46] A.H. Nayfeh, B. Balachandran, *Applied Nonlinear Dynamics*, John Wiley & Sons, New York, 1995.

[47] R.E. Mickens, *Truly Nonlinear Oscillations: Harmonic Balance, Parametric Expansions, Iteration, and Averaging Methods*, World Scientific, Singapore, 2010.

[48] R.H. Rand, Lecture Notes on Nonlinear Vibrations, version 52, (http://audiophile.tam.cornell.edu/randdocs/nlvibe52.pdf), Accessed 29 March 2010.

[49] Y. Ueda, *The Road to Chaos*, Aerial Press, Inc, Santa Cruz, 1992.

[50] S.H. Strogatz, *Nonlinear Dynamics and Chaos*, Addison-Wesley, New York, 1994.

[51] J.M.T. Thompson, H.B. Stewart, *Nonlinear Dynamics and Chaos*, John Wiley and Sons, Chichester, 2002.

[52] F.C. Moon, *Chaotic Vibrations*, John Wiley and Sons, New York, 2004.

[53] R.C. Hilborn, *Chaos and Nonlinear Dynamics, an Introduction for Scientists and Engineers*, 2nd edn, Oxford University Press, Oxford, 2006.

[54] H.J. Korsch, H-J Jodl, T. Hartmann, *Chaos, a Program Collection for the PC*, Third edn, Springer Verlag, Berlin, 2008.

[55] K. Magnus, *Vibrations*, Blackie & Son Ltd, Glasgow, 1965.

[56] K. Worden and G.R. Tomlinson, *Nonlinearity in Structural Dynamics, Detection, Identification and Modelling*, Institute of Physics Publishing, Bristol and Philadelphia, 2001.

[57] L.N. Virgin, *Introduction to Experimental Nonlinear Dynamics*, Cambridge University Press, Cambridge, 2000.

[58] J.J. Thomsen, *Vibrations and Stability, Advanced Theory, Analysis, and Tools*, 2nd edn, Springer, Berlin, 2003.

[59] D. Wagg and S. Neild, *Nonlinear Vibration with Control, for Flexible and Adaptive Structures*, Springer, Dordrecht, 2010.

2

Examples of physical systems described by the Duffing equation

Michael J. Brennan[1] and Ivana Kovacic[2]
[1]*University of Southampton, Institute of Sound and Vibration Research, United Kingdom*
[2]*University of Novi Sad, Faculty of Technical Sciences, Serbia*

2.1 Introduction

The Duffing equation in its various forms is used to describe many nonlinear systems. Although most physical systems cannot be described accurately in this way for a wide range of operating conditions, such as frequency and amplitude of excitation, in many cases it is possible to use this equation as an approximate description so that their behaviour can be studied qualitatively. In some situations, quantitative analysis can be conducted for small amplitudes of excitation. In many cases, it is the first step in moving from a linear to a nonlinear system.

In this chapter, several systems which can be represented by various forms of the Duffing equation, are described. They range from a pendulum, which was the main focus of Duffing's book, to the nonlinear behaviour of some isolators, beams, cables and electrical circuits. The emphasis is on the physical aspects of each system, and several different characteristics are described. In later chapters the starting point is the nondimensional equation of motion, so the aim here is to describe some of the

The Duffing Equation: Nonlinear Oscillators and their Behaviour, First Edition. Edited by Ivana Kovacic and Michael J. Brennan. Published 2011 by John Wiley & Sons, Ltd.

physical systems which can be modelled by the Duffing equation. The list of examples is not exhaustive; the literature survey reported in Chapter 1 has shown that many systems can be modelled approximately by the Duffing equation. Besides those considered below, among them are plates, shells, woofers, optical fibres, vessel-like or arched structures, micromechanical structures, nanomechanical resonators, rotors, prisms in fluid flow, guinea-pig cochlea, flight motor of an insect, ultrasonic cutting systems, a motorised momentum exchange tether on a circular earth orbit, piezo-ceramics under strong electric fields, etc. In Chapter VI of his book, Duffing gave three practical examples. They relate to an electrical circuit containing a saturating inductor, a capacitor and resistor, a pendulum – on which he also conducted his experiments, and three-phase electrical generators. The relevant sections of Duffing's book have been translated from German to English and are given in the Appendix to this book.

The interested reader can find several physical realisations of the Duffing equation by way of laboratory-based experimental models in Virgin's book [1]. However, the aim in this chapter is not to start with the equation and then represent it by a mechanical system, rather it is to show how and why several real physical systems can be represented approximately by the various forms of the Duffing equation.

Throughout this chapter, damping is neglected because the stiffness is the cause of the nonlinearity in the physical systems considered. Although damping can have an important effect on the dynamic behaviour, and it is discussed in many of the later chapters, it is not included in the equations of motion here for the sake of clarity. All the equations derived are for harmonic excitation (*forced vibration*); the equations for *free vibration* can be found by setting this force to zero. It should be noted that for all the nondimensionalised equations of motion presented, the symbol '~' is used for nondimensional variables such as the magnitude of the harmonic force, displacement and time. However, in later chapters, when these nondimensional equations of motion are considered, this symbol is omitted for brevity.

2.2 Nonlinear stiffness

The source of the nonlinearity in a mechanical system that results in its dynamic behaviour being modelled by the Duffing equation is the stiffness, which is shown in Figure 2.1a. In the case considered here, the stiffness is generally a function of position. This means that the force applied to the spring F_s, and the resulting displacement y have a nonlinear relationship. If the system is symmetric, i.e., the stiffness characteristic is the same when the spring is in compression or in tension, then the restoring force can be approximated as a series in y in which the exponents of y are odd integers. If this series is truncated after the first two terms then the force–deflection relationship is given by

$$F_s = k_1 y \pm k_3 y^3 \tag{2.2.1a}$$

If the cubic term is positive (negative), the spring is said to be hardening (softening) because the spring becomes stiffer (softer) as the displacement

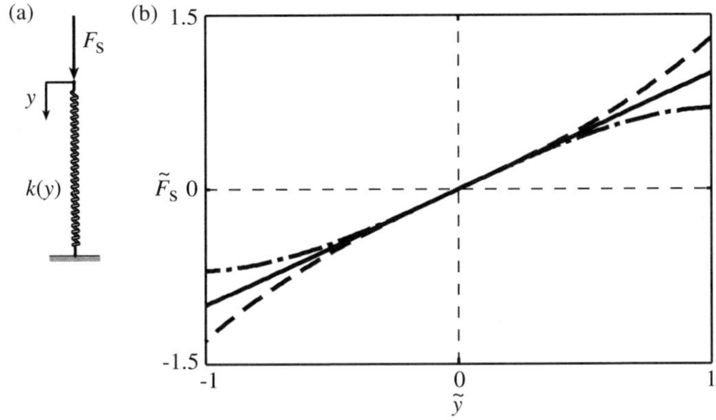

Figure 2.1 Characteristics of a nonlinear hardening and softening spring described by Equation (2.2.1b). (a) A nonlinear spring in which the stiffness is a function of the displacement; (b) Force deflection characteristics for a linear spring (solid line), hardening spring with $\gamma = 0.3$ (dashed line) and softening spring with $\gamma = 0.3$ (dashed-dotted line).

increases. For a spring of length l, Equation (2.2.1a) can be written in nondimensional form as

$$\tilde{F}_s = \tilde{y} \pm \gamma\tilde{y}^3 \tag{2.2.1b}$$

where $\tilde{F}_s = F_s/k_1 l$, $\tilde{y} = y/l$ and $\gamma = k_3 l^2/k_1$. In some of the later chapters this equation has been nondimensionalised in a different way so that there is a nondimensional coefficient, α for the linear term as well. In this book, the variable α is known as the linear stiffness parameter, but is referred to as the linear stiffness for brevity. The variable γ is known as the nonlinear stiffness parameter, but is referred to as nonlinear stiffness for brevity. Equation (2.2.1b) is plotted in Figure 2.1b to illustrate the nonlinear characteristics of the spring. The stiffness of the system is given by differentiating Equation (2.2.1a) to give $dF_s/dy = k_1 \pm 3k_3y^2$. It can be seen that the linear stiffness term k_1 is independent of position and that the nonlinear stiffness term $\pm 3k_3y^2$ is a function of displacement. It can also be seen that the stiffness is symmetric about $y = 0$, and that the cubic term in the force–deflection characteristic only starts to become important for about $\tilde{y} > 0.4$ (see Figure 2.1b).

If the stiffness is not symmetric about $y = 0$ then the force–deflection characteristic requires terms in the series of y, in which the exponents of y are integers of even order. Thus, Equation (2.2.1a) becomes

$$F_s = k_1 y \pm k_2 y^2 \pm k_3 y^3 \tag{2.2.2a}$$

which can be written in nondimensional form as

$$\tilde{F}_s = \tilde{y} \pm \beta\tilde{y}^2 \pm \gamma\tilde{y}^3 \tag{2.2.2b}$$

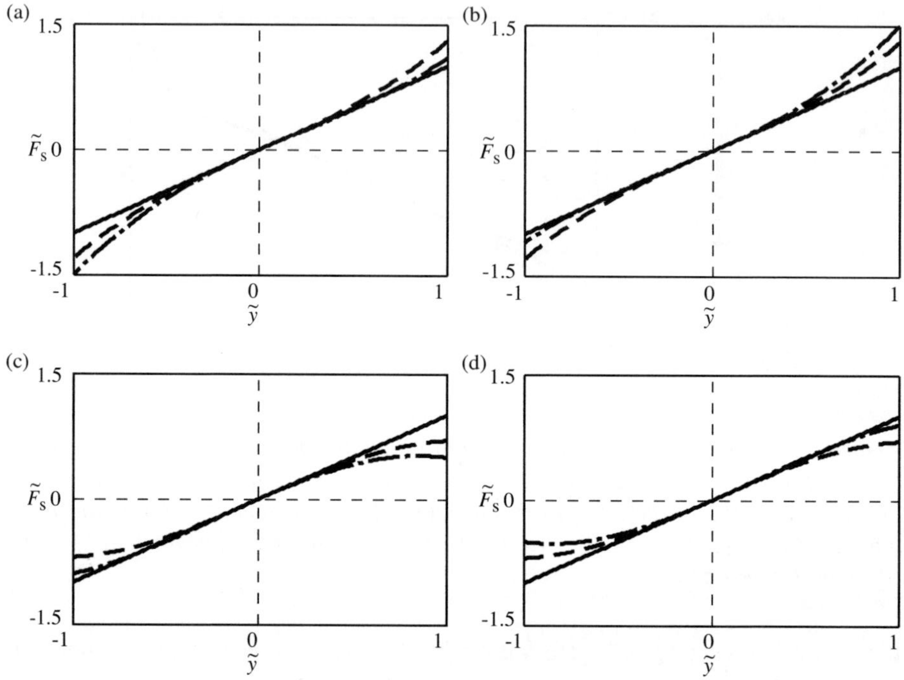

Figure 2.2 The effect of the quadratic term on the force deflection characteristic (a) $\tilde{F}_s = \tilde{y} - \beta\tilde{y}^2 + \gamma\tilde{y}^3$; (b) $\tilde{F}_s = \tilde{y} + \beta\tilde{y}^2 + \gamma\tilde{y}^3$; (c) $\tilde{F}_s = \tilde{y} - \beta\tilde{y}^2 - \gamma\tilde{y}^3$; (d) $\tilde{F}_s = \tilde{y} + \beta\tilde{y}^2 - \gamma\tilde{y}^3$. The linear case when $\beta = \gamma = 0$ (solid line), the case when $\beta = 0$ is for the pure hardening or softening case (dashed line), and the case when $\beta \neq 0$ there is an asymmetric stiffness (dashed-dotted line) for $\beta = 0.2$, $\gamma = 0.3$.

where $\beta = k_2 l / k_1$ is a nonlinear stiffness parameter. The stiffness due to the quadratic term in Equation (2.2.2a) is given by $2k_2 y$ and so varies linearly with y. It can be seen that when k_2 is positive (negative) then this will have a hardening (softening) effect when y is positive and vice versa. To visualise this effect, the nondimensional force–deflection characteristic is plotted in Figure 2.2. Examining this figure it can be seen that the quadratic term has the effect of hardening or softening the system depending on the sign of the term and whether the displacement is negative or positive. Note that in Figures 2.2(c) and (d), the stiffness becomes negative for some values of y. This means that the system will be unstable for large displacements.

2.3 The pendulum

The pendulum is the archetypal dynamical system studied in nonlinear dynamics. Its nonlinear characteristics were studied by Rayleigh, and more-or-less the whole of Duffing's book was devoted to it. The equation of motion describing the

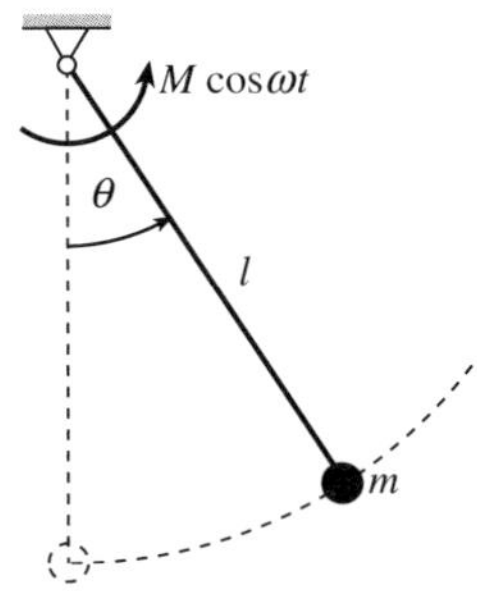

Figure 2.3 A harmonically excited pendulum.

angular displacement θ, of the harmonically excited pendulum shown in Figure 2.3 is given by

$$ml^2 \frac{d^2\theta}{dt^2} + mgl \sin\theta = M \cos\omega t \tag{2.3.1}$$

where l is the length of the pendulum, g is the acceleration due to gravity, and M is the amplitude of the applied moment. The angular displacement can be approximated by the series expansion

$$\sin\theta = \theta - \frac{\theta^3}{3!} + \frac{\theta^5}{5!} - \cdots \tag{2.3.2}$$

If this series is truncated to third order and substituted into Equation (2.3.1), the equation of motion becomes

$$ml^2 \frac{d^2\theta}{dt^2} + mgl\left(\theta - \frac{1}{6}\theta^3\right) = M \cos\omega t \tag{2.3.3a}$$

It can be seen that the pendulum has a softening stiffness characteristic because of the negative cubic term in the stiffness moment. Equation (2.3.3a) can be written in nondimensional form as

$$\ddot{\theta} + \theta - \frac{1}{6}\theta^3 = \tilde{M} \cos\Omega\tilde{t} \tag{2.3.3b}$$

where $\Omega = \omega/\omega_{\rm n}$ is nondimensional frequency, $\omega_{\rm n} = \sqrt{g/l}$ is the linear natural frequency, $\tilde{t} = \omega_{\rm n} t$ is nondimensional time, $\tilde{M} = M/\left(ml^2\omega_{\rm n}^2\right)$ is the normalised amplitude of the excitation moment. The overdots denote differentiation with respect to nondimensional time $\tilde{t}$; this notation is used throughout this book.

2.4 Example of geometrical nonlinearity

A simple system where the stiffness is a function of the displacement is shown in Figure 2.4. The distance d, is equal to the length of the spring when the system is at rest and d_0 is the length of the unstretched spring (original length).

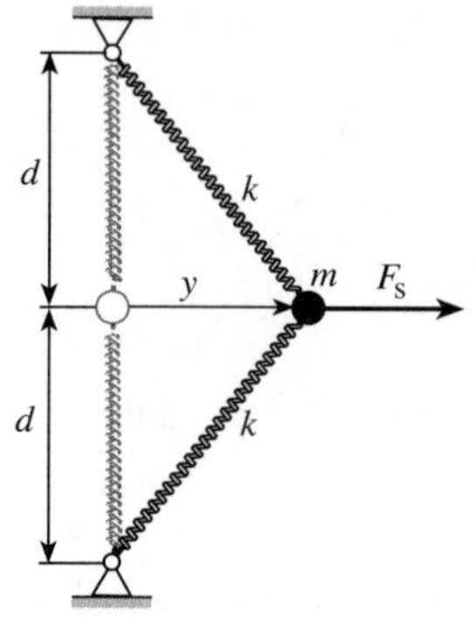

Figure 2.4 A simple example of a geometrically nonlinear system consisting of two linear springs and a mass.

When the mass moves in the y direction, the springs incline to accommodate the motion and it is this change in their length that is the cause of the nonlinearity. The relationship between the applied static force F_s and the resulting displacement y is given by

$$F_s = 2ky\left(1 - \frac{d_0}{\sqrt{y^2 + d^2}}\right) \tag{2.4.1a}$$

where $d_0 \leq d$ in this case. Equation (2.4.1a) can be written in nondimensional form as

$$\tilde{F}_s = \tilde{y}\left(1 - \frac{\tilde{d}}{\sqrt{\tilde{y}^2 + 1}}\right) \tag{2.4.1b}$$

where $\tilde{F}_s = F_s/2kd$, $\tilde{y} = y/d$ and $\tilde{d} = d_0/d$. Using the Taylor-series expansion to the third order for small y, Equation (2.4.1a) can be written as

$$F_s \approx k_1 y + k_3 y^3 \tag{2.4.2a}$$

where $k_1 = 2k(1 - d_0/d)$ and $k_3 = kd_0/d^3$. Equation (2.4.2a) can also be written in nondimensional form as

$$\tilde{F}_s = \alpha\tilde{y} + \gamma\tilde{y}^3 \tag{2.4.2b}$$

where $\alpha = 1 - \tilde{d}$ and $\gamma = \tilde{d}/2$. Equation (2.4.1b) and its approximation given by Equation (2.4.2b) are illustrated in Figure 2.5, for the particular case when $d_0/d = 0.9$.

It can be seen that for a displacement y, less than 40% of the length d, the percentage error between Equation (2.4.1b) and Equation (2.4.2b) is less than 5%. Furthermore, this error decreases for decreasing values of d_0/d. If a harmonic excitation force is applied to the mass in Figure 2.4, the resulting equation of motion

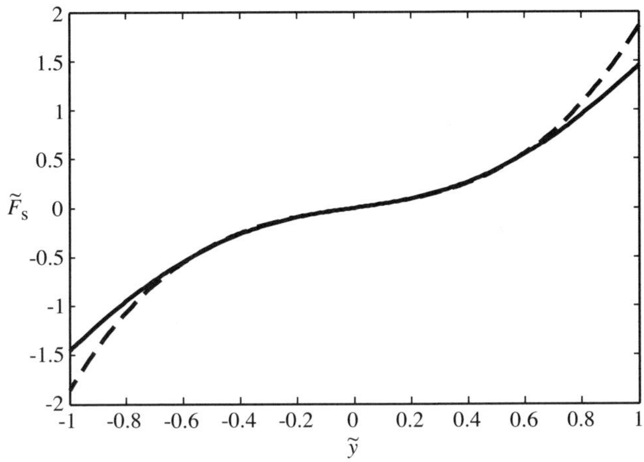

Figure 2.5 The nondimensional force–deflection characteristic of the spring system in Figure 2.4 for $d_0/d = 0.9$ given by Equation (2.4.1b) (solid line), approximate expression defined by Equation (2.4.2b) (dashed line).

for the system is given by

$$m\frac{d^2y}{dt^2} + k_1y + k_3y^3 = F\cos\omega t \tag{2.4.3a}$$

which can be written in nondimensional form as

$$\ddot{\tilde{y}} + \tilde{y} + \gamma\tilde{y}^3 = \tilde{F}\cos\Omega\tilde{t} \tag{2.4.3b}$$

where $\Omega = \omega/\omega_n$, $\omega_n = \sqrt{k_1/m}$, $\tilde{t} = \omega_n t$, $\tilde{y} = \sqrt{y/y_0}$, $y_0 = F/k_1|_{\omega=0}$, $\tilde{F} = F/\left(my_0\omega_n^2\right)$, $\gamma = y_0^2k_3/k_1$, and the overdots denote differentiation with respect to nondimensional time.

Note that the linear stiffness force can be set to zero in this system, if d_0 is set to be equal to d. Note also that this system is undamped. Clearly, in a real system damping would exist. It could, for example, be included as linear viscous damping as a term proportional to velocity in the equation of motion if required. The influence of this type of damping, *Coloumb damping* and other types of nonlinear damping on the response of this system is discussed in Chapter 6.

2.5 A system consisting of the pendulum and nonlinear stiffness

If the springs shown in Figure 2.4 are connected to the pendulum shown in Figure 2.3 in the way depicted in Figure 2.6, the system will exhibit both hardening and softening stiffness nonlinearities. The resulting equation of motion for small oscillations is given by

$$ml^2\frac{d^2\theta}{dt^2} + mgl\left(\theta - \frac{1}{6}\theta^3\right) + l_1\left(k_1l_1\theta + k_3l_1^3\theta^3\right) = M\cos\omega t \tag{2.5.1}$$

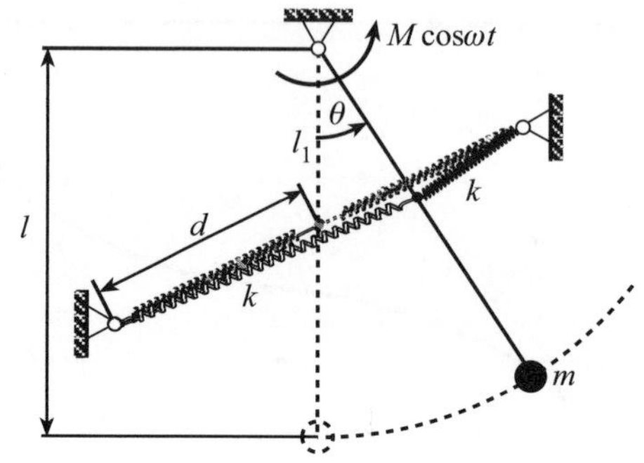

Figure 2.6 A system of springs that has a hardening stiffness characteristic connected to a pendulum that has a softening stiffness characteristic.

It can be seen that if the nonlinear stiffness term is chosen such that $k_3 = mgl/(6l_1^4)$, which means that that $k = mgld^3/(6l_1^4 d_0)$, then the nonlinearities cancel each other and the system acts as a linear system. An arrangement of this type can be incorporated into the design of physical systems, where nonlinearity is undesirable and needs to be eliminated.

2.6 Snap-through mechanism

The system shown in Figure 2.4 and described by Equation (2.4.1a) exhibits a hardening characteristic provided $d_0 \leq d$. If $d_0 > d$, then the characteristics of the system change dramatically. The system now exhibits negative stiffness when the springs are almost in line, and has three static equilibrium positions, two of them being stable and one being unstable. When the displacement is large, however, the system has a positive stiffness that tends to $2k$. The force–deflection characteristics described by Equation (2.4.1b) are plotted in Figure 2.7(a) to show the relationship between the nondimensional force and the nondimensional displacement as $\tilde{d}$ changes. It can be seen that the negative stiffness characteristic (negative gradient) occurs when $\tilde{d} > 1$. This can also be seen in Equation (2.4.2b). The cubic nonlinear stiffness term, however, is always positive and the system with such characteristics is called subsequently the system with a negative linear-positive nonlinear (cubic) stiffness. When connected to a mass, this system can potentially have very complex dynamics because of the snap-through behaviour. This is discussed fully in Chapter 7.

When $\tilde{d} < 1$, the gradient is always positive, so the linear stiffness is always positive. It can also be seen that as the displacement increases then the stiffness increases as expected due to the hardening characteristic of the spring as discussed above. The nondimensional potential energy $\tilde{V} = V/(2kd^2)$ can be determined by integrating Equation (2.4.2b) with respect to $\tilde{y}$ to give

$$\tilde{V} = \frac{1}{2}\left(1-\tilde{d}\right)\tilde{y}^2 + \frac{1}{8}\tilde{d}\tilde{y}^4 \tag{2.6.1}$$

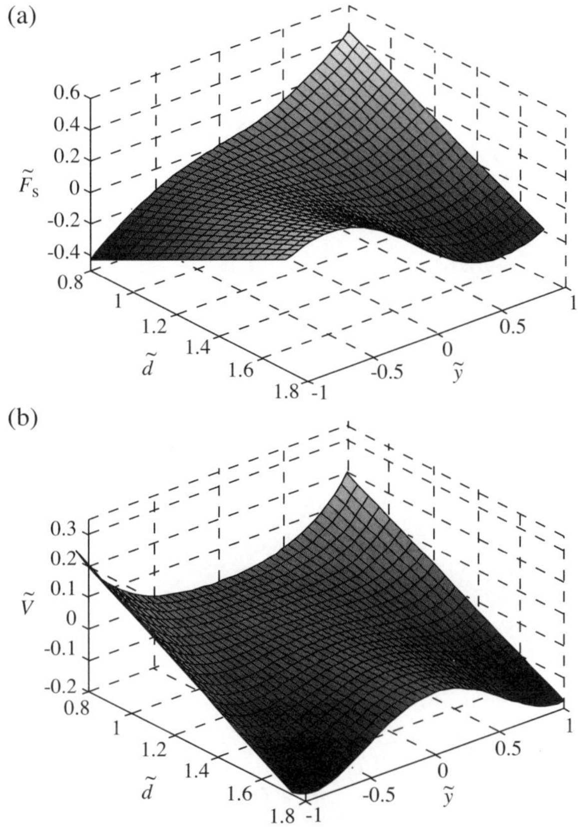

Figure 2.7 Characteristics of the two-spring system shown in Figure 2.4: (a) Nondimensional force–deflection characteristic as the original length of each spring is varied; (b) Nondimensional potential energy as a function of nondimensional displacement as the original length of each springs is varied.

and this is plotted in Figure 2.7(b) as $\tilde{d}$ is varied. The characteristic *single-well*, associated with a positive linear-positive nonlinear (cubic) stiffness, and the *double/two/twin-well potentia*l energy behaviour, associated with a negative linear-positive nonlinear (cubic) stiffness, can clearly be seen in this figure.

A famous example of a double-well system is the 'Moon beam' [2], a drawing of which is shown in Figure 2.8. This consists of a cantilever steel beam hung vertically, with the free end attracted by two magnets as shown in the figure. Assuming that a single-mode approximation can be made for the motion, the nondimensional equation of motion (note that here damping is neglected) is given by [2]

$$\ddot{\tilde{y}} - \alpha\tilde{y} + \gamma\tilde{y}^3 = \tilde{F}\cos\Omega\tilde{t} \tag{2.6.2}$$

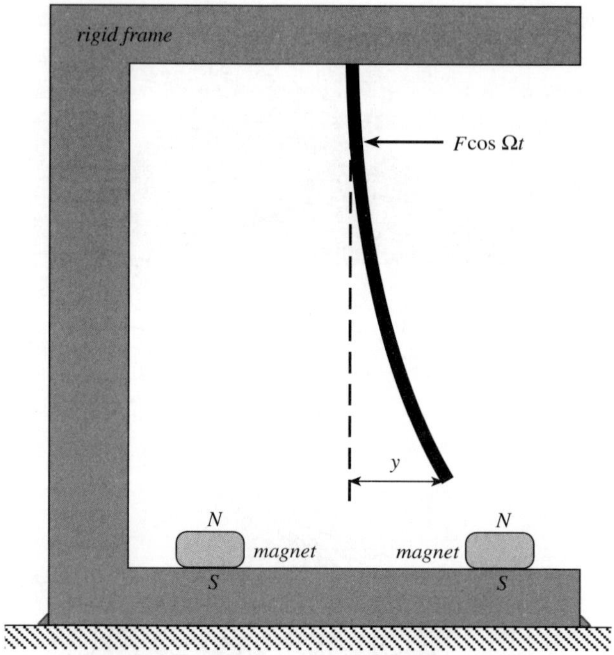

Figure 2.8 An example of a beam and magnet system, arranged such that the system has two stable equilibrium points. One of these points is shown in the figure, the other being when the beam is bent to the magnet on the left. This system has become known as the 'Moon beam' after Francis Moon, who first described this system with Holmes in [2].

where $\alpha = 1/2$ and $\gamma = 1/2$. This has been discussed at length in the literature with particular regard to chaotic behaviour. Instructions to make a 'chaotic toy', for practical investigations based on the 'Moon beam' are given in [3].

2.7 Nonlinear isolator

Nonlinear isolators are often used in engineering systems to isolate vibrating sources from their surroundings. The interested reader is referred to the textbook [4] and to a comprehensive review paper [5] for more details. Many nonlinear isolators have a force–deflection characteristic that is similar to that shown in Figure 2.5 If the static equilibrium position is set to $\tilde{y} = 0$ by an appropriate choice of stiffness and mass, then the stiffness at this position can be adjusted so that it is very low without having a large static deflection as with a linear isolator. A simple model of a nonlinear isolator is shown in Figure 2.9. It consists of the pair of springs shown in Figure 2.4 (but rotated so they are now horizontal) connected in parallel with a linear spring of stiffness k_v [6]. The

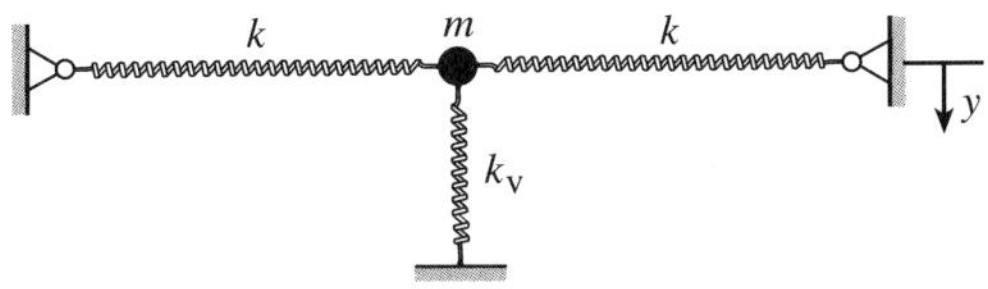

Figure 2.9 A nonlinear isolator loaded with a mass so that it is in the static equilibrium position.

force–deflection characteristic is, therefore, given by

$$F_s = k_v + 2ky\left(1 - \frac{d_0}{\sqrt{y^2 + d^2}}\right) \tag{2.7.1}$$

Note that a mass of a certain size is required to ensure that the springs with stiffness k in Figure 2.8 are in the horizontal position. If the mass is removed then the vertical spring will exert a vertical force on these springs so that they become oblique. The initial deflection due to the mass is called the static deflection and is given by $y_s = \sqrt{(d_0^2 - d^2)}$. More details on this can be found in [7]. Using the Taylor-series expansion to the third order for small y, Equation (2.7.1) can be written as

$$F_s \approx k_1 y + k_3 y^3 \tag{2.7.2}$$

where $k_1 = k_v + 2k(1 - d_0/d)$ and $k_3 = kd_0/d^3$. If $d_0/d > 1$, then the effect of the horizontal springs is to soften the isolator such that its stiffness is less than the vertical spring alone. If $k = k_v$, then d_0/d must be less than or equal to $3/2$ in order that the stiffness of the isolator does not become negative and have a snap-through characteristic. If a harmonic force is applied to the mass supported by the undamped nonlinear isolator system shown in Figure 2.9, the equation of motion is given by

$$m\frac{d^2y}{dt^2} + k_1 y + k_3 y^3 = F\cos\omega t \tag{2.7.3}$$

which can be written in nondimensional form as

$$\ddot{\tilde{y}} + \alpha\tilde{y} + \gamma\tilde{y}^3 = \tilde{F}\cos\Omega\tilde{t} \tag{2.7.4}$$

where the normalised displacement is given by $\tilde{y} = y/y_s$, in which y_s is the static defection of the isolator discussed above. The natural frequency of the system with the horizontal springs removed is given by $\omega_n = \sqrt{k_v/m}$, $\alpha = k_1/k_v$, $\gamma = k_3 y_s^2/k_v$, $\Omega = \omega/\omega_n$, $\tilde{t} = \omega_n t$, $\hat{F} = F/(k_v y_s)$.

2.7.1 Quasi-zero stiffness isolator

It is possible to choose the geometry of the springs in the model of the isolator such that the linear stiffness term is zero so that it has zero natural frequency, and the resulting system is purely cubic [7]. An example of such an isolator is the bubble mount. A photograph of this is shown in Figure 2.10(a) and the associated force–deflection curves in Figure 2.10(b).

If $k = k_v$, and $\tilde{d} = 3/2$ then Equation (2.7.4) becomes

$$\ddot{\tilde{y}} + \gamma\tilde{y}^3 = \tilde{F}\cos\Omega\,\tilde{t} \tag{2.7.5}$$

This *quasi-zero stiffness* characteristic relies heavily on the system being adjusted so that the lateral springs are horizontal when the system is in its static equilibrium position (tuned system). If this is not the case, then the dynamics of

(a)

(b)

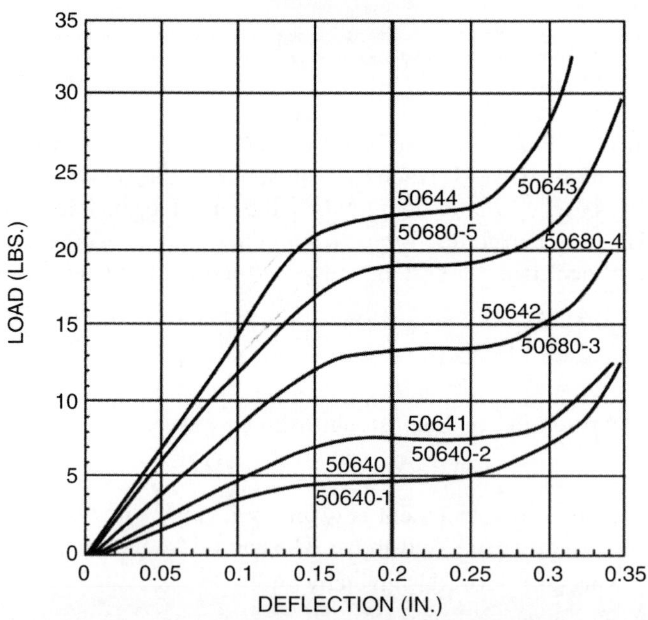

Figure 2.10 (a) Picture of the bubble mount; (b) Typical force–deflection characteristics of a bubble mount. Note the quasi-zero stiffness characteristic of the isolators at a certain deflection. (Image courtesy of Tech Products Corporation, http://www.novibes.com/Products&productId=45, Accessed 15 June 2010.)

the system can change considerably. To illustrate the effects that a mistuned system can have on the equation of motion a constant force F_0 is applied to the system so that

$$\ddot{\tilde{y}} + \gamma\tilde{y}^3 = \tilde{F}_0 + \tilde{F}\cos\Omega\,\tilde{t} \tag{2.7.6}$$

where $\hat{F}_0 = F_0/k_v y_s$. The deviation of the mass from the tuned static equilibrium position is equal to this force divided by the weight of the suspended mass. This detuning could also occur because of the steady acceleration of an aircraft or a vehicle, for example, in which the isolation system is situated. The effect of this force is to make the system asymmetric, such that the stiffness is not the same either side of the static equilibrium position. The resulting equation of motion in terms of the new variable z whose origin is the new static equilibrium is

$$z'' + z + \beta z^2 + \hat{\gamma} z^3 = \hat{F}\cos\hat{\Omega}\,\hat{t} \tag{2.7.7}$$

where

$$z = y - \sqrt[3]{\frac{\tilde{F}_0}{\gamma}}, \quad \beta = \sqrt[3]{\frac{\gamma}{\tilde{F}_0}}, \quad \hat{\gamma} = \sqrt[3]{\frac{\gamma^2}{27\tilde{F}_0^2}}, \quad \hat{F} = \sqrt[3]{\frac{\tilde{F}^3}{27\gamma\,\tilde{F}_0^2}},$$

$$\hat{\Omega} = \frac{\Omega}{\sqrt[6]{27\gamma\,\tilde{F}_0^2}}, \quad \hat{t} = \sqrt[6]{27\gamma\,\tilde{F}_0^2}\; t$$

and now primes denote differentiation with respect to $\hat{t}$. It can be seen that the system oscillates about $z = 0$ and the system now has a linear term as well as a cubic term. It also has a quadratic term because of the asymmetry in the system due to the constant force. To illustrate these effects, the stiffness of the system is plotted as a function of displacement in Figure 2.11. The symmetry of the system at the original equilibrium position shown in Figure 2.9 can be seen. When the static force is applied there is a new static equilibrium position, and the asymmetric stiffness characteristics about this position can also be seen.

A similar asymmetric characteristic stiffness occurs with a vibrating cable as discussed later in this chapter. Asymmetric oscillators described by Equation (2.7.6) are discussed in detail in Chapter 8.

2.8 Large deflection of a beam with nonlinear stiffness

Consider the pinned-pinned beam of length l shown in Figure 2.12(a). The left-hand support is fixed, but the right-hand support is free to slide to prevent inplane stretching of the beam when it is vibrating. Consider an element of the beam of length Δs excited along its length by force per unit length $f(x, t)$ as shown in Figure 2.12(b).

It is assumed that the motion of the beam in the x direction can be ignored. It is further assumed that the rotary inertia of the beam can be neglected. The equations of

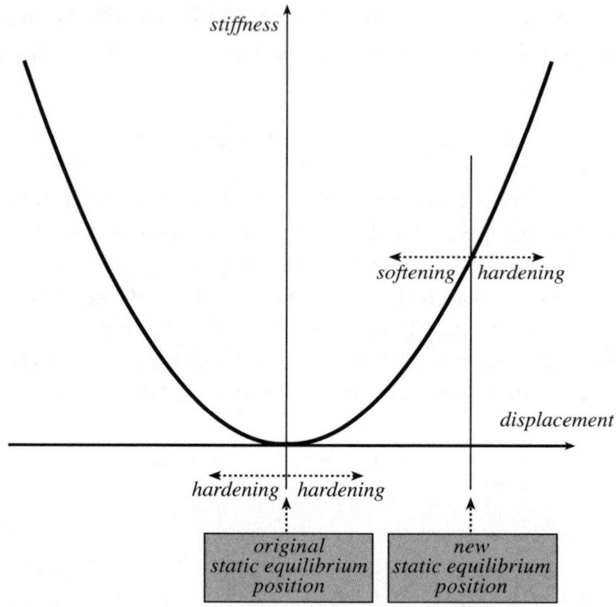

Figure 2.11 The stiffness characteristic of the quasi-zero stiffness isolator. The effects of mistuning the system so that the static equilibrium position is shifted is illustrated.

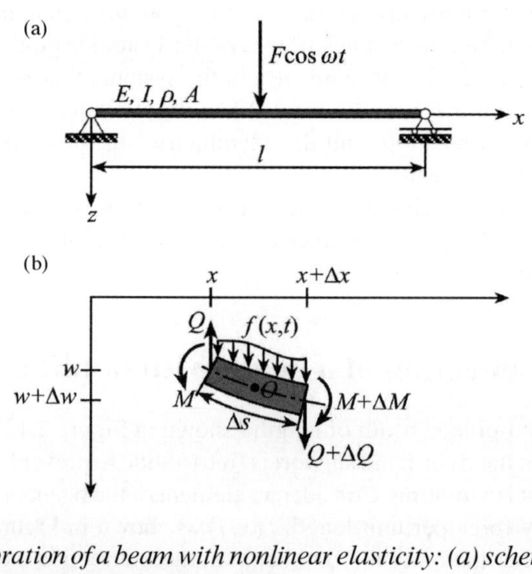

Figure 2.12 Vibration of a beam with nonlinear elasticity: (a) schematic of the beam; (b) section of the beam of length Δs.

dynamic equilibrium of the element of the vibrating beam can be determined by first by taking moments about point O, to give

$$(M + \Delta M) - M + Q\Delta x = 0 \tag{2.8.1a}$$

in which it has been assumed that the product $\Delta Q\Delta x$ is negligibly small, and secondly by summing forces in the z direction to give

$$(Q + \Delta Q) - Q + f(x, t) = \rho A \Delta s \frac{\partial^2 w}{\partial t^2} \tag{2.8.1b}$$

where the shear force and bending moment are given by Q and M, respectively; the density and cross-sectional area of the beam are given by ρ and A, respectively, so that ρA is the beam mass per unit length. Taking the limit as $\Delta x \to 0$, Equations (2.8.1a) and (2.8.1b) can be written as

$$\frac{\partial M}{\partial x} + Q = 0 \tag{2.8.2a}$$

$$\rho A \frac{\partial s}{\partial x} \frac{\partial^2 w}{\partial t^2} - \frac{\partial Q}{\partial x} = f(x, t) \tag{2.8.2b}$$

For large rotations of the beam the bending moment and curvature are related by [8],

$$M = EI \frac{\frac{\partial^2 w}{\partial x^2}}{\left(1 + \left(\frac{\partial w}{\partial x}\right)^2\right)^{\frac{3}{2}}} \tag{2.8.3a}$$

where E is the Young's modulus and I the second moment of area of the beam, respectively. If $(\partial w/\partial x)^2 \ll 1$, then Equation (2.8.3a) can be approximated by

$$M = EI \left(1 - \frac{3}{2}\left(\frac{\partial w}{\partial x}\right)^2\right) \frac{\partial^2 w}{\partial x^2} \tag{2.8.3b}$$

Note that $(\partial w/\partial x)^2$ is neglected in the denominator of Equation (2.8.3a) then $M = EI\partial^2 w/\partial x^2$ which is the linear relationship between the bending moment and curvature. Differentiating Equation (2.8.2a) with respect to x, substituting for M from Equation (2.8.3b) and combining this with Equation (2.8.2b) results in

$$\rho A \frac{\partial s}{\partial x} \frac{\partial^2 w}{\partial t^2} + EI \frac{\partial^4 w}{\partial x^4} - \frac{3}{2} EI \frac{\partial^2}{\partial x^2} \left(\left(\frac{\partial w}{\partial x}\right)^2 \frac{\partial^2 w}{\partial x^2} \right) = f(x, t) \tag{2.8.4}$$

Note that the first two terms on the left-hand side of Equation (2.8.4) are related to the linear behaviour of the Euler–Bernoulli beam, and the third term is due to the nonlinear relationship between the bending moment and the curvature of the beam. To determine the approximate equation of motion for the first mode of the beam, its displacement is written as the product of its mode shape $\varphi(x)$ and the modal amplitude $q(t)$, i.e., $w(x,t) = \varphi(x)q(t)$, where

$$\varphi(x) = \sin\left(\frac{\pi x}{L}\right) \tag{2.8.5}$$

in which L is the distance between the beam supports. The term $f(x,t)$ is the time-dependent force per unit length acting on the beam. If the applied force is assumed to be a time harmonic point force acting at the centre then $f(x,t)$ can be written as $f(x,t) = F\delta(x - l/2)\cos\omega t$, where δ is the Dirac delta function, which has units of m^{-1} in this case. Equation (2.8.4) is then multiplied by Equation (2.8.5) and integrated over the distance between the beam supports L, to give the approximate equation of motion for the first mode of vibration in terms of L. It is given by

$$m\frac{d^2q}{dt^2} + k_1 q - k_3 q^3 = F\cos\omega t \tag{2.8.6a}$$

where $m = \rho Al/2$, $k_1 = EI\pi^4/(2L^3)$, $k_3 = 3\pi^6 EI/\left(16L^5\right)$. However, note that L is not constant. It is, in fact, a function of the amplitude of the displacement and is given approximately by $L/l = \frac{1}{2}\left(1 + \left(1 - (q\pi/2l)^2\right)^{\frac{1}{2}}\right)$ [9]. Substituting this into Equation (2.8.6a) and neglecting powers of q greater than 3 results in coefficients of $k_1 = EI\pi^4/(2l^3)$, $k_3 = -3\pi^6 EI/\left(16l^5\right)$.

Thus, because the cubic term is now positive, the beam exhibits hardening stiffness for large deflections. Equation (2.8.6a) can be nondimensionalised by normalising the displacement q by the length of the beam l so that $\tilde{y} = q/l$ and by letting $\tilde{t} = \omega_n t$ to give

$$\ddot{\tilde{y}} + \tilde{y} + \gamma\tilde{y}^3 = \tilde{F}\cos\Omega\tilde{t} \tag{2.8.6b}$$

in which $\Omega = \omega/\omega_n$, $\omega_n = \sqrt{k_1/m}$, $\tilde{F} = F/(k_1 l)$, $\gamma = l^2 k_3/k_1$.

2.9 Beam with nonlinear stiffness due to inplane tension

Consider the beam of length l shown in Figure 2.13a. An initial tension (compression) T_s is applied to the beam which results in the right-hand support of the beam being shifted to the right (left) by Δl. An element of the beam of length Δs is shown in Figure 2.13b in which the shear force and bending moment are given by Q and M, respectively, and $T = T_s + T_d$, where $T_s = EA\Delta l/l$ and T_d is the inplane tension due to the vibration of the beam. As before, it is assumed that deflection of the beam $w(x,t)$ is small so that $\Delta s \approx \Delta x$ and motion of the beam in the x direction can be ignored. It is further assumed that the rotary inertia of the beam can be ignored.

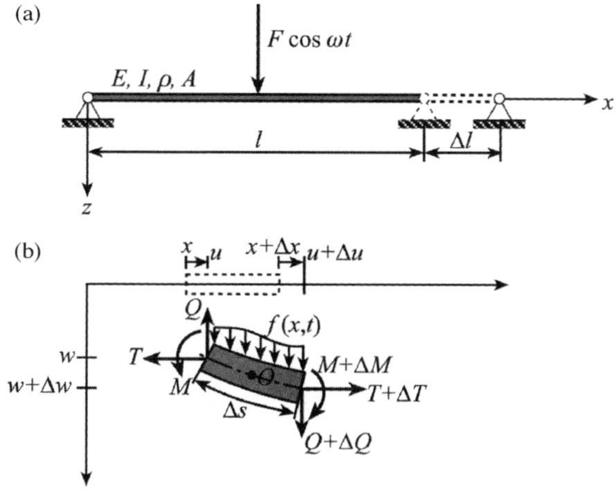

Figure 2.13 Vibrations of a beam with inplane tension: (a) schematic of the beam, (b) section of the beam of length Δs.

The equations of dynamic equilibrium of the element of the vibrating beam can be determined by first taking moments about point O, to give

$$(M+\Delta M)-M+Q(\Delta x+\Delta u)-T\Delta w=0 \tag{2.9.1a}$$

in which it has been assumed that $\Delta T\Delta w$ is negligibly small, and then summing forces in the z direction to give

$$(Q+\Delta Q)-Q+f(x,t)=\rho A\Delta x\frac{\partial^2 w}{\partial t^2} \tag{2.9.1b}$$

As before, the density and cross-sectional area of the beam are given by ρ and A, respectively, so that ρA is the beam mass per unit length. Finally, summing forces in the x direction gives

$$(T+\Delta T)-T=0 \tag{2.9.1c}$$

Taking the limit as $\Delta x\rightarrow 0$, Equations (2.9.1a–c) can be written as

$$\frac{\partial M}{\partial x}+Q-(T_{\text{s}}+T_{\text{d}})\frac{\partial w}{\partial x}=0 \tag{2.9.2a}$$

$$\frac{\partial Q}{\partial x}=\rho A\frac{\partial^2 w}{\partial t^2}-f(x,t) \tag{2.9.2b}$$

$$\frac{\partial T}{\partial x}=0 \tag{2.9.2c}$$

Note that it is assumed that $\partial u/\partial x \ll 1$ and so is neglected. Differentiating Equation (2.9.2a) with respect to x, noting that $M=EI\partial^2 w/\partial x^2$, where E is the Young's modulus and I the second moment of area of the beam respectively, and combining this with Equation (2.9.2b), gives

$$\rho A \frac{\partial^2 w}{\partial t^2} + EI \frac{\partial^4 w}{\partial x^4} - T \frac{\partial^2 w}{\partial x^2} = f(x, t) \tag{2.9.3}$$

Note that the first two terms on the left-hand side of Equation (2.9.3) are related to the linear behaviour of the Euler–Bernoulli beam, and the third term is due to the inplane tension. The inplane tension due to the inplane strain ε is given by

$$T = EA\varepsilon \tag{2.9.4}$$

where

$$\varepsilon = \frac{\sqrt{(\Delta x + \Delta u)^2 + \Delta w^2} - \Delta x}{\Delta x} \tag{2.9.5a}$$

which in the limit as $\Delta x \to 0$ becomes

$$\varepsilon \approx \frac{\partial u}{\partial x} + \frac{1}{2}\left(\frac{\partial w}{\partial x}\right)^2 \tag{2.9.5b}$$

Note that although $\partial u/\partial x$ was neglected compared to unity, it cannot be neglected compared with $(\partial w/\partial x)^2$. The term containing $(\partial w/\partial x)^2$ is the one that makes the resulting equation of motion nonlinear. Combining Equations (2.9.4) and (2.9.5b) gives

$$\frac{\partial u}{\partial x} = \frac{T}{EA} - \frac{1}{2}\left(\frac{\partial w}{\partial x}\right)^2 \tag{2.9.6}$$

Integrating this over the length of the beam and applying the end condition gives

$$u(l, t) = \Delta l = \frac{Tl}{EA} - \frac{1}{2}\int_0^l \left(\frac{\partial w}{\partial x}\right)^2 dx \tag{2.9.7}$$

from which the tension in the beam can be determined as

$$T = T_s + \frac{EA}{2l}\int_0^l \left(\frac{\partial w}{\partial x}\right)^2 dx \tag{2.9.8}$$

Substituting Equation (2.9.8) into Equation (2.9.3) gives the partial differential equation of motion for the beam:

$$\rho A \frac{\partial^2 w}{\partial t^2} + EI \frac{\partial^4 w}{\partial x^4} - \left(T_s + \frac{EA}{2l}\int_0^l \left(\frac{\partial w}{\partial x}\right)^2 dx\right) \frac{\partial^2 w}{\partial x^2} = f(x, t) \tag{2.9.9}$$

To determine the approximate equation of motion for the first mode of the beam the displacement of the beam is written as the product of its mode shape $\varphi(x)$ and the modal amplitude $q(t)$, i.e., $w(x,t) = \varphi(x)q(t)$, where

$$\varphi(x) = \sin\left(\frac{\pi x}{l}\right) \tag{2.9.10}$$

If it is assumed that the beam is excited by a point harmonic force at the centre such that $f(x,t) = F\delta(x - l/2)\cos\omega t$, and if Equation (2.9.9) is multiplied by Equation (2.9.10) and then integrated over the length of the beam, the approximate equation of motion for the first mode can be determined. It is given by

$$m\frac{d^2q}{dt^2} + k_1 q + k_3 q^3 = F\cos\omega t \tag{2.9.11a}$$

where $m = \rho A l/2$, $k_1 = (1 + T_s l^2/(EI\pi^2))EI\pi^4/(2l^3)$, $k_3 = \pi^4 EA/(8l^3)$.

Note that the static inplane tension only affects the linear stiffness, which also has a component due to bending of the beam. The coefficient of the cubic nonlinear term is due only to inplane stretching as a result of large out-of-plane vibrations. It is positive, which means that the beam behaves as a system with hardening stiffness. Equation (2.9.11a) can be nondimensionalised by normalising the displacement q by the length of the beam so that $\tilde{y} = q/l$ and by letting $\tilde{t} = \omega_n t$ to give

$$\ddot{\tilde{y}} + \tilde{y} + \gamma\tilde{y}^3 = \tilde{F}\cos\Omega\tilde{t} \tag{2.9.11b}$$

where $\Omega = \omega/\omega_n$, $\omega_n = \sqrt{k_1/m}$, $\tilde{F} = F/(k_1 l)$, $\gamma = l^2 k_3/k_1$. If the inplane force is compressive rather than tensile, then this has a softening effect on the linear stiffness terms as can be seen in Equation (2.9.11a), such that the linear natural frequency will be reduced. However, the large amplitude vibrations will effectively increase the dynamic inplane tension, and will therefore, have a hardening effect. This affects the cubic term in Equation (2.9.11a).

2.10 Nonlinear cable vibrations

Consider the uniform cable hanging in its static equilibrium state as shown in Figure 2.14(a).

The equations of static equilibrium of the element of cable of length Δs shown in Figure 2.14(b) are given by

$$(T_s + \Delta T_s)\cos(\theta + \Delta\theta) - T_s\cos\theta = 0 \tag{2.10.1a}$$

$$\rho A g\Delta s + (T_s + \Delta T_s)\sin(\theta + \Delta\theta) - T_s\sin\theta = 0 \tag{2.10.1b}$$

where ρ and A are cable density and cross-sectional area, respectively, so that ρA is the cable mass per unit length, g is the acceleration due to gravity, T_s is the static cable tension. Expanding Equations (2.10.1a,b), noting that $\cos(\Delta\theta) \approx 1$, $\sin(\Delta\theta) \approx \Delta\theta$, $\Delta x/\Delta s = \cos\theta$ and $\Delta y/\Delta s = \sin\theta$ neglecting $\Delta T_s \Delta\theta$ and taking the limit as $\Delta s \to 0$, Equations (2.10.1a) and (2.10.1b) can be written as

$$\frac{d}{ds}\left(T_s\frac{dx}{ds}\right) = 0 \tag{2.10.2a}$$

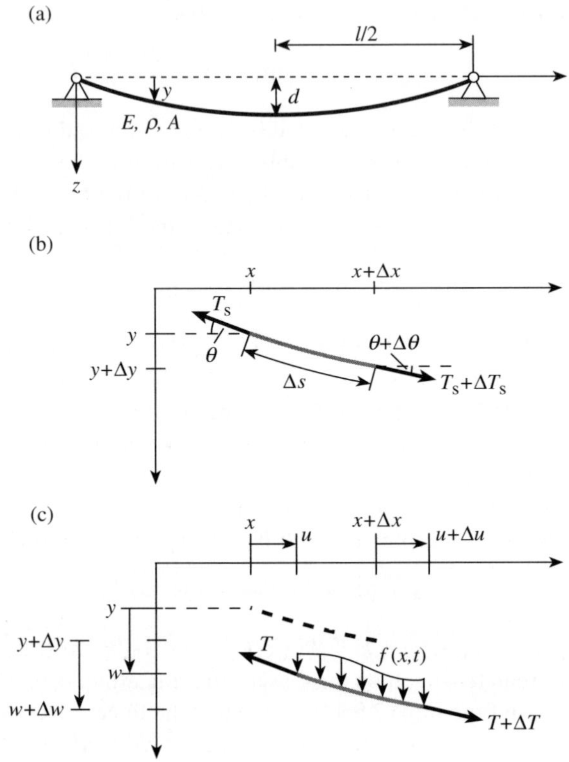

Figure 2.14 Nonlinear vibrations of a cable: (a) schematic of the cable in its static equilibrium position; (b) section of the cable of length Δs in the static equilibrium position; (c) section of the cable in dynamic equilibrium.

$$\rho Ag + \frac{d}{ds}\left(T_s \frac{dy}{ds}\right) = 0 \tag{2.10.2b}$$

It is also assumed that $d/l \ll 1$, so the profile of the cable is parabolic and is given by [10]

$$y = 4d\left(\frac{x}{l} - \left(\frac{x}{l}\right)^2\right) \tag{2.10.3}$$

where l is the span between the supports and d is the cable sag as shown in Figure 2.14(a). Given this assumption, the tension in the cable is approximately equal to its component the horizontal direction so $T_s \approx \rho Agl^2/8d$. Consider now, the cable when it is perturbed from the static equilibrium position, also shown in Figure 2.14(c), where $T = T_s + T_d$ in which T_d is the tension in the cable due to vibration in the vertical direction. The resulting equations obtained from dynamic equilibrium are

$$\frac{\partial}{\partial s}\left((T_s + T_d)\left(\frac{dx}{ds} + \frac{\partial u}{\partial s}\right)\right) = 0 \tag{2.10.4a}$$

$$\frac{\partial}{\partial s}\left((T_s+T_d)\left(\frac{dy}{ds}+\frac{\partial w}{\partial s}\right)\right)+\rho A g+f(x,t)=\rho A\frac{\partial^2 w}{\partial t^2} \tag{2.10.4b}$$

in which it assumed that the inertia force in the x direction is negligible. The components of displacement and the additional tension are functions of both position and time. Considering Equation (2.10.2b), Equation (2.10.4b) becomes

$$\rho A\frac{\partial^2 w}{\partial t^2}-T_s\frac{\partial^2 w}{\partial s^2}-\frac{\partial}{\partial s}\left(T_d\left(\frac{dy}{ds}+\frac{\partial w}{\partial s}\right)\right)=f(x,t) \tag{2.10.5}$$

Now, the additional tension T_d is related to the inplane strain ε in the cable by Hooke's law and is given by

$$T_d=EA\varepsilon \tag{2.10.6}$$

where E and A are the Young's modulus and cross-sectional area of the cable, respectively, and in which

$$\varepsilon=\frac{\sqrt{(dx+\partial u)^2+(dy+\partial w)^2}-\partial s}{\partial s}\approx\frac{\partial u}{\partial x}+\frac{dy}{dx}\frac{\partial w}{\partial x}+\frac{1}{2}\left(\frac{\partial w}{\partial x}\right)^2 \tag{2.10.7}$$

where is has been assumed that $dx/ds\approx 1$. It has also been assumed that $(dy/ds)^2\ll 1$ and $(\partial u/\partial s)^2\ll 1$, so can therefore be neglected. Following a similar procedure to that for the beam with inplane tension, Equations (2.10.6) and (2.10.7) can be combined to give

$$\frac{\partial u}{\partial x}=\frac{T_d}{EA}-\left(\frac{dy}{dx}\frac{\partial w}{\partial x}+\frac{1}{2}\left(\frac{\partial w}{\partial x}\right)^2\right) \tag{2.10.8}$$

Integrating over the length of the cable gives

$$u(l,t)=0=\frac{T_d l}{EA}-\int_0^l\left(\frac{dy}{dx}\frac{\partial w}{\partial x}+\frac{1}{2}\left(\frac{\partial w}{\partial x}\right)^2\right)dx \tag{2.10.9}$$

From which the tension in the cable due to the vibration is given by

$$T_d=\frac{EA}{l}\int_0^l\left(\frac{dy}{dx}\frac{\partial w}{\partial x}+\frac{1}{2}\left(\frac{\partial w}{\partial x}\right)^2\right)dx \tag{2.10.10}$$

which can be substituted into Equation (2.10.5) to give

$$\rho A\frac{\partial^2 w}{\partial t^2}-T_s\frac{\partial^2 w}{\partial x^2}-\frac{EA}{l}\left(\frac{d^2y}{dx^2}+\frac{\partial^2 w}{\partial x^2}\right)\left(\int_0^l\left(\frac{dy}{dx}\frac{\partial w}{\partial x}+\frac{1}{2}\left(\frac{\partial w}{\partial x}\right)^2\right)dx\right)=f(x,t) \tag{2.10.11}$$

Noting the relationship between y and x given in Equation (2.10.3) means that $dy/dx = 4d/l - 8dx/l^2$ and $d^2y/dx^2 = -8d/l^2$, so Equation (2.10.11) becomes

$$\rho A\frac{\partial^2 w}{\partial t^2} - T_s\frac{\partial^2 w}{\partial x^2} - \frac{EA}{l}\left(-\frac{8d}{l^2} + \frac{\partial^2 w}{\partial x^2}\right)\left(\int_0^l\left(\left(\frac{4d}{l} - \frac{8d}{l^2}x\right)\frac{\partial w}{\partial x} + \frac{1}{2}\left(\frac{\partial w}{\partial x}\right)^2\right)dx\right) = f(x,t) \tag{2.10.12}$$

Note that because the cable is fixed at each end $w(0,t) = w(l,t) = 0$, which means that

$$\int_0^l \frac{\partial^2 w}{\partial x^2} w dx = -\int_0^l \left(\frac{\partial w}{\partial x}\right)^2 dx, \quad \int_0^l x\frac{\partial w}{\partial x} dx = -\int_0^l w dx, \quad \int_0^l \frac{\partial w}{\partial x} dx = 0, \text{ so}$$

Equation (2.10.12) can be written as

$$\rho A\frac{\partial^2 w}{\partial t^2} - T_s\frac{\partial^2 w}{\partial x^2} + \frac{64EAd^2}{l^5}\int_0^l w dx - \frac{8EAd}{l^3}\left(\frac{\partial^2 w}{\partial x^2}\int_0^l w dx - \frac{1}{2}\int_0^l \left(\frac{\partial w}{\partial x}\right)^2 dx\right)$$

$$-\frac{EA}{2l}\frac{\partial^2 w}{\partial x^2}\int_0^l \left(\frac{\partial w}{\partial x}\right)^2 dx = f(x,t) \tag{2.10.13}$$

It is now assumed that the cable is excited harmonically at frequency ω such that it responds predominantly in its first mode of vibration. The displacement of the cable can thus be written as the product of its mode shape $\varphi(x)$ and the modal amplitude $q(t)$, i.e., $w(x,t) = \varphi(x)q(t)$. Following the introduction of the nondimensional parameter $\tilde{x} = x/l$, Equation (2.10.13) is multiplied by $\varphi(\tilde{x})$and then integrated over the nondimensional length of the cable to give

$$m\frac{d^2q}{dx^2} + k_1q + k_2q^2 + k_3q^3 = F\cos\omega t \tag{2.10.14}$$

where

$$m = \rho Al\int_0^1 \phi^2 d\tilde{x},$$

$$k_1 = \frac{T_s}{l}\left(\int_0^1 \left(\frac{d\phi}{d\tilde{x}}\right)^2 d\tilde{x} + 64\mu\tilde{d}^2\left(\int_0^1 \phi d\tilde{x}\right)^2\right),$$

$$k_2 = \frac{12EAd}{l^3}\int_0^1 \left(\frac{d\phi}{d\tilde{x}}\right)^2 d\tilde{x}\int_0^1 \phi d\tilde{x},$$

$$k_3 = \frac{EA}{2l^3}\left(\int_0^1 \left(\frac{d\phi}{d\tilde{x}}\right)^2 d\tilde{x}\right)^2,$$

$$F = l\int_0^1 p(\tilde{x}, t))\phi d\tilde{x}$$

and $64\mu\tilde{d}^2$ is Irvine's parameter (normally given the symbol λ^2), where $\mu = EA/T_s$, $\tilde{d} = d/l$ and $\tilde{d} \ll 1$ [10,11]. Note that because of the asymmetry of the cable, in that it is stiffer when it is forced downwards compared to when it is forced upwards, there is a quadratic term in the restoring force as well as the linear and the cubic terms. The square of the linear natural frequency of the cable is given by

$$\omega_n^2 = \frac{k_1}{m} = \frac{T_s}{\rho A l^2}\left(\frac{\int_0^1 \left(\frac{d\phi}{d\tilde{x}}\right)^2 d\tilde{x} + 64\mu\tilde{d}^2\left(\int_0^1 \phi d\tilde{x}\right)^2}{\int_0^1 \phi^2 d\tilde{x}}\right) \tag{2.10.15}$$

which can also be determined from the transcendental equation [10]

$$\tan\left(\frac{\pi}{2}\frac{\omega_n}{\omega_0}\right) = \left(\frac{\pi}{2}\frac{\omega_n}{\omega_0}\right) - \frac{1}{16\mu\tilde{d}^2}\left(\frac{\pi}{2}\frac{\omega_n}{\omega_0}\right)^3 \tag{2.10.16}$$

where the square of the fundamental natural frequency of a taught linear string with the same tension T_s is given by

$$\omega_0 = \frac{\pi^2 T_s}{\rho A l^2} \tag{2.10.17}$$

It is clear that the tension has an effect on the linear natural frequency directly through the term T_s. It also has an effect through the term $\mu\tilde{d}^2$ which is proportional to T_s. Equation (2.10.16) is plotted in Figure 2.15, which can be used to determine the linear natural frequency of the cable. It can be seen that when $\mu\tilde{d}^2 \ll 1$ such that the inplane tension dominates, the cable has a similar natural frequency to a taught string.

When $\mu\tilde{d}^2 \to 1$, the inplane stiffness of the cable becomes increasingly dominant, and has a profound effect on the natural frequency of the cable. The mode shape is given by [11]

$$\phi(x) = \frac{1}{1-\sec\left(\frac{\pi\omega_n}{2\omega_o}\right)}\left(1-\tan\left(\frac{\pi\omega_n}{2\omega_o}\right)\sin\left(\frac{\pi\omega_n}{\omega_o}x\right)-\cos\left(\frac{\pi\omega_n}{\omega_o}x\right)\right) \tag{2.10.18}$$

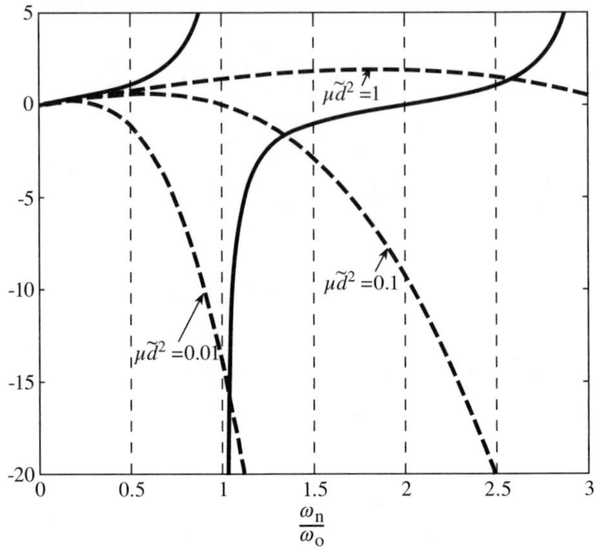

Figure 2.15 Graphical solution to the transcendental equation given in Equation (2.10.16). The solid lines correspond to the left-hand side of this equation, and the dashed lines correspond to the right-hand side of this equation for a range of values of $\mu\tilde{d}^2$.

The effect of $\mu\tilde{d}^2$ on the mode shape corresponding to the fundamental natural frequency can be seen in Figure 2.16. The parameter $\tilde{d}$ has been fixed at 0.1, and the static tension is varied compared to the inplane stiffness. It can be seen that when the tension is dominant, the mode shape is similar to that for a taught string, which is described by a simple sine function. When the effect of the tension diminishes, the mode shape changes considerably, as shown in Figure 2.16.

Equation (2.10.14) can be nondimensionalised two ways; one is by normalising the displacement q by the length of the cable l, so that $\tilde{y} = q/l$ and by letting $\tilde{t} = \omega_n t$ to give

$$\ddot{\tilde{y}} + \tilde{y} + \beta\tilde{y}^2 + \gamma\tilde{y}^3 = \tilde{F}\cos\Omega\tilde{t} \tag{2.10.19}$$

where

$$\beta = \frac{12}{\pi^2}\mu\tilde{d}\left(\frac{\omega_o}{\omega_n}\right)^2\frac{I_2}{I_m}, \quad \gamma = \frac{1}{2\pi^2}\mu\left(\frac{\omega_o}{\omega_n}\right)^2\frac{I_3}{I_m}, \quad \tilde{F} = \frac{F}{k_1 l}$$

in which

$$I_2 = \int_0^1 \phi d\tilde{x} \int_0^1 \left(\frac{\partial\phi}{\partial\tilde{x}}\right)^2 d\tilde{x}, \; I_3 = \left(\int_0^1 \left(\frac{\partial\phi}{\partial\tilde{x}}\right)^2 d\tilde{x}\right)^2, \; I_m = \int_0^1 \phi^2 d\tilde{x}$$

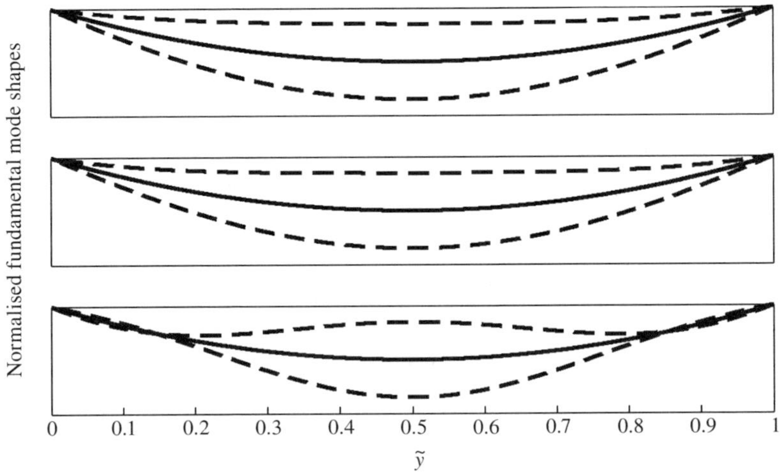

Figure 2.16 Shape of the cable when it is vibrating in its fundamental mode, for various values of $\mu\tilde{d}^2$: static position of the cable (solid line); mode shape of the cable (dashed), $\tilde{d} = 0.1$: a) $\mu = 1$; b) $\mu = 10$; c) $\mu = 100$.

Another way to nondimensionalise is to normalise the displacement q to the sag d, i.e., $\tilde{y} = q/d$. In this case

$$\beta = \frac{12}{\pi^2}\mu\tilde{d}^2\left(\frac{\omega_o}{\omega_n}\right)^2\frac{I_2}{I_m}, \quad \gamma = \frac{1}{2\pi^2}\mu\tilde{d}^2\left(\frac{\omega_o}{\omega_n}\right)^2\frac{I_3}{I_m}, \quad \tilde{F} = \frac{F}{k_1 d}$$

It is clear that it is more convenient to nondimensionalise the equation the second way as Irvine's parameter $64\mu\tilde{d}^2$ appears in both the quadratic and the cubic terms. However, this form of the equation does not reduce to a taught (straight) cable because in this case $d \to 0$. The values of the integrals given above, and the nonlinear parameters are given in Table 2.1, for values of $\mu\tilde{d}^2 = 0.01$, 1 and 10. The values when the mode shape is approximated by a Sine function are also given for reference.

The characteristics of the nonlinearities associated with cables, i.e., the quadratic term due to the asymmetry of the system and the cubic term due to the inplane stiffness

Table 2.1 Evaluation of the integrals and nondimensional coefficients for the quadratic and cubic terms in the Helmholtz–Duffing equation describing the vibration of a cable.

$\frac{\omega_n}{\omega_o}$	$\mu\tilde{d}^2$	I_m	I_2	I_3	β	γ
1	1 (approx)	$\frac{1}{2}$	π (3.141)	$\frac{\pi^4}{4}$ (24.35)	$\frac{6}{25\pi}$ (0.0764)	$\frac{\pi^2}{400}$ (0.0247)
1.05	0.01	0.4978	3.118	24.15	0.0808	0.0261
1.35	1	0.4803	2.945	22.71	1.146	0.3683
2.60	10	0.2661	2.669	47.27	57.93	42.75

also occur in curved beams and shells. Amabali has presented comprehensive theoretical and practical results for thin shells [12]. The same characteristics can also be found in micro- and nanoresonators commonly used in sensing devices [13]. In these micro- and nanosystems, the quadratic term is often due to the asymmetrical electromechanical coupling, and the cubic term is due to geometrical nonlinearity.

Dynamic behaviour of the asymmetric oscillators modelled by Equation (2.10.19) is investigated in detail in Chapter 8.

2.11 Nonlinear electrical circuit

The electrical circuit of interest is shown in Figure 2.17(a). It consists of a nonlinear inductor and a linear capacitor; the circuit would also have some inherent resistance and hysteresis, but this is neglected here because the focus is on nonlinearity. This circuit was studied many years ago before Duffing wrote his book, by Martienssen [14], Biermanns [15] and later by Hayashi [16] and Ueda [17]. It was also included in Duffing's book, the relevant part of which has been translated in the Appendix. When the inductor is considered to be linear, the relationship between the current i, and the magnetic flux ϕ is given by $i = \phi/L$, in which L is the inductance. However, when the inductor exhibits nonlinear behaviour the inductance is not a constant but is a function of

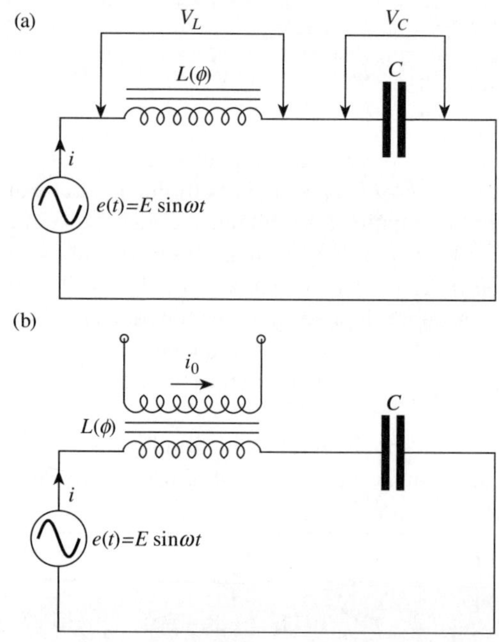

Figure 2.17 Electrical circuits containing a saturating nonlinear inductor and a linear capacitor: (a) with nonlinear inductance that has a symmetric characteristic; (b) with nonlinear inductance that has an asymmetric characteristic.

the current (in a similar way to stiffness being a function of displacement in a mechanical system). The potential difference V_L, across the inductor is given by [16]

$$V_L = N\frac{d\phi}{dt} \tag{2.11.1}$$

where N is the number of turns of the inductor coil and ϕ is the magnetic flux in the inductor core. The potential difference across the capacitance, V_C is given by

$$V_C = \frac{q}{C} \tag{2.11.2}$$

where C is capacitance and $i = dq/dt$ is electrical charge on the capacitor. The sum of the potential differences across the inductor and capacitor must be equal to the applied voltage $e(t)$, so that

$$N\frac{d\phi}{dt} + \frac{q}{C} = e(t) \tag{2.11.3}$$

The nonlinear relationship between the current and the magnetic flux can be represented by several different functions. However, a simple representation is as a power series in ϕ as in Biermanns [15] and Hayashi [16]

$$i = a_1\phi + a_3\phi^3 + a_5\phi^5 + \cdots \tag{2.11.4}$$

Assuming a harmonic supply voltage of the form $e(t) = E\sin\omega t$ and differentiating Equation (2.11.3) with respect to time gives

$$N\frac{d^2\phi}{dt^2} + \frac{i}{C} = \omega E\cos\omega t \tag{2.11.5}$$

Truncating Equation (2.11.4) to the third order, and substituting for the current in Equation (2.11.5) from Equation (2.11.4) gives

$$\frac{d^2\phi}{dt^2} + \frac{1}{L_1C}\phi + \frac{1}{L_3C}\phi^3 = \frac{\omega E}{N}\cos\omega t \tag{2.11.6}$$

where $1/L_n = a_n/N$, $n = 1, 3$. Note that the linear natural frequency of the system is given by $\omega_n = \sqrt{1/(L_1C)}$. Equation (2.11.6) can be written in nondimensional form as

$$\ddot{\tilde{y}} + \tilde{y} + \gamma\tilde{y}^3 = \Omega\cos\Omega\tilde{t} \tag{2.11.7}$$

where $\Omega = \omega/\omega_n$, $\tilde{t} = \omega_n t$, $\tilde{y} = \phi/\Phi$, $\gamma = \Phi^2 L_1/L_3$, $\Phi = E/(N\omega_n)$.

2.11.1 The electrical circuit studied by Ueda

The circuit shown in Figure 2.17(a) can be modified so that it can be described by the equation studied by Yoshisuki Ueda, who discovered *chaos* in a certain type of Duffing oscillator (Some related facts about this are given in Section 1.6.2). Here, damping (resistance) is again neglected for clarity.

If the inductor has a secondary coil with a biasing DC current i_0 passed through it as shown in Figure 2.17(b), the relationship between the current and the magnetic flux can be approximated by

$$i = a_1\phi + a_2\phi^2 + a_3\phi^3 \tag{2.11.8}$$

Note that this is now asymmetric. The resulting nondimensional equation describing the system can be determined in a similar way to that for Equation (2.11.7), to give

$$\ddot{\tilde{y}} + \tilde{y} + \beta\tilde{y}^2 + \gamma\tilde{y}^3 = \Omega\cos\Omega\tilde{t} \tag{2.11.9}$$

where $\gamma = \Phi L_1/L_2$ in which $1/L_2 = a_2/N$. If the system parameters are adjusted so that $\beta = \sqrt{3\gamma}$ and the coordinate shifted so that $x = y + \beta/(3\gamma)$ then Equation (2.11.9) can be written as

$$\ddot{x} + \gamma x^3 = \frac{1}{3\sqrt{3\gamma}} + \Omega\cos\Omega\tilde{t} \tag{2.11.10}$$

which is an oscillator excited by a constant and a harmonic term. Note that it has a pure cubic nonlinearity, and is similar in form to the equation describing the QZS isolator discussed in Section . If γ is set to unity, Ω is sufficiently large and set to $\tilde{B}$, and a damping term included that is proportional to the rate of change of x then Equation (2.11.10) can be written as

$$\ddot{x} + c\dot{x} + x^3 = \tilde{B}\cos\Omega\tilde{t} \tag{2.11.11}$$

which is the equation having the form of the one that Ueda studied when he discovered chaos in the system [17,18]. Some of these results are reported also in Section 5.4.3.

2.12 Summary

In this chapter, several physical examples that can be modelled by the Duffing equation have been described. The physical systems have included simple springs that are connected in such a way that the nonlinearity is due to geometry. They have also included vibration isolators and the vibration of the first mode of distributed parameters systems such as beams and cables. A simple electrical nonlinear electrical system has also been described.

The systems presented illustrate the physical phenomena that gives rise to several types of geometric nonlinearity. These comprise systems with hardening nonlinearity, corresponding to positive linear and positive nonlinear stiffness; softening nonlinearity, corresponding to positive linear and negative nonlinear stiffness; positive linear-negative cubic stiffness nonlinearity, which corresponds to a system with a double/two/twin-well potential; finally, it includes the case of pure nonlinearity, when there is only a nonlinear stiffness term, with no linear stiffness term.

Although the physical systems discussed in this chapter can be represented by various forms of the Duffing equation, it should be noted that the nondimensional coefficients in the equations are related to quite different physical properties in the systems they model.

References

[1] L.N. Virgin, *Introduction to Experimental Nonlinear Dynamics: A Case Study in Mechanical Vibration*, Cambridge University Press, Cambridge, 2000.

[2] F.C. Moon and P.I. Holmes, A magnetoelastic strange attractor. *Journal of Sound and Vibration*, 65, 275–296, 1979.

[3] F.C. Moon, *Chaotic Motions: An Introduction for Applied Scientists and Engineers*, John Wiley & Sons, New York, 1987.

[4] E.I. Rivin, *Passive Vibration Isolation*, ASME Press, New York, 2001.

[5] R.A. Ibrahim, Recent advances in nonlinear passive vibration isolators. *Journal of Sound and Vibration*, 314, 371–452, 2008.

[6] P. Alabuzhev, A. Gritchin, L. Kim, G. Migirenko, V. Chon, P. Stepanov, *Vibration Protecting and Measuring Systems with Quasi-Zero Stiffness*, Hemisphere Publishing, New York, 1989.

[7] A. Carrella, M.J. Brennan, T.P. Waters, Static analysis of a passive vibration isolator with Quasi-Zero Stiffness Characteristic. *Journal of Sound and Vibration*, 301, 678–689, 2007.

[8] D. Wagg, S. Neild, *Nonlinear Vibration with Control for Flexible and Adaptive Structures*, Springer, Dordrecht, Netherlands, 2009.

[9] A. Abolfathi, M.J. Brennan and T.P. Waters, On the large deflection of a simply supported beam, *ISVR Technical Memorandum*, University of Southampton, Southampton, 998, 2010.

[10] H.M. Irvine, T.K Caughey, The linear theory of free vibrations of a suspended cable. *Proceedings of the Royal Society A*, 341, 299–315, 1974.

[11] A. Luongo, G. Rega, F. Vestroni, Planar non-linear free vibrations of an elastic cable. *International Journal of Non-Linear Mechanics*, 19, 39–52, 1984.

[12] M. Amabili, *Nonlinear Vibrations and Stability of Shells and Plates*, Cambridge University Press, Cambridge, 2008.

[13] J.F. Rhoads, S.W. Shaw, K.L. Turner, Nonlinear dynamics and its applications in Micro- and nanoresonators. *Journal of Dynamic Systems, Measurement, and Control*, 132/ 034001, 1–13, 2010.

[14] V.O. Martienssen, Über neue, resonanzerscheinungen in wechselstromkreisen. *Physik Zeitschrift – Leipz*, 11, 448–460, 1910.

[15] J. Biermanns, Der schwingungskreis mit heisenhaltiger induktivität. *Archiv für Elektrotechnik*, 345–353, 1915.

[16] C. Hayashi, *Nonlinear Oscillations in Physical Systems*, McGraw-Hill Book Company, New York, 1964.

[17] Y. Ueda, Random phenomena resulting from nonlinearity in the system described by Duffing's equation. *International Journal of Non-linear Mechanics*, 20, 481–491, 1985.

[18] Y. Ueda, Randomly transitionally phenomena in the system governed by Duffing's equation. *Journal of Statistical Physics*, 20, 181–196, 1979.

3

Free vibration of a Duffing oscillator with viscous damping

Hiroshi Yabuno
Faculty of Science and Technology, Keio University, Japan

3.1 Introduction

In this chapter, *free vibration* of the Duffing oscillator described by the following nondimensional equation

$$\ddot{y}+2\zeta\dot{y}+\mu\dot{y}^3+\alpha y+\gamma y^3=0 \tag{3.1.1}$$

is investigated, where y, ζ, μ, α and γ are the displacement, damping ratio, nonlinear damping parameter, linear stiffness parameter and nonlinear (cubic) stiffness parameter (for brevity, the last two parameters are subsequently referred to as the linear stiffness and the nonlinear stiffness, respectively). This system exhibits many qualitatively different phenomena, depending on the linear and nonlinear parameters. Even in the linearised system

$$\ddot{y}+2\zeta\dot{y}+\alpha y=0 \tag{3.1.2}$$

the system undergoes dynamic and static instability through *local bifurcations*. Divergence (buckling) is produced when the linear stiffness α changes from being positive to negative (which is hereafter called negative stiffness). Furthermore,

The Duffing Equation: Nonlinear Oscillators and their Behaviour, First Edition. Edited by Ivana Kovacic and Michael J. Brennan. Published 2011 by John Wiley & Sons, Ltd.

self-excited oscillation occurs when the damping ratio ζ changes from being positive to negative (which is hereafter called negative damping).

The nonlinear stiffness γ affects the postbuckling behaviour, i.e., the existence of nontrivial *fixed points* and their *stability*. Also, the nonlinear damping parameter μ influences the existence of the steady-state response for the self-excited oscillation and determines the magnitude of the steady-state amplitude and its stability. In particular if $\gamma = 0$, Equation (3.1.1) corresponds to so-called Rayleigh's equation [1] given by

$$\ddot{y} - \varepsilon\left(\dot{y} - \dot{y}^3\right) + \omega_0^2 y = 0 \tag{3.1.3}$$

where $2\zeta = -\varepsilon$, $\mu = \varepsilon$, and $\alpha = \omega_0^2$. Further transformation $v = \sqrt{3}\dot{y}$ yields the van der Pol equation [1] as

$$\ddot{v} - \varepsilon\left(1 - v^2\right)\dot{v} + \omega_0^2 v = 0 \tag{3.1.4}$$

Equations (3.1.3) and (3.1.4) have been analysed for many years as they are important models for self-excited oscillations [2].

As for static instability, the case with nontrivial fixed points is very interesting. In particular, negative damping induces complex dynamics, i.e., *global bifurcations* and some types of *periodic orbits* that enclose one fixed point or three fixed points. By using *phase portraits* of the unperturbed *Hamiltonian system*, qualitative analysis is performed to investigate the global nonlinear dynamics.

The reminder of this chapter is organised as follows. Local analysis of the fixed points is conducted in Sections 3.2 and 3.3. The variation of the fixed points with the nonlinear stiffness is examined, and the static and dynamic instability in the neighbourhood of each fixed point is analysed. The nonlinear characteristics of the local bifurcations near the fixed points are also examined. Using a relationship between the local bifurcations, the global dynamics are analysed from a qualitative point of view in Sections 3.4 and 3.5. The variation of dynamics depending on the system parameters is studied and it is shown that periodic orbits surrounding the nontrivial and trivial fixed points are produced through global bifurcations (*homoclinic and heteroclinic bifurcations*).

3.2 Fixed points and their stability

In this section, the fixed points are determined and their stability is examined by performing local analysis. First, Equation (3.1.1) is expressed in terms of state variables $y_1 = y$, $y_2 = \dot{y}$ as

$$\dot{y}_1 = y_2 \tag{3.2.1}$$

$$\dot{y}_2 = -\alpha y_1 - 2\zeta y_2 - \gamma y_1^3 - \mu y_2^3 \tag{3.2.2}$$

which can also be written down as

$$\frac{d\mathbf{y}}{dt} = \mathbf{G}(\mathbf{y}) \tag{3.2.3}$$

$$\mathbf{y} = \begin{bmatrix} y_1 \\ y_2 \end{bmatrix}, \quad \mathbf{G}(\mathbf{y}) = \begin{bmatrix} g_1(y_1, y_2) \\ g_2(y_1, y_2) \end{bmatrix} \tag{3.2.4}$$

where

$$g_1(y_1, y_2) = y_2, \quad g_2(y_1, y_2) = -\alpha y_1 - 2\zeta y_2 - \gamma y_1^3 - \mu y_2^3 \tag{3.2.5}$$

Putting $\dot{y}_1 = \dot{y}_2 = 0$ gives the *equilibrium* equations

$$0 = y_{2\mathrm{st}},\ 0 = -\alpha y_{1\mathrm{st}} - 2\zeta y_{2\mathrm{st}} - \gamma y_{1\mathrm{st}}^3 - \mu y_{2\mathrm{st}}^3 \tag{3.2.6}$$

where $(y_{1\mathrm{st}}, y_{2\mathrm{st}})$ denotes fixed points. Therefore, at the fixed points $y_{2\mathrm{st}} = 0$ and y_{1st} satisfies

$$y_{1\mathrm{st}}\left(\alpha + \gamma y_{1\mathrm{st}}^2\right) = 0 \tag{3.2.7}$$

In the case of $\alpha\gamma > 0$, there is only a trivial fixed point $(y_{1\mathrm{st}}, y_{2\mathrm{st}}) = (0, 0)$. In the case of $\alpha\gamma < 0$, there are two nontrivial fixed points in addition to the trivial one. The nontrivial fixed points are $(y_{1\mathrm{st}}, y_{2\mathrm{st}}) = (y_{\mathrm{st}+}, 0)$ and $(y_{1\mathrm{st}}, y_{2\mathrm{st}}) = (y_{\mathrm{st}-}, 0)$, where $y_{\mathrm{st}+} = \sqrt{-\alpha/\gamma}$ and $y_{\mathrm{st}-} = -\sqrt{-\alpha/\gamma}$.

The local stability of these fixed points is now examined. Substituting $y_1(t) = y_{1\mathrm{st}} + \Delta y_1(t)$ $(|\Delta y_1| << 1)$ and $y_2(t) = y_{2\mathrm{st}} + \Delta y_2(t)$ $(|\Delta y_2| << 1)$ into Equations (3.2.1) and (3.2.2), gives

$$\Delta\dot{y}_1 = \Delta y_2,\ \Delta\dot{y}_2 = -(\alpha + 3\gamma y_{1\mathrm{st}}^2)\Delta y_1 - \zeta\,\Delta y_2 \tag{3.2.8}$$

or

$$\frac{d\Delta\mathbf{y}}{dt} = \mathbf{J_G}\Delta\mathbf{y} \tag{3.2.9}$$

where

$$\Delta\mathbf{y} = \begin{bmatrix} \Delta y_1 \\ \Delta y_2 \end{bmatrix}, \quad \mathbf{J_G} = \begin{bmatrix} 0 & 1 \\ -(\alpha + 3\gamma y_{1\mathrm{st}}^2) & -2\zeta \end{bmatrix} \tag{3.2.10}$$

$\mathbf{J_G}$ is the *Jacobian* of $\mathbf{G}$ at $\mathbf{y} = \mathbf{y}_{\mathrm{st}} = [y_{1\mathrm{st}}\ 0]^T$, where T denotes the transpose. Equation (3.2.8) can be rewritten as the second-order system of the variable Δy_1

$$\Delta\ddot{y}_1 + 2\zeta\Delta\dot{y}_1 + (\alpha + 3\gamma y_{1\mathrm{st}}^2)\Delta y_1 = 0 \tag{3.2.11}$$

and the characteristic equation is

$$\lambda^2 + 2\zeta\lambda + \alpha + 3\gamma y_{1\mathrm{st}}^2 = 0 \tag{3.2.12}$$

The solution of this equation gives the eigenvalues λ_1 and λ_2 that determine the stability of the fixed points. In the case when $\zeta > 0$ $(\zeta < 0)$, Δy_1 has the dynamics of a second-order system with positive (negative) damping. For $\alpha + 3\gamma y_{1\mathrm{st}}^2 > 0$ and

$\alpha + 3\gamma y_{1st}^2 < 0$, Δy_1 has the dynamics of a second-order system with positive and negative stiffness, respectively.

3.2.1 Case when the nontrivial fixed points do not exist ($\alpha\gamma > 0$)

In this case, only the trivial fixed point $(y_{1st}, y_{2st}) = (0, 0)$ exists and the characteristic equation is expressed as

$$\lambda^2 + 2\zeta\lambda + \alpha = 0 \tag{3.2.13}$$

3.2.1.1 Stability of the trivial fixed point in the case of positive linear and positive nonlinear stiffness ($\alpha > 0, \gamma > 0$)

As the damping ratio ζ changes, the eigenvalues of the characteristic equation change, too. Setting $\alpha = \omega_0^2$, the dependence of the stability on ζ is examined:

1. $\zeta < -\omega_0$: There are two positive eigenvalues and the trivial fixed point is an *unstable node*, as shown in Figure 3.1(a).
2. $\zeta = -\omega_0$: There are multiple positive eigenvalues and the trivial fixed point is an *unstable inflected node*.
3. $-\omega_0 < \zeta < 0$: There are complex conjugate eigenvalues with a positive real part and because the trivial fixed point is an unstable *focus*, depicted in Figure 3.1(b), self-excited oscillation is produced.

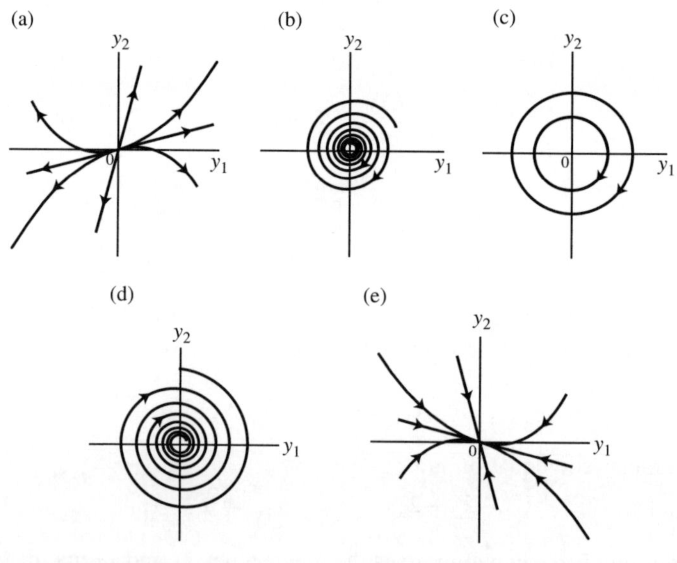

Figure 3.1 Phase portraits for the case $\alpha = \omega_0^2 > 0, \gamma > 0$: (a) $\zeta < -\omega_0$; (b)$-\omega_0 < \zeta < 0$; (c) $\zeta = 0$; (d) $0 < \zeta < \omega_0$ (e) $\zeta > \omega_0$.

4. $\zeta = 0$: There are pure imaginary complex conjugate eigenvalues and since the trivial fixed point is a *centre*, illustrated in Figure 3.1(c), y is oscillatory with no damping.

5. $0 < \zeta < \omega_0$: There are complex conjugate eigenvalues with a negative real part and because the trivial fixed point is a stable focus, shown in Fig 3.1(d), y is oscillatory and damped.

6. $\zeta = \omega_0$: There are multiple negative eigenvalues and the trivial fixed point is a *stable inflected node*.

7. $\omega_0 < \zeta$: There are two negative eigenvalues and the trivial fixed point is a *stable node*, which is plotted in Figure 3.1(e).

3.2.1.2 Stability of the trivial fixed point in the case of negative linear and negative nonlinear stiffness ($\alpha < 0, \gamma < 0$)

In this case, the characteristic equation has one positive and one negative eigenvalue that are independent of ζ. The trivial fixed point is a saddle, which is unstable. This fixed point is shown in Figure 3.2.

3.2.2 Case when the nontrivial fixed points exist ($\alpha\gamma < 0$)

In this case, there is one trivial and two nontrivial fixed points. Depending on the sign of α (a positive and a negative sign denote positive and negative stiffness, respectively), the stability of the trivial and nontrivial fixed points changes.

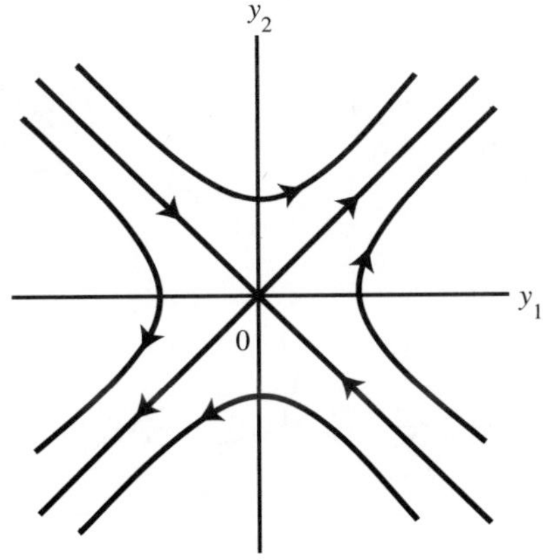

Figure 3.2 Phase portrait for the case $\alpha < 0, \gamma < 0$.

3.2.2.1 Case of positive linear and negative nonlinear stiffness ($\alpha > 0, \gamma < 0$)

As in the case $\alpha\gamma > 0$, the stability of the trivial fixed point is dependent on ζ when $\alpha > 0, \gamma < 0$ as is illustrated in Figures 3.3(a)–(e). For the nontrivial fixed points, $(y_{1st+}, 0)$ and $(y_{1st-}, 0)$, where $y_{st+} = \sqrt{-\alpha/\gamma}$ and $y_{st-} = -\sqrt{-\alpha/\gamma}$, Equations (3.2.11) and (3.2.12) become

$$\Delta\ddot{y}_1 + 2\zeta\,\Delta\dot{y}_1 - 2\alpha\Delta y_1 = 0 \tag{3.2.14}$$

and

$$\lambda^2 + 2\zeta\lambda - 2\alpha = 0 \tag{3.2.15}$$

Therefore, the nontrivial fixed points are saddles independent of ζ, as illustrated in Figure 3.3.

3.2.2.2 Case of negative linear and positive nonlinear stiffness ($\alpha < 0, \gamma > 0$)

In this case, the trivial fixed point is a *saddle point* independent of ζ, as shown in Figure 3.4.

Setting $\alpha = -\omega_0^2$ and using Equation (3.2.14), it can be seen that the stability of the nontrivial fixed points changes depending on ζ as follows:

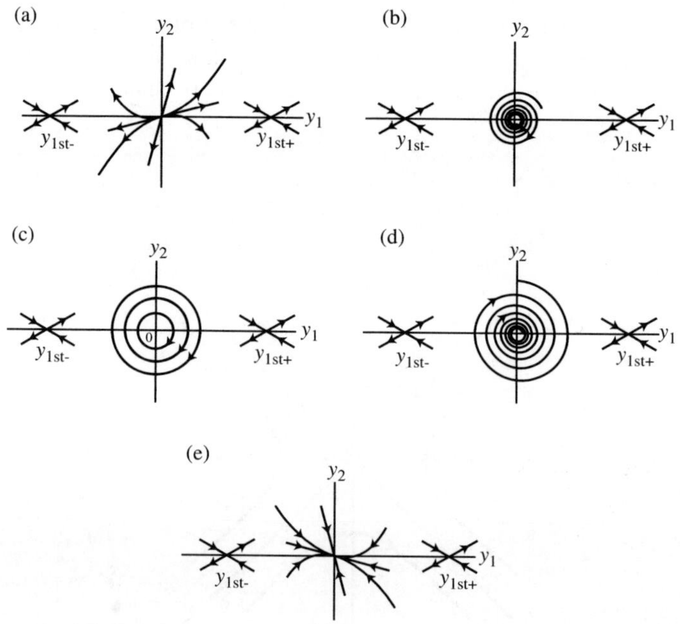

Figure 3.3 Phase portraits for the case $\alpha = \omega_0^2 > 0, \gamma < 0$: (a) $\zeta < -\omega_0$; (b) $-\omega_0 < \zeta < 0$; (c) $\zeta = 0$; (d) $0 < \zeta < \omega_0$ (e) $\zeta > \omega_0$.

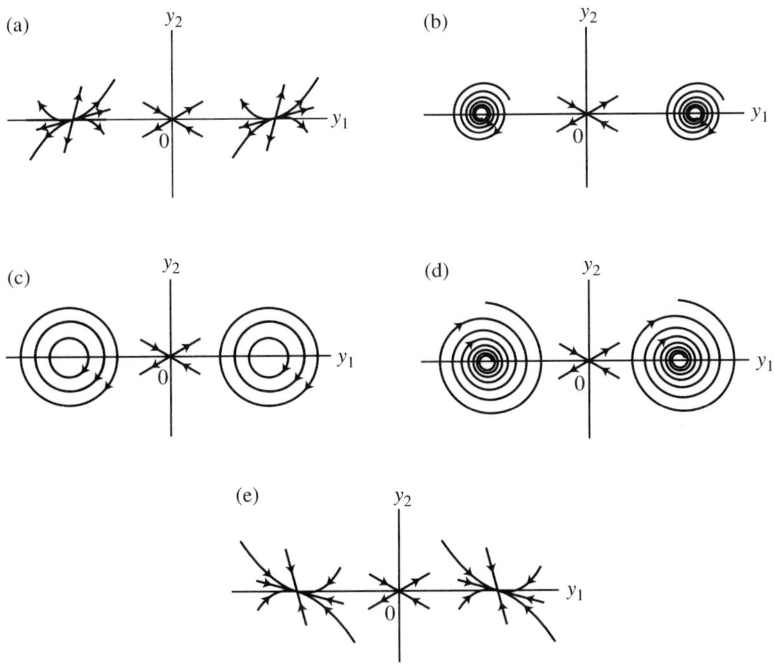

Figure 3.4 *Phase portraits for the case* $\alpha = -\omega_0^2 < 0, \gamma > 0$*: (a)* $\zeta < -\sqrt{2}\omega_0$*; (b)* $-\sqrt{2}\omega_0 < \zeta < 0$*; (c)* $\zeta = 0$*; (d)* $0 < \zeta < \sqrt{2}\omega_0$ *(e)* $\zeta > \sqrt{2}\omega_0$.

1. $\zeta < -\sqrt{2}\omega_0$: There are two positive eigenvalues and the nontrivial fixed point is an unstable node as shown in Figure 3.4(a).
2. $\zeta = -\sqrt{2}\omega_0$: There are multiple positive eigenvalues and the nontrivial fixed point is an unstable inflected node.
3. $-\sqrt{2}\omega_0 < \zeta < 0$: There are complex conjugate eigenvalues with a positive real part and the nontrivial fixed point is an unstable focus given in Figure 3.4(b). Self-excited oscillation is produced. The displacement is oscillatory and grows with time.
4. $\zeta = 0$: There are pure imaginary complex conjugate eigenvalues and the nontrivial fixed point is a centre, illustrated in Figure 3.4(c). The displacement is oscillatory without damping.
5. $0 < \zeta < \sqrt{2}\omega_0$: There are complex conjugate eigenvalues with a negative real part and the nontrivial fixed point is a stable focus as shown in Figure 3.4(d).
6. $\zeta = \sqrt{2}\omega_0$: There are multiple negative eigenvalues and the nontrivial fixed point is a stable inflected node.
7. $\sqrt{2}\omega_0 < \zeta$: There are two negative eigenvalues and the nontrivial fixed point is a stable node shown in Figure 3.4(e).

3.2.3 Variation of phase portraits depending on linear stiffness and linear damping

By using the previous linear analytical results near the fixed points, a general overview can be given of the phase portraits, depending on the linear stiffness and linear damping. Figures 3.5 and 3.6 show typical phase portraits that correspond to certain combinations of these two parameters for the system with softening nonlinear stiffness ($\gamma < 0$) and with hardening nonlinear stiffness ($\gamma > 0$), respectively.

3.3 Local bifurcation analysis

3.3.1 Bifurcation from trivial fixed points

In this section, local bifurcation near the trivial fixed point for the case when $\alpha > 0$ ($\alpha = \omega_0^2$) is investigated. Of interest is the bifurcation produced when the sign of ζ is changed and the effects of the nonlinear stiffness and nonlinear damping on this bifurcation are examined.

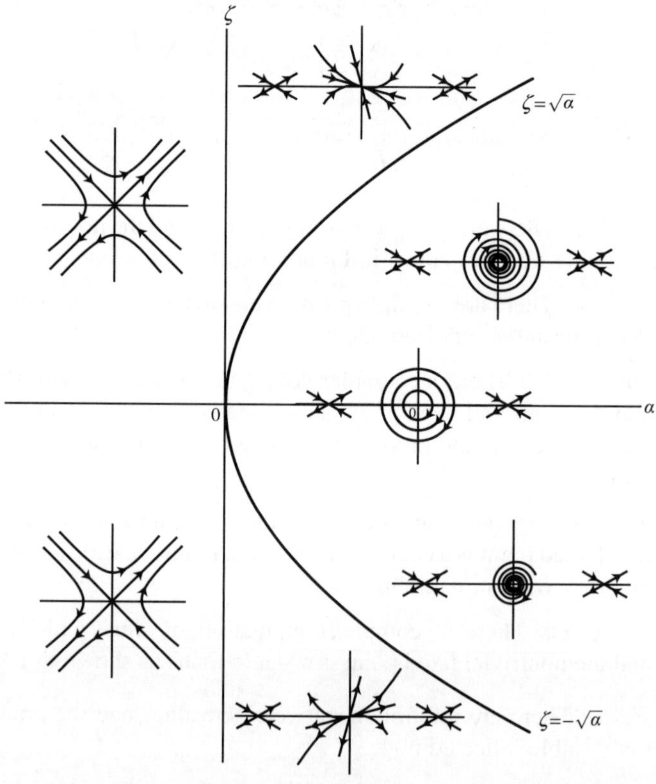

Figure 3.5 Phase portraits by local analysis corresponding to $\gamma < 0$.

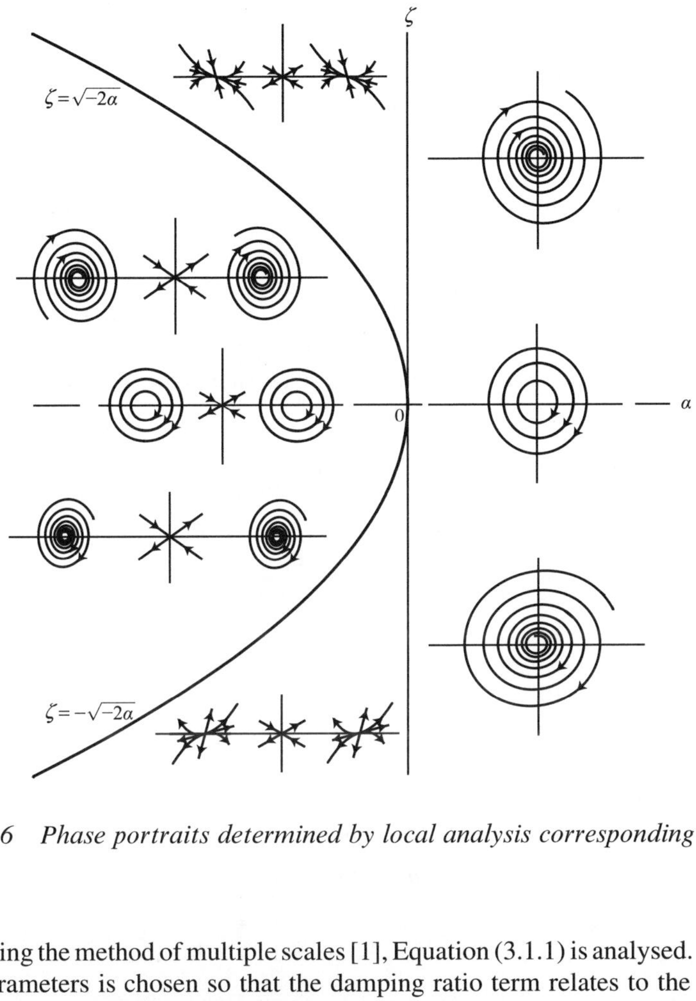

Figure 3.6 Phase portraits determined by local analysis corresponding to $\gamma > 0$.

By using the method of multiple scales [1], Equation (3.1.1) is analysed. The order of the parameters is chosen so that the damping ratio term relates to the nonlinear stiffness and damping terms as follows:

$$\zeta = O(\varepsilon^2),\ \mu = O(1),\ \gamma = O(1) \tag{3.3.1}$$

where ε is an order parameter. Then, ζ can be expressed as:

$$\zeta = \varepsilon^2\,\hat{\zeta} \quad (\hat{\zeta} = O(1)) \tag{3.3.2}$$

The solution of $y_1 = y$ is expanded approximately as

$$y_1 = \varepsilon y_{11} + \varepsilon^3\, y_{13} + \cdots \tag{3.3.3}$$

and the multiple *timescales*: $T_0 = t,\ T_2 = \varepsilon^2\, t$ are introduced. Substituting Equations (3.3.2) and (3.3.3) into Equation (3.1.1) and equating the same powers of ε yields

$$D_0^2 y_{11} + \omega_0^2 y_{11} = 0 \tag{3.3.4}$$

$$D_0^2 y_{13} + \omega_0^2 y_{13} = -2D_0 D_2 y_{11} - 2\hat{\zeta} D_0 y_{11} - \mu(D_0 y_{11})^3 - \gamma y_{11}^3 \tag{3.3.5}$$

where $D_0 = \partial/\partial T_0$ and $D_2 = \partial/\partial T_2$.

The solution of Equation (3.3.4) is

$$y_{11} = A(T_2)e^{j\omega_0 T_0} + A^*(T_2)e^{j\omega_0 T_0} = A(T_2)e^{j\omega_0 T_0} + \text{c.c.} \tag{3.3.6}$$

where c.c. denotes a complex conjugate of the preceding term. Then, Equation (3.3.5) can be written as

$$\begin{aligned} D_0^2 y_{13} + \omega_0^2 y_{13} = &-\left(2j\omega_0 D_2 A + 2j\hat{\zeta}\omega_0 A + 3j\mu\omega_0^3 |A|^2 A + 3\gamma |A|^2 A\right) e^{j\omega_0 T_0} \\ &+ NST + \text{c.c.} \end{aligned} \tag{3.3.7}$$

where $|A| = AA^*$ and NST denotes the terms that do not produce *secular terms* for y_3. The secular term is set to zero to give

$$D_2 A + \hat{\zeta} A + \frac{3\mu\omega_0^2}{2} |A|^2 A - j\frac{3\gamma}{2\omega_0} |A|^2 A = 0 \tag{3.3.8}$$

The variable A can be written in polar form as

$$A = \frac{\hat{a}(T_2)}{2} e^{j\theta(T_2)} \tag{3.3.9}$$

Then, noting that $T_2 = \varepsilon^2 t$ and $\varepsilon^2 \hat{\zeta} = \zeta$, Equation (3.3.8) can be separated into its real and imaginary parts as

$$\frac{da}{dt} + \zeta a + \frac{3\mu\omega_0^2}{8} a^3 = 0 \tag{3.3.10}$$

$$a\frac{d\theta}{dt} - \frac{3\gamma}{8\omega_0} a^3 = 0 \tag{3.3.11}$$

where $a = \varepsilon\hat{a}$.

Using Equation (3.3.6), Equation (3.3.3) can be written as

$$y_1 = a(t)\cos(\omega_0 t + \theta(t)) + O(\varepsilon^3) \tag{3.3.12}$$

where the time variation of a and θ in Equation (3.3.12) is governed by Equations (3.3.10) and (3.3.11). By setting $da/dt = 0$ for Equation (3.3.10), the steady-state amplitude a_{st} is sought which satisfies

$$a_{\text{st}}\left(\zeta + \frac{3\mu\omega_0^2}{8} a_{\text{st}}^2\right) = 0 \tag{3.3.13}$$

Thus, there is a trivial steady-state amplitude $a_{\text{st}} = 0$ for any ζ. On the other hand, a nontrivial steady-state amplitude exists for the case $\zeta\mu < 0$

$$a_{\text{st}} = 0, \sqrt{\frac{-8\zeta}{3\mu\omega_0^2}} \ (\zeta\mu < 0); \quad a_{\text{st}} = 0 \ (\zeta\mu > 0) \tag{3.3.14}$$

Next, the stability of the steady-state amplitude is examined. Substituting $a(t) = a_{\text{st}} + \Delta a(t)$ into Equation (3.3.10) and considering Equation (3.3.13), yields

$$\frac{d\Delta a}{dt} + \left(\zeta + \frac{9}{8}\mu\omega_0^2 a_{\text{st}}^2\right)\Delta a = 0 \tag{3.3.15}$$

From the time evolution of Δa, the stability of the trivial and nontrivial steady-state amplitude a_{st} can be determined. For the trivial steady-state amplitude $a_{\text{st}} = 0$, Equation (3.3.15) is rewritten as

$$\frac{d\Delta a}{dt} + \zeta\Delta a = 0 \tag{3.3.16}$$

Its solution is

$$\Delta a = C_1 e^{-\zeta t} \tag{3.3.17}$$

where C_1 is a constant that can be calculated from the initial condition. Therefore, the trivial steady-state amplitude ($y_1 = y_2 = 0$) is stable for $\zeta > 0$ and unstable for $\zeta < 0$, as shown in Figure 3.7.

Furthermore, for the nontrivial steady-state amplitude of $a_{\text{st}} = \sqrt{-8\zeta/(3\mu\omega_0^2)}$, Equation (3.3.15) can be expressed as

$$\frac{d\Delta a}{dt} - 2\zeta\Delta a = 0 \tag{3.3.18}$$

The solution is

$$\Delta a = C_2 e^{2\zeta t} \tag{3.3.19}$$

where C_2 is a constant that can be determined from the initial condition. The steady-state amplitude for $\zeta < 0$ is stable and for $\zeta > 0$ is unstable.

As a result, when $\zeta < 0$ (negative damping), the trivial fixed point is unstable and the small disturbance Δa grows, oscillating with time. The oscillation is called a self-excited oscillation. If $\mu > 0$, the stable steady-state oscillation in y undergoes a *supercritical Hopf bifurcation*, as illustrated in Figure 3.7(a). The situation is described by the arrow (A) in Figures 3.8 and 3.11, which is discussed later, while the amplitude continues to grow with time if $\mu < 0$.

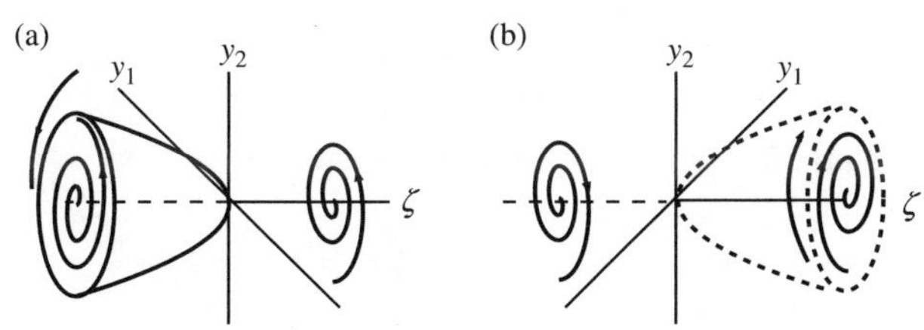

Figure 3.7 Hopf bifurcation: (a) supercritical; (b) subcritical.

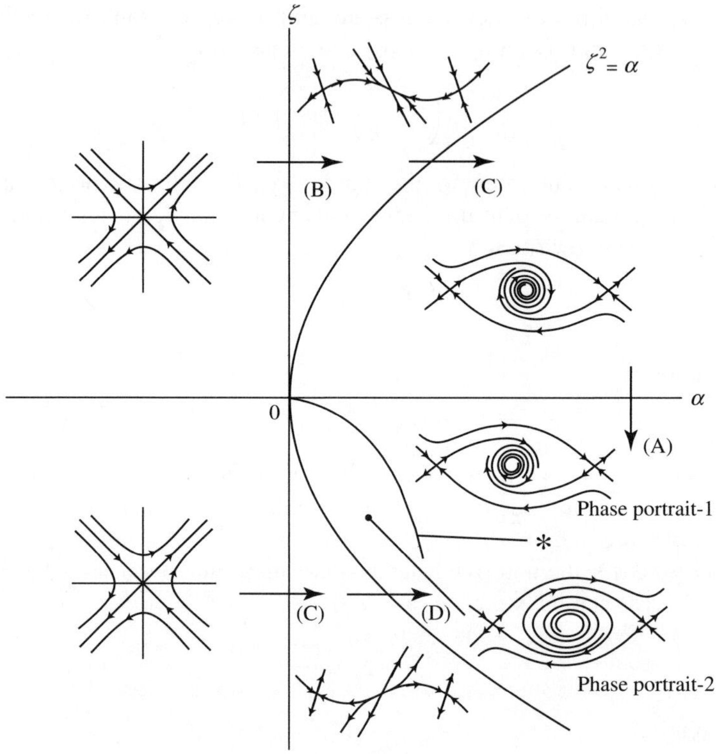

Figure 3.8 Phase portraits by global analysis for the case $\gamma < 0, \mu > 0$.

On the other hand, in the case when $\zeta > 0$, the trivial fixed point is stable for small disturbances, and Δa decays with time. There is an unstable *limit cycle* denoted by the dotted line in Figure 3.7(b) for the case $\mu < 0$ and when the disturbance is outside the limit cycle, it grows with time. Hence, with a large disturbance, self-excited oscillation occurs even if $\zeta > 0$ (positive damping). This self-excited oscillation cannot be predicted by linear analysis. From the above analysis, it is noted that nonlinear damping can produce a limit cycle, but the nonlinear stiffness is not related to its creation.

Finally, by setting $d\theta/dt = 0$ in Equation (3.3.11), θ_{st} is in the steady state when

$$\theta_{st} = \frac{3\gamma}{8\omega_0} a_{st}^2 t + \theta_0 \tag{3.3.20}$$

Then, the nontrivial steady-state oscillation is expressed as

$$y_1 = a_{st} \cos\left((\omega_0 + \frac{3\gamma}{8\omega_0} a_{st}^2) t + \theta_0 \right) + O(\varepsilon^3) \tag{3.3.21}$$

where θ_0 is a constant defined by the initial condition. Thus, in the steady-state, the response frequency is different from the linear natural frequency ω_0 and depends on the steady-state amplitude. For the hardening-type nonlinear stiffness ($\gamma > 0$), the response frequency is greater than the linear natural frequency, and for the softening-type nonlinear stiffness ($\gamma < 0$), the response frequency is less than the linear natural frequency.

3.3.2 Bifurcation from nontrivial fixed points

In this section, the local bifurcation near the nontrivial fixed points in the case $\alpha < 0$ $(2\alpha = -\omega_1^2)$ and $\gamma > 0$ is analysed. The nontrivial fixed points follow from the result derived in Section 3.2

$$y_{1\mathrm{st}} = \pm\sqrt{-\alpha/\gamma} \tag{3.3.22}$$

Letting

$$y = y_{\mathrm{st}} + \Delta y^1 \tag{3.3.23}$$

Equation (3.1.1) leads to

$$\Delta\ddot{y}_1 + 2\zeta\Delta\dot{y}_1 + \mu\Delta\dot{y}_1^3 + \omega_1^2\Delta y_1 + 3\gamma y_{\mathrm{st}}\Delta y_1^2 + \gamma\Delta y_1^3 = 0 \tag{3.3.24}$$

where $\omega_1^2 = -2\alpha$. By using the method of multiple scales in a way similar to that in Section 3.3.1, Equation (3.3.24) is analysed. The order of the parameters is chosen so that the damping ratio term relates to the nonlinear stiffness and damping terms as follows:

$$\zeta = O(\varepsilon^2),\ \mu = O(1),\ \gamma = O(1) \tag{3.3.25}$$

where ε is an order parameter. Then, ζ can be expressed as:

$$\zeta = \varepsilon^2\hat{\zeta}\quad (\hat{\zeta} = O(1)) \tag{3.3.26}$$

The solution can be written approximately as

$$\Delta y_1 = \varepsilon\Delta y_{11} + \varepsilon^2\Delta y_{12} + \varepsilon^3\Delta y_{13} + \cdots \tag{3.3.27}$$

and the multiple timescales: $T_0 = t,\ T_2 = \varepsilon^2 t$ are introduced. Substituting Equations (3.3.26) and (3.3.27) into Equation (3.3.24) and equating the same powers of ε, yields

$$D_0^2\Delta y_{11} + \omega_1^2\Delta y_{11} = 0 \tag{3.3.28}$$

$$D_0^2\Delta y_{12} + \omega_1^2\Delta y_{12} = -3y_{st}\gamma\Delta y_{11}^2 \tag{3.3.29}$$

$$\begin{aligned} D_0^2\Delta y_{13} + \omega_1^2\Delta y_{13} = \; & -2D_0D_2\Delta y_{11} - 2\hat{\zeta}D_0\Delta y_{11} - 6\gamma y_{\mathrm{st}}\Delta y_{11}\Delta y_{12} \\ & -\gamma\Delta y_{11}^3 - \mu(D_0\Delta y_{11})^3 \end{aligned} \tag{3.3.30}$$

where $D_0 = \partial/\partial T_0$ and $D_2 = \partial/\partial T_2$. The solution of Equation (3.3.28) is

$$\Delta y_{11} = A(T_2)e^{j\omega_1 T_0} + \mathrm{c.c.} \tag{3.3.31}$$

A particular solution of Equation (3.3.29) is

$$\Delta y_{12} = \frac{y_{st}\gamma}{\omega_1^2}\left(A^2 e^{2j\omega_1 T_0} - 3|A|^2\right) + \text{c.c.} \tag{3.3.32}$$

So, Equation (3.3.30) can be rewritten as

$$D_0^2 y_{13} + \omega_1^2 y_{13} = -\Big(2j\hat{\zeta}\omega_1 D_2 A + 2j\hat{\zeta}\omega_1 A + 3j\mu\omega_1^3 |A|^2 A$$

$$+ \frac{3}{\omega_1^2}\left(\omega_1^2\gamma - 10y_{st}^2\gamma^2\right)|A|^2 A)\Big) e^{j\omega_1 T_0} + \text{NST} + \text{c.c.} \tag{3.3.33}$$

The secular term is set to zero to give

$$D_2 A + \hat{\zeta} A + \frac{3\mu\omega_1^2}{2}|A|^2 A - j\frac{3\gamma}{2\omega_1^3}\left(\omega_1^2 - 10y_{st}^2\gamma\right)|A|^2 A = 0 \tag{3.3.34}$$

The variable A can be written in polar form as

$$A = \frac{\hat{a}(T_2)}{2} e^{j\theta(T_2)} \tag{3.3.35}$$

Noting that $T_2 = \varepsilon^2 t$ and $\varepsilon^2\hat{\zeta} = \zeta$, Equation (3.3.34) can be separated into its real and imaginary parts:

$$\frac{da}{dt} + \zeta a + \frac{3\mu\omega_1^2}{8} a^3 = 0 \tag{3.3.36}$$

$$a\frac{d\theta}{dt} + \frac{3\gamma}{2\omega_1} a^3 = 0 \tag{3.3.37}$$

where $a = \varepsilon\hat{a}$.

Using Equation (3.3.31), Equation (3.3.27) can be written as

$$\Delta y_1 = a(t)\cos(\omega_1 t + \theta(t)) + O(\varepsilon^3) \tag{3.3.38}$$

where the time variation of a and θ in Equation (3.3.38) is governed by Equations (3.3.36) and (3.3.37). Similar to the discussion in Section 3.1, the supercritical and *subcritical Hopf bifurcations* are produced for the case of $\mu > 0$ (Figure 3.7(a)) and $\mu < 0$ (Figure 3.7(b)), respectively.

The nonlinear analysis in Section 3.3 is local to each fixed point. A global bifurcation analysis is performed in the next section.

3.4 Global analysis for softening nonlinear stiffness ($\gamma < 0$)

3.4.1 Phase portraits

Based on the results of Sections 3.2 and 3.3, the *bifurcation diagram* and associated phase portraits for $\mu > 0$ are plotted in Figure 3.8. Changing the damping ratio ζ from positive to negative for $\alpha > 0$, as indicated by the arrow (A) from the first quadrant to the fourth quadrant, a supercritical Hopf bifurcation occurs and a stable limit cycle is produced around the trivial fixed point. The system undergoes the subcritical *Pitchfork bifurcation* across $\alpha = 0$ along the arrow (B) and along the arrow (C), the trivial fixed point becomes a stable focus across the curve $\zeta^2 = \alpha$ where the trivial fixed point is a stable inflected node. The nontrivial fixed points emerge across $\alpha = 0$ along the arrow (C). In this case, the trivial fixed point is an unstable node and along the arrow (D), the trivial fixed point becomes an unstable focus across the curve $\zeta^2 = \alpha$ where the trivial fixed point is an unstable inflected node. Then, because Phase portrait-1 and Phase portrait-2 are not *homeomorphic*, a global bifurcation between the phase portraits is expected at the part of '*' in the fourth quadrant in Figure 3.8. The global bifurcation is analysed in the next section.

3.4.2 Global bifurcation analysis

3.4.2.1 Hamiltonian

The linearisation of Equations (3.2.1) and (3.2.2) about the trivial fixed point leads to

$$\dot{y}_1 = y_2,\ \dot{y}_2 = -\alpha y_1 - 2\zeta y_2 \tag{3.4.1}$$

or

$$\frac{d\Delta\mathbf{y}}{dt} = \mathbf{J_G}(0)\Delta\mathbf{y} \tag{3.4.2}$$

where

$$\mathbf{J_G}(0) = \begin{bmatrix} 0 & 1 \\ -\alpha & -2\zeta \end{bmatrix} \tag{3.4.3}$$

In particular, in the case $\alpha = \zeta = 0$, i.e., no stiffness and no damping, the Jacobian is

$$\mathbf{J_G}(0) = \begin{bmatrix} 0 & 1 \\ 0 & 0 \end{bmatrix} \tag{3.4.4}$$

which has multiple zero eigenvalues. This singularity is called the Bogdanov–Takens singularity [2,3]. Here, the same technique as Carr [4] and Guckenheimer and Holmes [2] is followed to study the global dynamics [5]. A rescaling transformation is first performed. Setting y_1, y_2, α, ζ, and t as

$$y_1 = \varepsilon^{1/3}\bar{y}_1, y_2 = \varepsilon^{2/3}\bar{y}_2, \alpha = \varepsilon^{2/3}\bar{\alpha}, \zeta = \varepsilon^{4/3}\bar{\zeta}, t = \varepsilon^{-1/3}\bar{t} \tag{3.4.5}$$

results in

$$\dot{\bar{y}}_1 = \bar{y}_2,\ \dot{\bar{y}}_2 = -\bar{\alpha}\bar{y}_1 - \gamma\bar{y}_1^3 - \varepsilon\left(2\bar{\zeta}\bar{y}_2 + \mu\bar{y}_2^3\right) \tag{3.4.6}$$

where the overdot denotes the derivative with respect to $\bar{t}$.

If the small terms including ε are neglected, Equation (3.4.6) becomes an integrable Hamiltonian system

$$\dot{\bar{y}}_1 = \bar{y}_2, \quad \dot{\bar{y}}_2 = -\bar{\alpha}\bar{y}_1 - \gamma\bar{y}_1^3 \tag{3.4.7}$$

with the *Hamiltonian*

$$H = \frac{1}{2}\bar{y}_2^2 + \frac{\bar{\alpha}}{2}\bar{y}_1^2 + \frac{\gamma}{4}\bar{y}_1^4 \tag{3.4.8}$$

Then, along the solutions of Equation (3.4.6), H is varied with time as

$$\frac{dH}{d\bar{t}} = \frac{\partial H}{\partial \bar{y}_1}\frac{d\bar{y}_1}{d\bar{t}} + \frac{\partial H}{\partial \bar{y}_2}\frac{d\bar{y}_2}{d\bar{t}} = -\varepsilon(2\zeta\bar{y}_2^2 + \mu\bar{y}_2^4) \tag{3.4.9}$$

Allowing a linear rescaling of time, μ can be set to 1 without loss of generality and the behaviour for $\mu = -1$ is obtained by reversing the time in the result for $\mu = 1$. On the other hand, γ can be set to 1 or -1 without loss of generality, but the system has qualitatively different dynamics in these cases. Therefore, two cases: $\gamma = 1$ and $\gamma = -1$ are analysed for $\mu = 1$.

3.4.2.2 Parameter values for heteroclinic orbits

For the case $\gamma < 0$, the conditions for the existence of a *heteroclinic orbit* are investigated: $\bar{\alpha}$ can be set to 1 and γ to –1 without loss of generality. Then, the system of Equation (3.4.7) has fixed points $\bar{y}_{\mathrm{st}} = 0$, $\bar{y}_{\mathrm{st}+} = 1$, and $\bar{y}_{\mathrm{st}-} = -1$, and H is expressed as

$$H = \frac{1}{2}\bar{y}_2^2 + \frac{1}{2}\bar{y}_1^2 - \frac{1}{4}\bar{y}_1^4 \tag{3.4.10}$$

The corresponding potential energy and phase portrait are plotted in Figure 3.9(a).

There are heteroclinic orbits connecting the saddle points (nontrivial fixed points) at $(\bar{y}_1, \bar{y}_2) = (\pm 1, 0)$, which correspond to $H(\bar{y}_1, \bar{y}_2) = 1/4$. The locus is expressed as

$$y_2 = \pm\sqrt{\frac{1}{2}(1 - 2\bar{y}_1^2 + \bar{y}_1^4)} \tag{3.4.11}$$

Let S and U be the *stable and unstable manifolds* of the nontrivial fixed point $(\bar{y}_1, \bar{y}_2) = (-1, 0)$. Let $H^+(\bar{y}_1, \bar{y}_2)$ and $H^-(\bar{y}_1, \bar{y}_2)$ be H for U and S, respectively. Let $H(+)$ be the value of H^+ when U hits $\bar{y}_1 > 0$ and $\bar{y}_2 = 0$ and let $H(-)$ be the value of H^- when S hits $\bar{y}_1 > 0$ and $\bar{y}_2 = 0$. Equation (3.1.1) has a heteroclinic orbit if

$$H(+) = H(-) \tag{3.4.12}$$

Hence,

$$\int_0^{\infty} \frac{dH^+}{d\bar{t}}d\bar{t} = \int_0^{-\infty} \frac{dH^-}{d\bar{t}}d\bar{t} = -\int_0^{\infty} \frac{dH^-}{d\bar{t}}d\bar{t} \tag{3.4.13}$$

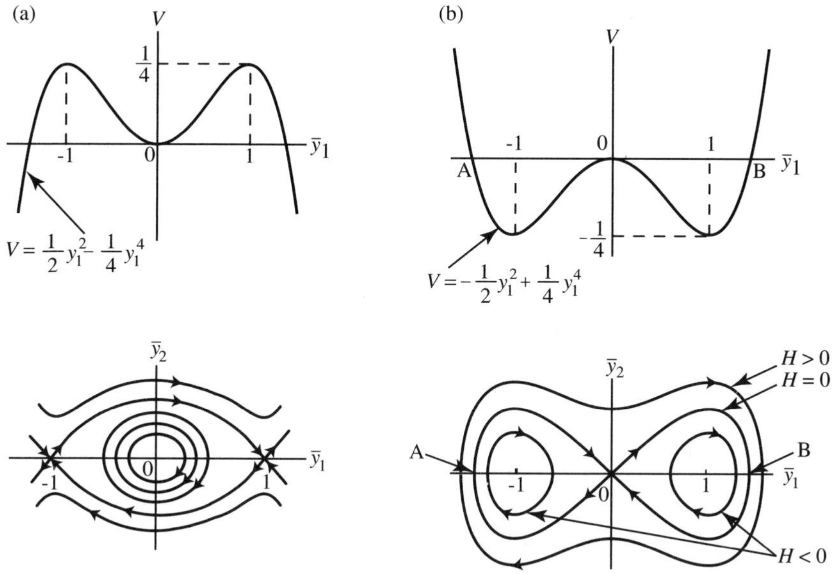

Figure 3.9 Potential-energy plots and phase portraits for $\varepsilon = 0$: *(a)* $\bar{\alpha} = 1,\ \ \gamma = -1$; *(b)* $\bar{\alpha} = -1,\ \ \gamma = 1$.

By considering $H^+(\bar{y}_1, \bar{y}_2) = H^-(\bar{y}_1, -\bar{y}_2)$, the above condition can be rewritten as

$$\int_0^\infty \frac{dH^+}{d\bar{t}} d\bar{t} = 0 \tag{3.4.14}$$

Using Equation (3.4.9), Equation (3.4.14) can be expressed as

$$\int_0^\infty 2\bar{\zeta}\bar{y}_2^2 + \bar{y}_2^4 dt = 0 \tag{3.4.15}$$

where μ is set to 1.

Furthermore, considering Equations (3.2.1) and (3.4.11), gives

$$\bar{\zeta} = -\frac{\int_{-1}^{1} \bar{y}_2^3 d\bar{y}_1}{2\int_{-1}^{1} \bar{y}_2 d\bar{y}_1} = -\frac{\int_{-1}^{1} \left(\frac{1}{2}(1-2\bar{y}_1^2+\bar{y}_1^4)\right)^{3/2} d\bar{y}_1}{2\int_{-1}^{1} \sqrt{\frac{1}{2}(1-2\bar{y}_1^2+\bar{y}_1^4)} d\bar{y}_1} \tag{3.4.16}$$

From Equation (3.4.5), the condition for the heteroclinic orbit shown in Figure 3.10 is expressed in terms of the original parameters $\alpha(=\varepsilon^{2/3}\bar{\alpha})$ and

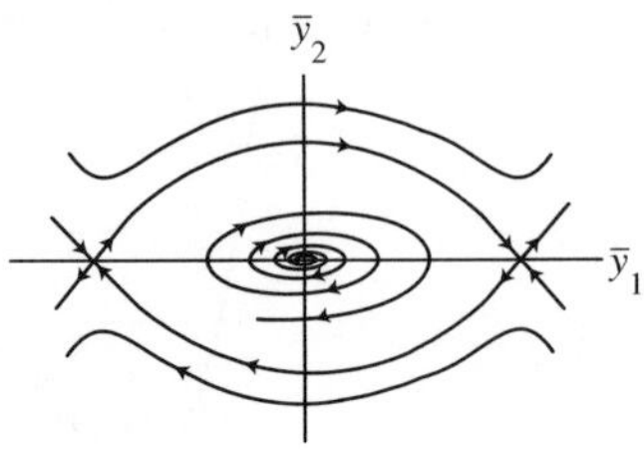

Figure 3.10 Heteroclinic orbit.

$\zeta(=\varepsilon^{4/3}\bar{\zeta})$ in Equation (3.1.1) as

$$\zeta = -\frac{\displaystyle\int_{-1}^{1}\left(\tfrac{1}{2}(1-2\bar{y}_1^2+\bar{y}_1^4)\right)^{3/2}d\bar{y}_1}{2\displaystyle\int_{-1}^{1}\sqrt{\tfrac{1}{2}(1-2\bar{y}_1^2+\bar{y}_1^4)}d\bar{y}_1}\alpha^2 \tag{3.4.17}$$

This is a *bifurcation set* and it is shown in Figure 3.8 in $\alpha-\zeta$ plane for $\mu = 1$ as a curve marked by '*'.

3.5 Global analysis for hardening nonlinear stiffness ($\gamma > 0$)

3.5.1 Phase portraits

Using the results obtained in Sections 3.2 and 3.3, the bifurcation diagram and associated phase portraits for $\mu > 0$ can be plotted as shown in Figure 3.11. Using the local bifurcation analysis in Sections 3.3, the following *Hopf bifurcation* is predicted. Changing the parameter ζ from positive to negative with $\alpha > 0$, i.e., from the first quadrant to the fourth quadrant, as shown by the arrow (A) in Figure 3.11, the supercritical Hopf bifurcation occurs and one stable limit cycle is produced around the trivial fixed point as seen from the results in Sections 3.3.1. Also, changing the parameter ζ from positive to negative with $\alpha < 0$, i.e., from the second quadrant to the third quadrant, as illustrated by the arrow (B), the supercritical Hopf bifurcation occurs and two stable limit cycles are produced around the nontrivial fixed points as seen from the result obtained in Sections 3.3.2. The system undergoes the supercritical pitchfork bifurcation across $\alpha = 0$ along the arrow (C) and along the arrow (C), the nontrivial fixed points becomes a stable focus across the curve $\zeta^2 = -2\alpha$ where the nontrivial fixed points are a stable inflected node. The nontrivial fixed points emerge across $\alpha = 0$ along the arrow (E). In this case the nontrivial fixed points are an unstable node and along the arrow (F), the nontrivial fixed points become an unstable focus across the curve $\zeta^2 = -2\alpha$ where the nontrivial fixed points are an unstable inflected node. Then, because Phase portrait-1 and Phase portrait-2 are not homeomorphic, global bifurcations between phase portraits are expected at the part of '*' in the third quadrant in Figure 3.11. The global bifurcations are analysed in the next section.

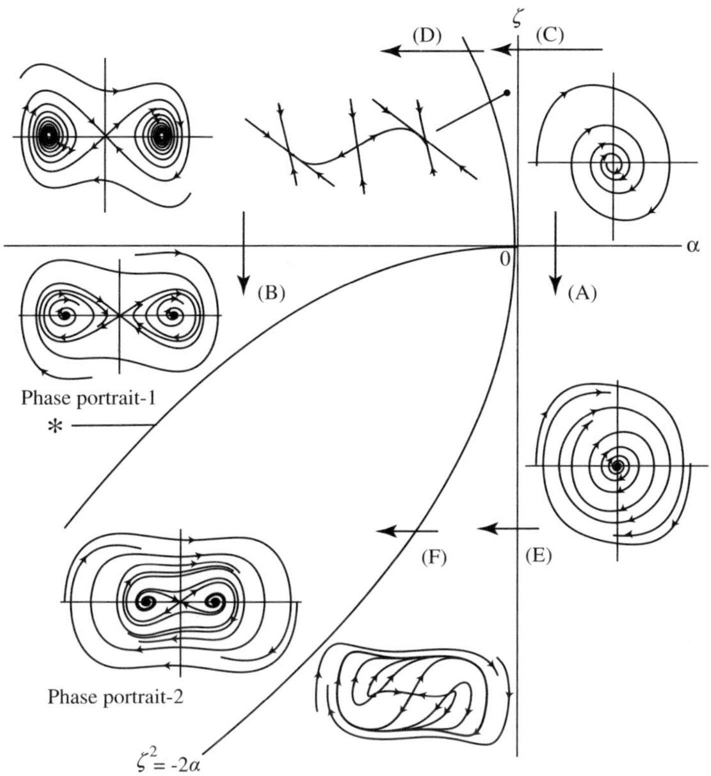

Figure 3.11 Phase portraits by global analysis for the case $\gamma > 0$, $\mu > 0$.

3.5.2 Global bifurcation analysis

In this section, global bifurcations are considered. First, note that, for $\zeta > 0$ and $\mu > 0$, no closed orbits exist, because $\partial g_1/\partial y_1 + \partial g_2/\partial y_2 = -2\zeta - 3\mu y_2^2 < 0$ and Bendixon's criterion [2] can be applied. On the other hand, in the transition between the phase portraits in the third quadrant in the region of $\zeta < 0$ in Figure 3.11, which are indicated by the local bifurcation analysis, additional bifurcations (global bifurcations) are expected.

3.5.2.1 Hamiltonian

The dynamics in the third quadrant of the bifurcation set plotted in Figure 3.11 are analysed. Setting $\bar{\alpha} = -1$ and $\gamma = 1$, the *two/double/twin-well potential* energy associated with Equation (3.4.8), and the corresponding phase portrait for $\varepsilon = 0$ are shown in Figure 3.9(b). When $\varepsilon = 0$, the total energy H is invariant from Equation (3.4.9). Let the total energy be b. The phase portraits are qualitatively different depending on the magnitude of b. When $H(\bar{y}_1, \bar{y}_2) = b$, there is the double homoclinic orbit if $b = 0$, and a single closed curve surrounding the three fixed points if $b > 0$. Also, if $-\alpha^2/(4\gamma) < b < 0$, $H(y_1, y_2) = b$ consists of two periodic curves surrounding each fixed point.

Setting $\alpha = -1$ and $\gamma = 1$, the relationships between (α, ζ) are sought such that Equation (3.1.1) has the above single periodic orbit, double homoclinic orbit, and two periodic orbits. The fixed points are $\bar{y}_{st} = 0$, $\bar{y}_{st+} = 1$, and $\bar{y}_{st-} = -1$, and H is expressed as

$$H = \frac{1}{2}\bar{y}_2^2 - \frac{1}{2}\bar{y}_1^2 + \frac{1}{4}\bar{y}_1^4 \tag{3.5.1}$$

for $\varepsilon = 0$. The curve in the phase portrait corresponding to $H(\bar{y}_1, \bar{y}_2) = b$ for $\varepsilon = 0$ represents a homoclinic orbit, one periodic orbit surrounding the three fixed points, and two periodic orbits surrounding each nontrivial fixed point, for $b = 0$, $b > 0$, and $-1/4 < b < 0$, respectively. The phase trajectory is expressed as a function of b as

$$\bar{y}_2 = \pm\sqrt{2b + \bar{y}_1^2 - \bar{y}_1^4/2} \tag{3.5.2}$$

These phase portraits are described in the region of $\bar{y}_1 > 0$ in Figure 3.12, where the homoclinic orbit hits the $\bar{y}_1$ axis at point B.

The points on the positive $\bar{y}_1$ axis where the periodic orbit hits the $\bar{y}_1$ axis are expressed as

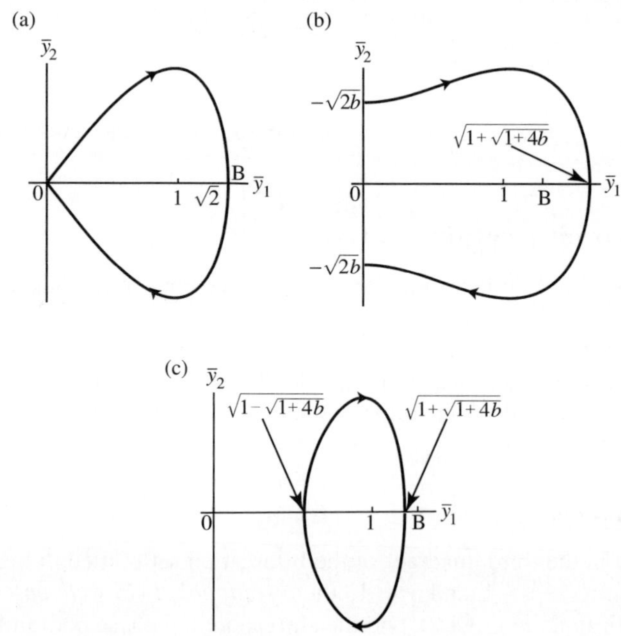

Figure 3.12 Periodic orbits for $\varepsilon = 0$ corresponding to the case $\bar{\alpha} = -1$, $\gamma = 1$: (a) $b = 0$;(b) $b > 0$;(c) $0 > b > -1/4$.

$$\bar{y}_1 = \sqrt{1 \pm \sqrt{1+4b}} \tag{3.5.3}$$

Therefore, as shown in Figure 3.12, the homoclinic orbit, existing for $b = 0$, passes the origin and $(\bar{y}_1, \bar{y}_2) = (\sqrt{2}, 0)$, the periodic orbit surrounding the three fixed points, occurring for $b > 0$, passes one point $(\bar{y}_1, \bar{y}_2) = (\sqrt{1+\sqrt{1+4b}}, 0)$, and a periodic orbit surrounding a nontrivial fixed point, existing for $-1/4 < b < 0$ passes two points, $(\sqrt{1-\sqrt{1+4b}}, 0)$ and $(\sqrt{1+\sqrt{1+4b}}, 0)$.

3.5.2.2 Homoclinic orbit

First, the parameter values for the existence of the homoclinic orbit for Equations (3.2.1) and (3.2.2) are sought. When $\varepsilon = 0$, the *level curve* for the homoclinic orbits is $H(\bar{y}_1, \bar{y}_2) = b = 0$ and it is expressed as

$$\bar{y}_2 = \pm\sqrt{\bar{y}_1^2 - \frac{1}{2}\bar{y}_1^4} \tag{3.5.4}$$

which is plotted in Figure 3.12(a) for the region $\bar{y}_1 > 0$.

Let S and U be the stable and unstable manifolds at the trivial fixed point (saddle). Let $H^+(\bar{y}_1, \bar{y}_2)$ and $H^-(\bar{y}_1, \bar{y}_2)$ be H for U and S, respectively. Let $H(+)$ be the value of H^+ when U hits $\bar{y}_1 > 0, \bar{y}_2 = 0$ at $\bar{t} = T$ and let $H(-)$ be the value of H^- when S hits $\bar{y}_1 > 0, \bar{y}_2 = 0$ at $\bar{t} = -T$. Equations (3.2.1) and (3.2.2) have a homoclinic orbit if

$$H(+) = H(-) \tag{3.5.5}$$

Hence,

$$\int_0^T \frac{dH^+}{d\bar{t}} d\bar{t} = \int_0^{-T} \frac{dH^-}{d\bar{t}} d\bar{t} = -\int_0^T \frac{dH^-}{d\bar{t}} d\bar{t} \tag{3.5.6}$$

By considering $H^+(\bar{y}_1, \bar{y}_2) = H^-(\bar{y}_1, -\bar{y}_2)$, the above condition can be rewritten as

$$\int_0^T \frac{dH^+}{d\bar{t}} d\bar{t} = 0 \tag{3.5.7}$$

Using Equation (3.4.9), Equation (3.5.7) is expressed as

$$\int_0^T (2\zeta\bar{y}_2^2 + \bar{y}_2^4) d\bar{t} = 0 \tag{3.5.8}$$

Furthermore, considering Equations (3.2.1) and (3.5.4), the following is obtained

$$\int_0^{\sqrt{2}} (2\zeta\bar{y}_2 + \bar{y}_2^3) d\bar{y}_1 = \int_0^{\sqrt{2}} \left(2\zeta\sqrt{\bar{y}_1^2 - \bar{y}_1^4/2} + (\bar{y}_1^2 - \bar{y}_1^4/2)^{3/2}\right) d\bar{y}_1 = 0 \tag{3.5.9}$$

$$\bar{\zeta} = -\frac{\int_0^{\sqrt{2}} \left((\bar{y}_1^2 - \bar{y}_1^4/2)^{3/2}\right)^{3/2} d\bar{y}_1}{2\int_0^{\sqrt{2}} \sqrt{\bar{y}_1^2 - \bar{y}_1^4/2} d\bar{y}_1} \tag{3.5.10}$$

Noting that $\alpha = \varepsilon^{2/3}\bar{\alpha}$ and $\zeta = \varepsilon^{4/3}\bar{\zeta}$, Equation (3.5.10) can be rewritten as

$$\zeta = -\frac{\int_0^{\sqrt{2}} \left((\bar{y}_1^2 - \bar{y}_1^4/2)^{3/2}\right)^{3/2} d\bar{y}_1}{2\int_0^{\sqrt{2}} \sqrt{\bar{y}_1^2 - \bar{y}_1^4/2}\, d\bar{y}_1}\alpha^2 \tag{3.5.11}$$

3.5.2.3 Orbit surrounding the trivial and nontrivial fixed points

Next, parameter values for the existence of the periodic orbit surrounding three fixed points are sought. When $\varepsilon = 0$, the phase trajectory for such a periodic orbit is

$$\bar{y}_2 = \pm\sqrt{2b + \bar{y}_1^2 - \bar{y}_1^4/2} \tag{3.5.12}$$

where $b > 0$ and it is given in Figure 3.12b for the region $\bar{y}_1 > 0$. Parameter values are sought such that the integration of dH/dt from $(\bar{y}_1, \bar{y}_2) = (0, \sqrt{2b})$ to $(\sqrt{1+\sqrt{1+4b}}, 0)$ is the same as that from $(\bar{y}_1, \bar{y}_2) = (0, -\sqrt{2b})$ to $(\sqrt{1+\sqrt{1+4b}}, 0)$. Similar to the analysis of the homoclinic orbit, the relationship between ζ and α is determined to be

$$\zeta = -\frac{\int_0^{\sqrt{1+\sqrt{1+4b}}} (2b + \bar{y}_1^2 - \bar{y}_1^4/2)^{3/2} d\bar{y}_1}{2\int_0^{\sqrt{1+\sqrt{1+4b}}} \sqrt{2b + \bar{y}_1^2 - \bar{y}_1^4/2}\, d\bar{y}_1}\alpha^2 \tag{3.5.13}$$

3.5.2.4 Orbit surrounding a nontrivial fixed point

When $\varepsilon = 0$, the phase trajectory for such a periodic orbit is given by

$$\bar{y}_2 = \pm\sqrt{2b + \bar{y}_1^2 - \bar{y}_1^4/2} \tag{3.5.14}$$

where $-1/4 < b < 0$ and it is shown in Figure 3.12(c) for the region $y_1 > 0$. Parameter values are sought such that the integration of dH/dt from $(\bar{y}_1, \bar{y}_2) = (\sqrt{1-\sqrt{1+4b}}, 0)$ to $(\sqrt{1+\sqrt{1+4b}}, 0)$ along the upper curve is the same as that from $(\bar{y}_1, \bar{y}_2) = (\sqrt{1-\sqrt{1+4b}}, 0)$ to $(\sqrt{1+\sqrt{1+4b}}, 0)$ along the lower curve. Similar to the analysis of the homoclinic orbit, the relationship between ζ and α is determined to be

$$\zeta = -\frac{\displaystyle\int_{\sqrt{1-\sqrt{1+4b}}}^{\sqrt{1+\sqrt{1+4b}}} \left(2b+\bar{y}_1^2-\bar{y}_1^4/2\right)^{3/2} d\bar{y}_1}{\displaystyle 2\int_{\sqrt{1-\sqrt{1+4b}}}^{\sqrt{1+\sqrt{1+4b}}} \sqrt{2b+\bar{y}_1^2-\bar{y}_1^4/2}\, d\bar{y}_1}\alpha^2 \tag{3.5.15}$$

3.5.2.5 Bifurcation sets

On the basis of Equations (3.5.11), (3.5.13) and (3.5.15), the relationship between ζ and α is uniformly expressed by using the total energy b as

$$\zeta = -P(b)\alpha^2 \tag{3.5.16}$$

The graph of $P(b)$ is plotted in Figure 3.13. Except for the narrow part marked with a dotted circle near $b = 0$, $P(b)$ monotonically increases with an increase of b. When the value $P(b) = \zeta/\alpha^2$ is less than $P(b) \approx 0.171411$, b is negative

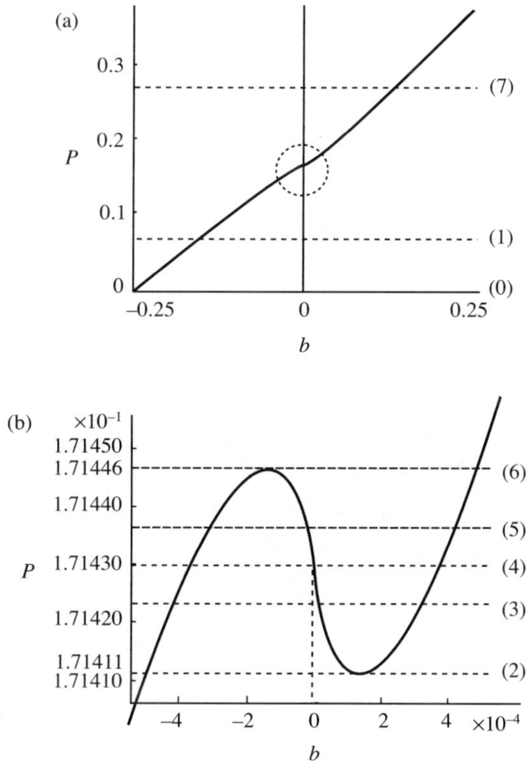

Figure 3.13 Graph of $P(b)$: (a) Total energy and linear damping effect;(b) Expansion near $b = 0$.

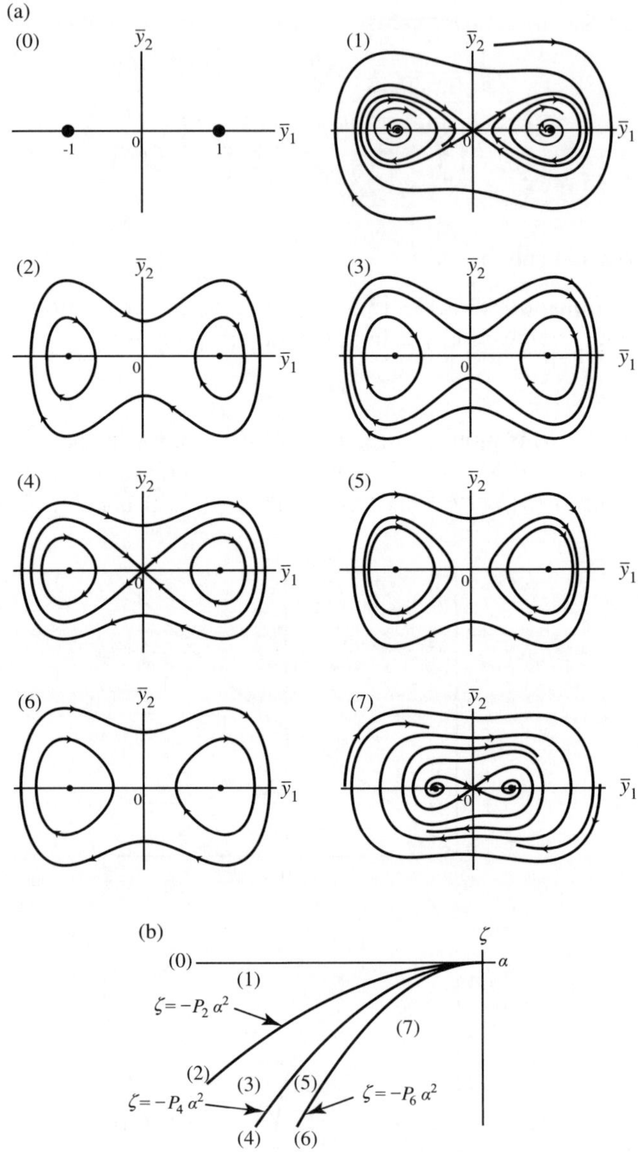

Figure 3.14 Phase portraits and global bifurcations: (a) Phase portraits; (b) Completion of the bifurcation set.

$(H = b < 0)$. Then, there are two periodic orbits surrounding a nontrivial fixed point and the phase portrait is plotted as the case (1) in Figure 3.14(a). When $P(b) = \zeta/\alpha^2$ is larger than about $P(b) \approx 0.171446$, b is positive. Then, there is one periodic orbit surrounding three fixed points and the phase portrait is shown as the case (7) in

Figure 3.14(a). The graph of $P(b)$ near $b \approx 0$ ($P(b) \approx 0.17143$) does not monotonically increase and in this part, some global bifurcations can be produced. By using Figure 3.13(b), the change of a phase portrait depending on $\zeta/(\alpha^2)$ is examined.

When $P(b) < 0$, there is no periodic orbit. When $P(b) = 0$, a Hopf bifurcation occurs at the nontrivial fixed points (this is the case (0) in Figure 3.13(a) and the associated phase portrait is labelled by (0) in Figure 3.14(a). The above mentioned periodic orbits are produced as the phase portrait (1) in Figure 3.14(a).

When $P(b)$ is increased until $P(b) \approx 0.171411$ (this is case (2) in Figure 3.13(b) and the associated phase portrait is labelled by (2) in Figure 3.14(a). At this point, a periodic orbit surrounding three fixed points is produced, and when $P(b)$ becomes larger than $P(b) = 0.171411$ (this is case (3) in Figure 3.13(b)), this periodic orbit bifurcates into two periodic orbits surrounding the three fixed points labelled by (3) in Figure 3.14(a).

For case (4) shown in Figure 3.13(b), the inner periodic orbit surrounding the three fixed points is connected at the origin and a homoclinic orbit appears because of $b = 0$ ($P(b) = 0.171430$). The phase portrait is shown as case (4) in Figure 3.14(a).

When $P(b)$ is further increased (case (5) in Figure 3.13(b)), the homoclinic orbit changes to an additional new periodic orbits surrounding each nontrivial fixed point, corresponding to the phase portrait (5) in Figure 3.14(a). When $P(b)$ is increased until $P(b) = 0.171446$ (this is case (6) in Figure 3.13(b) and the associated phase portrait is (6) in Figure 3.14(a), two periodic orbits surrounding each nontrivial fixed point are connected and only one periodic orbit surrounding each nontrivial fixed point exist. When $P(b)$ becomes larger than $P(b) = 0.171446$ (case (7) in Figure 3.13(b)), these periodic orbits disappear and only one periodic orbit surrounding the three fixed points remains as shown as case (7) in Figure 3.14(a). The phase portrait corresponds to Phase portrait-2 in the third quadrant in Figure 3.11.

From the above results, it can be seen that global bifurcations are produced in cases (2), (4), and (6) in Figure 3.13. Let the corresponding value of $P(b)$ be P_2, P_4, and P_6. Bifurcation sets for these bifurcations can be expressed as $\zeta = -P_2\alpha^2$, $\zeta = -P_4\alpha^2$, and $\zeta = -P_6\alpha^2$ and these are plotted in Figure 3.14(b). These bifurcation sets can be seen as one curve labelled by '*' in Figure 3.11.

3.6 Summary

In this chapter, the nonlinear characteristics of free vibrations of the Duffing oscillator with viscous damping have been discussed. First, by performing a local analysis, the fixed points have been determined and their stability examined. It has been shown that nonlinear stiffness and nonlinear damping influence postbuckling behaviour and the production of a limit cycle for self-excited oscillations. Then, using the knowledge of local dynamics, global analysis has been carried out from a qualitative point of view. For the case with a softening nonlinear stiffness ($\gamma < 0$), the system undergoes a Hopf bifurcation only at the trivial fixed point and the self-excited oscillation has a stable limit cycle. When the effect of negative linear damping becomes much larger, the stable limit cycle breaks though the heteroclinic bifurcation and there are no stable fixed points or limit cycles. The case with a hardening nonlinearity is more complicated. For the cases with positive and negative linear stiffness, the Hopf bifurcations occur for trivial and nontrivial fixed points, when the damping ratio changes from being positive to negative. In particular, the bifurcation at the nontrivial fixed points induces global bifurcations with an increase of negative damping, and through the global bifurcations, a periodic orbit including the trivial and two nontrivial fixed points is produced.

Acknowledgments

The author thanks Professor Kazuyuki Yagasaki at Niigata University for many useful suggestions and comments on the bifurcation analysis, and Mr. Yusuke Kanai and Mr. Hiroki Kato, graduate students at Keio University, for the help with drawing the figures.

References

[1] A. Nayfeh, D. T. Mook, *Non-linear Oscillations*, Wiley, New York, 1979.

[2] J. Guckenheimer, P. Holmes, *Non-linear Oscillations Dynamical Systems, and Bifurcations of Vector Fields*, Springer-Verlag, New York, 1990.

[3] P. Holmes, D. Rand, Phase portraits and bifurcations of the non-linear oscillator: $\ddot{x} + (\alpha + \gamma x^2)\dot{x} + \beta x + \delta x^3 = 0$. *International Journal of Non-linear Mechanics*, 15, 449–458, 1980.

[4] J. Carr, *Application of Center Manifold Theory*, Springer-Verlag, New York, 1981.

[5] H. Yabuno, Y. Kanai, Global bifurcations of Rayleigh-Duffing Oscillator compared with van der Pol oscillator, to be submitted to Nonlinear Dynamics.

4

Analysis techniques for the various forms of the Duffing equation

Livija Cveticanin
University of Novi Sad, Faculty of Technical Sciences, Serbia

4.1 Introduction

As discussed in Chapter 1 and seen in the Appendix, special attention in Duffing's book [1] was given to the analytical solution of the differential equation with cubic nonlinearity, which describes relatively small oscillations of a pendulum, given by the nondimensional equation

$$\ddot{y} + \alpha y - \gamma y^3 = F \cos \Omega t \tag{4.1.1}$$

where y is the displacement, α and γ are the linear and nonlinear stiffness parameters, respectively, and F and Ω are the amplitude and frequency of the harmonic excitation force respectively. Duffing presented an approximate analytical solution for the case when the nonlinearity is small, i.e., when $\gamma \ll \alpha$, and the excitation force is neglected, writing it down in the form of the Weierstrass elliptic integral $\wp$ [2]. This solution was used to analyse *free oscillations* of a symmetrical pendulum. The solution has some limitations: first, it is based on the assumption that the nonlinearity is small, and secondly, its form can be seen as complicated and not convenient for practical

The Duffing Equation: Nonlinear Oscillators and their Behaviour, First Edition. Edited by Ivana Kovacic and Michael J. Brennan. Published 2011 by John Wiley & Sons, Ltd.

applications. Later, many investigations have been carried out to find simpler, but sufficiently accurate approximate analytical solutions for the differential equation

$$\ddot{y} \pm \alpha y \pm \gamma y^3 = f(y, \dot{y}, \Omega t) \tag{4.1.2}$$

which represents the modified and more generalised version of Equation (4.1.1). This equation has an additional function *f*, which can consist of linear or nonlinear terms and also time-periodic functions that are not necessarily small.

For the case when there is a small cubic nonlinearity, the most widely applied methods are: the method of multiple scales [3], the Bogolubov–Mitropolski method [4], the Krylov–Bogolubov method [5], the straightforward expansion [3], Lindstedt–Poincaré method [3], etc. These methods have been modified for solving a system of two coupled differential equations with constant coefficients [6,7] and with parameters that slowly vary in time [8–12] by the author of this chapter. It is common for these methods to have a solution based on that of a linear differential equation, which is a trigonometric function.

The analytical procedures for solving the strong nonlinear Duffing equation given by Equation (4.1.2) are an extension of the methods mentioned previously. The methods of solution are based on the exact solution of the differential equation

$$\ddot{y} \pm \alpha y \pm \gamma y^3 = 0 \tag{4.1.3}$$

The exact solution of Equation (4.1.3) is given by *Jacobi elliptic functions* [13]. These functions represent a more general class of periodic functions, which include trigonometric functions as a particular case (see Appendix 4AI). In this chapter, the closed form analytic solution of Equation (4.1.3) is widely discussed for various initial conditions. Using the exact solution this equation in conjunction with perturbation and nonperturbation methods, the approximate solutions for various forms of Duffing equations are obtained, based on Jacobi elliptic functions.

It is believed that the Jacobi elliptic function was first used as a generating solution of the Duffing equation in 1969 by Barkham and Soudack [14]. Since then many approximate analytical solving methods have been adopted, taking into consideration the specific properties of Jacobi elliptic functions. In [15–17], modifications of the averaging method are presented, where the trial solution is the Jacobi elliptic function, rather than a trigonometric function. The elliptic-Krylov–Bogolubov method [18–22] uses a general expression for the time derivative of the amplitude and phase similar to those obtained by the conventional averaging method that uses a trigonometric function. In [23–27], the conventional harmonic balance method is modified into the elliptic harmonic balance method for solving the strongly nonlinear Duffing equation. The methods combining harmonic balancing and the Krylov–Bogolubov method [28], the elliptic-perturbation technique [29], the elliptic Lindstedt–Poincaré method [30], the elliptic multiple scales method [31], the generalised Galerkin method [32], the homotopy-perturbation techniques [33], etc., use also the Jacobi elliptic function as the basic solution of Equation (4.1.3).

The author of this chapter has developed an approximate analytical solution method for the Duffing oscillator that has strong cubic [34], and strong quadratic nonlinearities [35,36]. It has also been developed for the generalised nonlinear oscillator [37], where the order of a strong nonlinear term need not be cubic or quadratic but of any fractional-order. The methods have been extended for solving of a system of coupled Duffing equations [38–40], and in particular, for solving differential equations that have complex variables [41–52]. The approximate analytical solution methods have also been developed for a strongly nonlinear Duffing equation with parameters that vary slowly in time [53].

The large number of approximate analytical solution methods for the Duffing equation published in numerous papers is usually divided into two groups. The first group involves techniques that require small physical parameters. In the second group the methods do not have such restriction. The methods belonging to the first group which are presented in this section are: the method of straightforward expansion, the elliptic-averaging method, the elliptic Krylov–Bogolubov method and the elliptic Lindstedt–Poincaré method. Usually, the last two methods are called 'elliptic perturbation methods', as the parameters of the elliptic function are perturbed. The approximate methods: the elliptic harmonic balance method, the weighted residual methods (Galerkin procedure) and the homotopy techniques, which are shown in this section, do not require the existence of a small parameter and belong to the second group. These methods are commonly called 'nonperturbation methods'.

In this chapter the following analytic asymptotic procedures are shown: (1) the straightforward expansion method, (2) the parameter-expanding method, (3) the averaging method, (4) the parameter perturbation method, (5) the harmonic balance method, (6) the weighted residual method, (7) the homotopy perturbation method and (8) the homotopy analysis method. They are all based on the exact analytic solution of Equation (4.1.3). The theoretical considerations shown in the sections are applied to solve some examples. The analytically obtained approximate solutions are compared with numerically obtained 'exact' solutions.

In spite of the fact that numerous methods have been developed to solve the strongly nonlinear Duffing equation analytically, some asymptotic approaches still need to be investigated. All the existing asymptotic solving procedures have some disadvantages as well as advantages and these are also discussed in this chapter.

4.2 Exact solution for free oscillations of the Duffing equation with cubic nonlinearity

The differential equation (4.1.3) with initial conditions

$$y(0) = y_0, \quad \dot{y}(0) = \dot{y}_0 \tag{4.2.1a,b}$$

has an exact solution

$$y = Y \,\mathrm{ep}(\psi, k^2) \tag{4.2.2}$$

where ep denotes the general expression for the Jacobi elliptic sine (sn), cosine (cn) and delta (dn) functions [54], and ψ is the argument of the Jacobi elliptic function

$$\psi = \omega t + \theta \tag{4.2.3}$$

where ω is the frequency, k^2 is the modulus of the elliptic function and Y and θ are the amplitude and phase angle, which are constants that depend on the initial conditions (4.2.1).

Depending on the sign of the coefficients α and γ, the following type of equations can be distinguished (see [19]): hardening ($\alpha > 0$ and $\gamma > 0$); softening ($\alpha > 0$ and $\gamma < 0$); and, finally, the negative linear-positive cubic one ($\alpha < 0$ and $\gamma > 0$).

For the case of a strong cubic hardening nonlinearity, the Duffing equation

$$\ddot{y} + \alpha y + \gamma y^3 = 0 \tag{4.2.4}$$

has an exact analytical solution in the form of the Jacobi elliptic function

$$y = Y \mathrm{cn}(\omega t + \theta, k^2) \tag{4.2.5}$$

where cn is the Jacobi elliptic function (see Appendix 4AI and [54]) and:

$$\omega^2 = \alpha + \gamma Y^2, \quad k^2 = \frac{\gamma Y^2}{2(\alpha + \gamma Y^2)} \tag{4.2.6a,b}$$

Substituting Equation (4.2.5) and its time derivative into Equation (4.2.4), gives the amplitude Y

$$Y = \left(-\frac{\alpha}{\gamma} + \frac{\alpha + \gamma y_0^2}{\gamma} \sqrt{1 + \frac{2\dot{y}_0^2 \gamma}{\alpha}} \right)^{1/2} \tag{4.2.7}$$

and the phase angle θ, which satisfies the following relation

$$\frac{\mathrm{sn}(\theta, k^2)}{\mathrm{cn}(\theta, k^2)} \mathrm{dn}(\theta, k^2) = -\frac{\dot{y}_0}{y_0 \omega} \tag{4.2.8}$$

For the specific initial conditions

$$y(0) = y_0, \quad \dot{y}(0) = 0 \tag{4.2.9a,b}$$

the amplitude and phase angle are

$$Y = y_0, \quad \theta = 0 \tag{4.2.10a,b}$$

while for

$$y(0) = 0, \quad \dot{y}(0) = \dot{y}_0 \tag{4.2.11a,b}$$

they are given by

$$Y = \left(-\frac{\alpha}{\gamma} + \frac{\alpha}{\gamma} \sqrt{1 + \frac{2\dot{y}_0^2 \gamma}{\alpha}} \right)^{1/2}, \quad \theta = K(k) \tag{4.2.12a,b}$$

Table 4.1 The solutions of Equation (4.2.4) depending on the sign of the coefficients.

Type	Solution	Frequency	Modulus
$\alpha > 0, \gamma < 0$	$y = Y\,\mathrm{sn}(\omega t + \theta, k^2)$	$\omega^2 = \alpha - \dfrac{\gamma Y^2}{2} > 0$	$k^2 = \dfrac{\gamma Y^2}{2\omega^2}$
$\alpha < 0, \gamma > 0$	$y = Y\,\mathrm{cn}(\omega t + \theta, k^2)$	$\omega^2 = \gamma Y^2 - \alpha > 0$	$k^2 = \dfrac{\gamma Y^2}{2\omega^2}$

where $K(k)$ is the complete elliptic integral of the first kind [55]. The first term of the series expansion of Equation (4.2.12) gives the approximate amplitude $Y \approx \dot{y}_0$.

Using the aforementioned procedure, the solutions for the softening and negative linear-positive cubic Duffing equations can be determined and are given in Table 4.1.

4.2.1 The frequency and period of free oscillations of the Duffing oscillator

Using the properties of the Jacobi elliptic functions, given in Appendix 4AI, and the relations for ω and k^2 given above, the period of oscillations for the hardening, softening and negative linear-positive cubic Duffing equations can be determined. Namely, the frequency of the Jacobi elliptic functions ω differs from the frequency of vibration Ω. The period of the Jacobi elliptic function depends on two parameters: the frequency of the Jacobi elliptic function ω and the modulus k^2

$$T = \frac{4K(k)}{\omega} \tag{4.2.13}$$

On the other hand, due to the physical properties of free vibrating systems, the period of vibration is $T = 2\pi/\Omega$. Equating the period values, the frequency of vibration is given by

$$\Omega = \frac{2\pi}{T} = \frac{\pi\omega}{2K(k)} \tag{4.2.14}$$

For the hardening Duffing equation the period of vibration is

$$T = \frac{4K\left(\dfrac{\gamma Y^2}{2(\alpha + \gamma Y^2)}\right)}{\sqrt{\alpha + \gamma Y^2}} \tag{4.2.15}$$

and the frequency of vibration, calculated for a total phase angle of 2π is

$$\Omega = \frac{\pi\sqrt{\alpha + \gamma Y^2}}{2K\left(\dfrac{\gamma Y^2}{2(\alpha + \gamma Y^2)}\right)} \tag{4.2.16}$$

Using the development of the total elliptic integral of the first kind (see Equation (4AI.3)) and assuming that the cubic term γY^3 is small in comparison to the linear term αY, the approximate value of Ω^2, is obtained

$$\Omega^2 = (\alpha+\gamma Y^2)\left(1-\frac{\gamma Y^2}{4\alpha}\left(1-\frac{\gamma Y^2}{\alpha}\right)\right)+\ldots \approx \alpha+\frac{3}{4}\gamma Y^2+\frac{(\gamma Y^2)^3}{4\alpha^2}+\ldots \quad (4.2.17)$$

Neglecting the third and higher terms of the expansion as the small values of the third and higher order, the relation (4.2.17) simplifies to

$$\Omega = \sqrt{\alpha+\frac{3}{4}\gamma Y^2} \quad (4.2.18)$$

This form of the equation for the approximate frequency can also be obtained if the solution of Equation (4.2.4) is assumed in the form of a trigonometric function [5].

In accordance with Equations (4.2.13) and (4.2.14) and the expressions given in Table 4.1, the period of vibration of the softening oscillator is:

$$T = \frac{4K\left(\dfrac{\gamma Y^2}{2\alpha-\gamma Y^2}\right)}{\sqrt{\alpha-\dfrac{\gamma Y^2}{2}}} \quad (4.2.19)$$

and the frequency is

$$\Omega = \frac{\pi\sqrt{\alpha-\dfrac{\gamma Y^2}{2}}}{2K\left(\dfrac{\gamma Y^2}{2\alpha-\gamma Y^2}\right)} \quad (4.2.20)$$

Similarly, the period of oscillations of the negative linear-positive cubic oscillator is given by:

$$T = \frac{4K\left(\dfrac{\gamma Y^2}{2(\gamma Y^2-\alpha)}\right)}{\sqrt{\gamma Y^2-\alpha}} \quad (4.2.21)$$

the frequency of which is

$$\Omega = \frac{\pi\sqrt{\gamma Y^2-\alpha}}{2K\left(\dfrac{\gamma Y^2}{2(\gamma Y^2-\alpha)}\right)} \quad (4.2.22)$$

It is clear that the period and frequency of vibration for all types of the Duffing oscillator depend on the initial conditions, i.e., initial amplitude and velocity, and also on the coefficients of the linear and cubic terms of the system. For small values of Y, the frequency of the hardening (4.2.16) and softening cubic oscillator (4.2.20) tends to the natural frequency of the linear oscillator, i.e., $\Omega = \sqrt{\alpha}$. By increasing the initial displacement, the frequency of vibration of the oscillator with hard cubic nonlinearity increases, too, tending to infinity for an infinite value of initial displacement. The increase of the initial displacement in the softening oscillator is limited by $\gamma Y^2 \leq 2\alpha$. Besides, increasing the value of Y causes the decrease of the frequency of vibration for the softening-type oscillator. In the negative linear-positive cubic oscillator a real value of frequency exists for $\gamma Y^2 \geq \alpha$: the higher the value of Y, the higher the value of the frequency Ω.

4.2.2 Discussion

From the results obtained, the following conclusions can be drawn:

1. The frequency and the modulus of the Jacobi elliptic function strongly depend on the amplitude Y, i.e., the initial values y_0 and $\dot{y}_0$.

2. The period and the frequency of the vibration depend on the initial conditions, too. In spite of the fact that the period of vibration is equal to the corresponding value of the Jacobi elliptic function, the frequency of the Jacobi elliptic function and of the vibrating system have different numerical values.

3. According to Equation (4.2.5) and Equation (4.2.6), it is clear that the solution for the hardening oscillator exists for all values of α and γ, but for the softening and negative linear-positive cubic oscillators the motion is periodic only for some special relations between the parameters α and γ and the initial amplitude of Y, as given in Table 4.1.

4. Examining the values of k^2 for the types of oscillator discussed, it can be seen that the absolute value of the modulus is in the interval 0 to 1/2.

5. The amplitude Y and phase θ are calculated using the initial conditions (4.2.1), (4.2.9) or (4.2.11).

6. For the pure cubic equation, when $\alpha = 0$,

$$\ddot{y} + \gamma y^3 = 0 \tag{4.2.23}$$

   the modulus of the Jacobi elliptic function is constant $k^2 = 1/2$ and the frequency is $\omega = Y\sqrt{\gamma}$.

   The closed form solution is

$$y = Y \operatorname{cn}(Yt\sqrt{\gamma} + \theta, 1/2) \tag{4.2.24}$$

   where for the initial conditions given in Equation (4.2.1), the amplitude of vibration Y has the value

$$Y = \left(y_0^4 + \frac{2}{\gamma} \dot{y}_0^2 \right)^{1/4} \tag{4.2.25}$$

and the phase angle θ satisfies Equation (4.2.8). The period of the Jacobi elliptic function that corresponds to the period of vibration is

$$T = \frac{4K(1/2)}{Y\sqrt{\gamma}} = \frac{7.41630}{Y\sqrt{\gamma}} \tag{4.2.26}$$

and the frequency of vibration is

$$\Omega = \frac{2\pi}{4K(1/2)} Y\sqrt{\gamma} = 0.84721Y\sqrt{\gamma} \tag{4.2.27}$$

Using Equation (4.2.27) and the initial conditions (4.2.9) i.e., Equation (4.2.10), the approximate solution of Equation (4.2.23) is introduced in the form of a trigonometric cosine function

$$y = Y\cos(0.84721Y\sqrt{\gamma}t) \tag{4.2.28}$$

In Figure 4.1, the exact $y_e = \text{cn}(t, 1/2)$ and the approximate $y_a = \cos(0.84721t)$ solutions of the differential equation (4.2.23) for the initial conditions $y(0) = y_0 = 1$, $\dot{y}(0) = 0$ are compared. It can be concluded that, although the functions describing the approximate solutions are different, the solution given by Equation (4.2.28) is

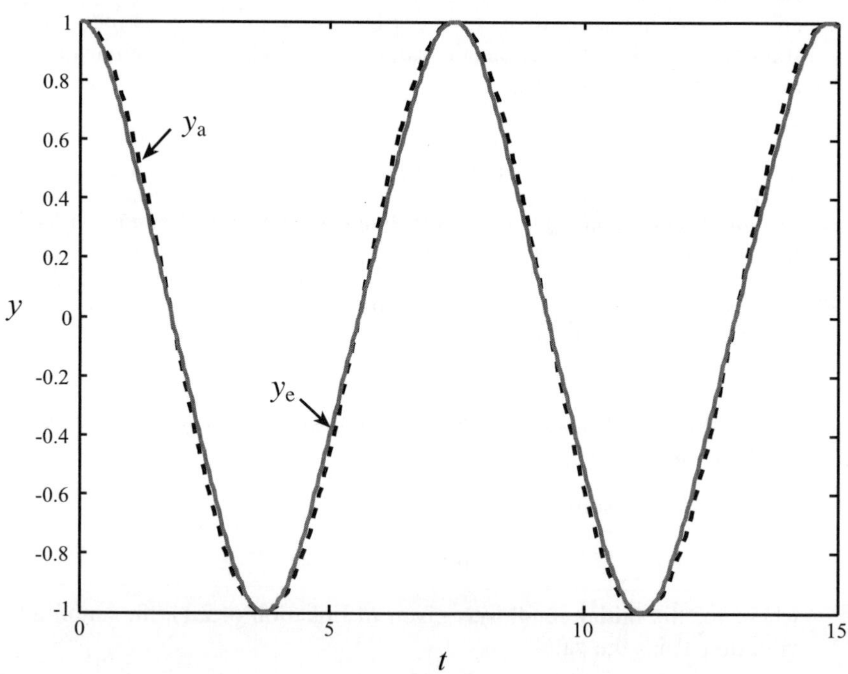

Figure 4.1 The exact y_e (grey solid line) and approximate y_a (black dashed line) solution for the pure cubic Duffing equation.

satisfactory as the maximal amplitudes and the periods of functions are equal to the ones given by the exact solution.

In the literature, the approximate value of the frequency of vibration Ω is often sought. Assuming the solution of Equation (4.2.23) in the form of a trigonometric function, the approximate value of the frequency is

$$\Omega_{\mathrm{a}} = \sqrt{\frac{3}{4}\gamma Y^2} = 0.86603 Y\sqrt{\gamma} \tag{4.2.29}$$

To compare the approximate value of frequency Ω_{a} (4.2.29) with the exact frequency given in Equation (4.2.27), the percentage error is calculated to be

$$\frac{\Omega_{\mathrm{a}} - \Omega}{\Omega} 100\% = 2.02\% \tag{4.2.30}$$

The difference between the approximate and exact frequency of vibration is small and constant, and does not depend on the parameters of the system and the initial amplitude.

It should be emphasised that for simplicity, most of the analysis in this chapter is carried out for the hardening Duffing oscillator described by

$$\ddot{y} + \alpha y + \gamma y^3 = f(y, \dot{y}), \text{ and } \ddot{y} + \alpha y + \gamma y^3 = f(\Omega t, y, \dot{y}) \tag{4.2.31a,b}$$

with the initial conditions given in Equation (4.2.1).

4.3 The elliptic harmonic balance method

The method of elliptic harmonic balance is one of the most widely used methods for solving the strong cubic Duffing equation (4.1.2), where the function f depends on the displacement y, the velocity $\dot{y}$ and the time-periodic function. The solution is assumed to be of the form of the Jacobi elliptic function (4.2.2). Substituting the assumed solution into the Duffing equation (4.2.4), gives the following algebraic equation

$$a_1(p)\mathrm{ep} + a_3(p)\mathrm{ep}^3 + \sum_i^N a_i(p)\mathrm{ep}^i = 0 \tag{4.3.1}$$

where $a_1, a_2, \ldots, a_i$ are functions of the unknown parameters p and N is an arbitrary integer. The elliptic harmonic balance method involves equating to zero all the terms in Equation (4.3.1) that have the same order of the elliptic function. The following system of coupled algebraic equations is obtained

$$a_1(p) = 0, \quad a_3(p) = 0, \ldots, \quad a_i(p) = 0 \tag{4.3.2}$$

Usually, the number of equations (4.3.2) is higher than the number of the parameters p. The number of equations is then chosen from the system of equations given in Equation (4.3.2) so that it corresponds to the number of the unknown parameters. By solving this subsystem of algebraic equations, the parameters p can be determined. The better the choice of equations, the more accurate the values of the parameters determined.

In *autonomous systems*, for example, very often only two parameters are unknown: the frequency and the modulus of the elliptic function. In accordance with

$$a_1(\omega, k^2) = 0, \quad a_3(\omega, k^2) = 0 \tag{4.3.3a,b}$$

the parameters ω and k^2 are determined.

The main advantage of the harmonic balance method is its simplicity. The main disadvantage of the method is that there are no strict rules for the assumption of the solution for the Duffing equation, and the accuracy of the method strictly depends on the choice of the number of harmonics in the solution.

In this section the elliptic harmonic balance method is applied to solve the Duffing equation with a strong quadratic term and with harmonic excitation, and the Duffing equation with strong damping.

4.3.1 The Duffing equation with a strong quadratic term

The Duffing equation with an additional strong quadratic nonlinearity is given by

$$\ddot{y} + \alpha y + \beta y|y| + \gamma y^3 = 0 \tag{4.3.4}$$

and the initial conditions are

$$y(0) = y_0, \quad \dot{y}(0) = 0 \tag{4.3.5a,b}$$

The solution for Equation (4.3.4) is assumed to be of the form Equation (4.2.5), where ω and k^2 are unknown parameters. Substituting Equation (4.2.5) into Equation (4.3.4) and using the series expansion of the function $\mathrm{cn}|\mathrm{cn}|$, as is given by Equation (4AII.10) in Appendix 4AII

$$\mathrm{cn}|\mathrm{cn}| \approx a_0\mathrm{cn} + a_1\mathrm{cn}^3 \tag{4.3.6}$$

the following approximate equation is obtained

$$-Y\omega^2\mathrm{cn}(1-2k^2 + 2k^2\mathrm{cn}^2) + \alpha Y\,\mathrm{cn} + \gamma Y^3\mathrm{cn}^3 + \beta Y|Y|(a_0\,\mathrm{cn} + a_1\,\mathrm{cn}^3) = 0 \tag{4.3.7}$$

Applying the harmonic balance procedure and equating the terms with the same order of the elliptic function cn, the following two coupled algebraic equations are obtained

$$\begin{aligned} &\mathrm{cn}: -\omega^2(1-2k^2) + \alpha + \beta|Y|a_0 = 0, \\ &\mathrm{cn}^3: -2k^2\omega^2 + \gamma Y^2 + \beta|Y|a_1 = 0 \end{aligned} \tag{4.3.8a,b}$$

Solving the system of equations (4.3.8), the parameters of the elliptic function are found to be

$$\omega^2 = \alpha + \beta|Y|(a_0 + a_1) + \gamma Y^2, \quad k^2 = \frac{\gamma Y^2 + \beta|Y|a_1}{2(\alpha + \beta|Y|(a_0 + a_1) + \gamma Y^2)} \tag{4.3.9a,b}$$

Examining Equations (4.3.9), it can be seen that the coefficient β of the quadratic term has a significant influence on both the parameters of the elliptic function. The frequency and the modulus of the elliptic function, i.e., the period of oscillations, given by

$$T = \frac{4K(k)}{\omega} = \frac{4K\left(\dfrac{\gamma Y^2 + \beta|Y|a_1}{2(\alpha + \beta|Y|(a_0 + a_1) + \gamma Y^2)}\right)}{\sqrt{\alpha + \beta|Y|(a_0 + a_1) + \gamma Y^2}} \tag{4.3.10}$$

depend on Y. Using Equations (4.2.5), (4.3.5) and (4.3.9), the approximate solution of Equation (4.3.4) is given by

$$y = y_0 \operatorname{cn}\left(t\sqrt{\alpha + \gamma y_0^2 + \beta|y_0|(a_0 + a_1)}, \frac{\gamma y_0^2 + \beta|y_0|a_1}{2(\alpha + \gamma y_0^2 + \beta|y_0|(a_0 + a_1))}\right) \tag{4.3.11}$$

The frequency and the modulus of the elliptic function do not depend on the sign of the initial displacement y_0. For $\alpha = \beta = \gamma = 1$, the initial conditions $y_0 = 1$ and $\dot{y}_0 = 0$, and the known parameter values $a_0 = 16/15\pi$ and $a_1 = 32/15\pi$ (see Appendix 4AII, Equation (4AII.14)), the approximate solution is

$$y_a = \operatorname{cn}(1.7374t, 0.27812) \tag{4.3.12}$$

In Figure 4.2, y_a is plotted together with the solution y_N, obtained by solving the differential equation (4.3.4) numerically using the Runge–Kutta method. Comparing the solutions, it can be seen that they are in good agreement despite the fact that the initial values are large and that the coefficients of the nonlinear terms β and γ are of the same order as the coefficient of the linear term α. In addition, it can be seen that the accuracy of the approximate solution does not deteriorate even after a long period of time.

4.3.2 The Duffing equation with damping

The elliptic harmonic balance method is applied to the damped Duffing equation

$$\ddot{y} + 2\delta\dot{y} + \alpha y + \gamma y^3 = 0 \tag{4.3.13}$$

where δ is the damping coefficient that does not need to be small. The approximate solution is assumed as

$$y = Ye^{-st}\operatorname{cn}(\omega t + \theta, k^2) \equiv Ye^{-st}\operatorname{cn} \tag{4.3.14}$$

where s, ω and k^2 are the unknown parameters that need to be calculated. It is assumed that the initial amplitude Y is y_0 and the phase θ is zero and that

$$y = y_0 e^{-st}\operatorname{cn}(\omega t, k^2) \equiv y_0 e^{-st}\operatorname{cn} \tag{4.3.15}$$

Substituting Equation (4.3.15) and its time derivatives into Equation (4.3.13), and separating the terms with the same order of the Jacobi elliptic function, the following system of algebraic equations is obtained

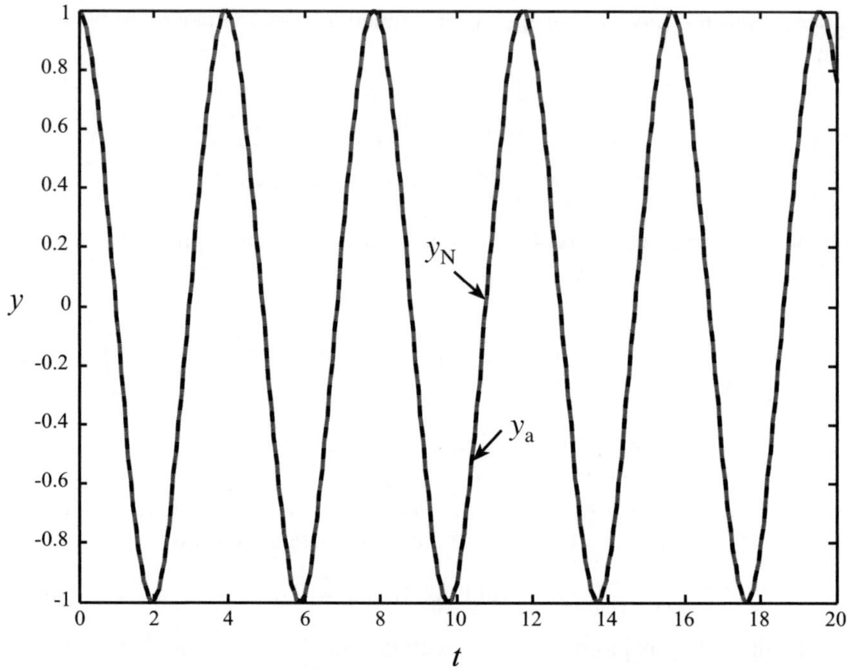

Figure 4.2 The time histories obtained analytically y_a (black dashed line) and numerically y_N (grey solid line) for $\alpha=\beta=\gamma=1$ and initial conditions $y_0=1$ and $\dot{y}_0=0$.

$$\begin{aligned}
&\text{cn} : s^2-\omega^2(1-2k^2)+\alpha-2\delta s=0,\\
&\text{sn dn} : 2(s-\delta)\omega=0, \qquad\qquad (4.3.16\text{a–c})\\
&\text{cn}^3 : -2k^2\omega^2+\gamma y_0^2=0
\end{aligned}$$

The constraint for relations (4.3.16) is that the initial displacement must satisfy the relation $Y=y_0<1$. Solving Equations (4.3.16) gives

$$s=\delta, \quad \omega^2=\alpha+\gamma y_0^2-\delta^2, \quad k^2=\frac{\gamma y_0^2}{2(\alpha+\gamma y_0^2-\delta^2)} \qquad (4.3.17\text{a–c})$$

Substituting Equation (4.3.17) into Equation (4.3.15), the following approximate solution is obtained

$$y=y_0\,\mathrm{e}^{-\delta t}\mathrm{cn}\left(t\sqrt{\alpha-\delta^2+\gamma y_0^2},\quad \frac{\gamma y_0^2}{2(\alpha-\delta^2+\gamma y_0^2)}\right) \qquad (4.3.18)$$

The frequency and the modulus of the Jacobi elliptic function depend not only on the coefficients of linear and cubic terms and initial conditions but also on the damping in the system. For $\sqrt{\alpha - \delta^2 + \gamma y_0^2} > 0$, the amplitude of vibration decreases exponentially, but the period of oscillation is constant

$$T = \frac{4K\left(\dfrac{\gamma y_0^2}{\alpha - \delta^2 + \gamma y_0^2}\right)}{\sqrt{\alpha - \delta^2 + \gamma y_0^2}}$$

To check the accuracy of the result, numerical examples are considered. For the oscillator

$$\ddot{y} + 0.2\dot{y} + y + y^3 = 0 \tag{4.3.19}$$

with the initial conditions

$$y_0 = 0.1, \quad \dot{y}_0 = -0.01 \tag{4.3.20a,b}$$

the approximate analytical solution is

$$y_a = 0.1e^{-0.1t}\text{cn}(t, 0.005) \tag{4.3.21}$$

The solution y_a is compared with the numerical solution y_N, obtained by solving Equation (4.3.13) by using the Runge–Kutta method. The corresponding solutions are plotted in Figure 4.3(a).

For the oscillator

$$\ddot{y} + \dot{y} + y + y^3 = 0 \tag{4.3.22}$$

with initial conditions

$$y_0 = 0.1. \quad \dot{y}_0 = -0.05 \tag{4.3.23a,b}$$

the analytical solution is

$$y_a = 0.1e^{-0.5t}\,\text{cn}(0.87178t, 0.0068) \tag{4.3.24}$$

which is plotted in Figure 4.3(b). It can be seen that the analytical solution and the numerical solution y_N are in good agreement.

4.3.3 The harmonically excited Duffing oscillator

The harmonically excited Duffing oscillator is described by

$$\ddot{y} + \alpha y + \gamma y^3 = F\cos\Omega t \tag{4.3.25}$$

with the initial conditions

$$y(0) = 0, \quad \dot{y}(0) = 0 \tag{4.3.26a,b}$$

where F and Ω are the amplitude and frequency of the excitation force, respectively. In spite of the fact that the principle of superposition does not work in nonlinear systems, it is assumed that the solution is a sum of the Jacobi elliptic function which

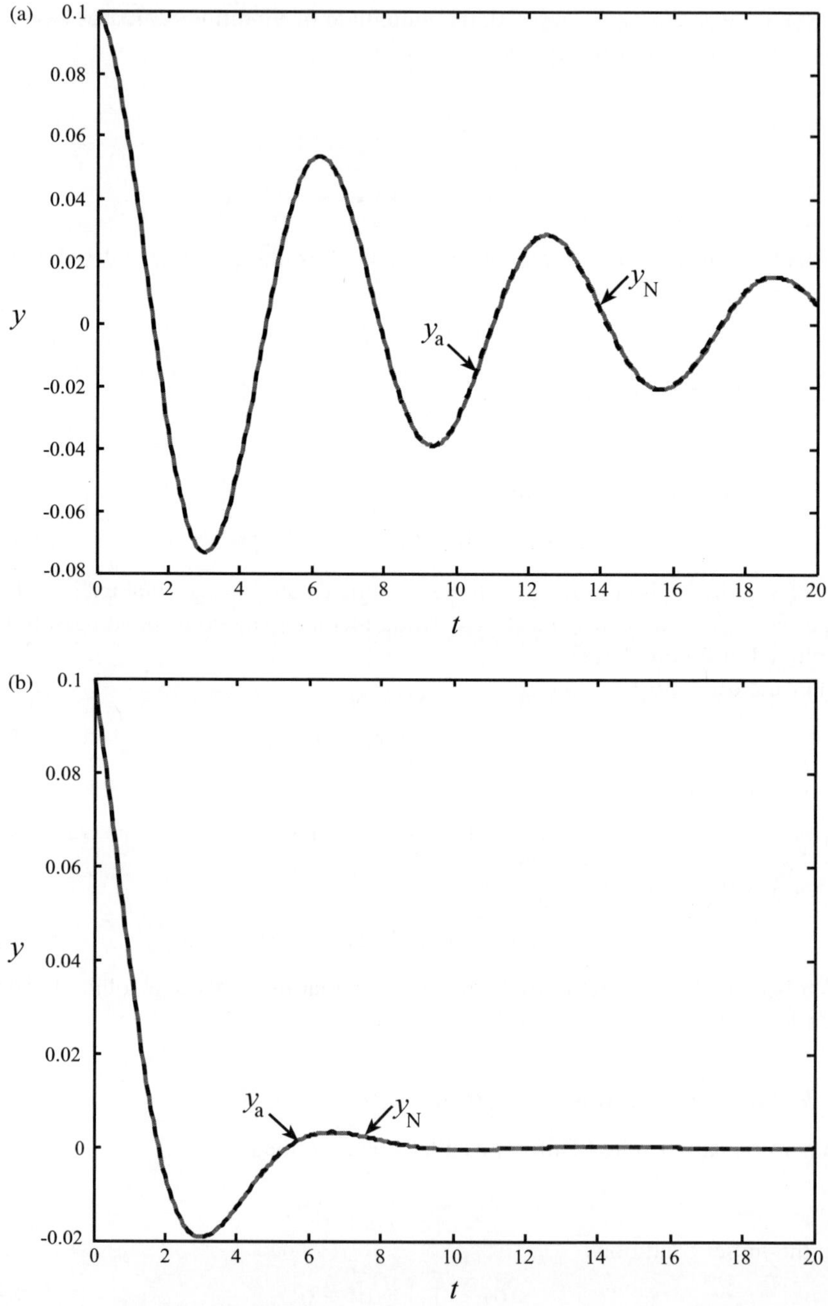

Figure 4.3 The time histories obtained analytically y_a (black dashed line) and numerically y_N (grey solid line) for: (a) $y_0 = 0.1$ and $\delta = 0.1$; (b) $y_0 = 0.1$ and $\delta = 0.5$.

has the form of Equation (4.2.5) and a cosine function with the frequency of excitation force Ω

$$y = Y\left(\mathrm{cn}(\Omega_1 t, k^2) - \cos\Omega t\right) \tag{4.3.27}$$

where Y, k^2 and Ω_1 are the unknown parameters which need to be calculated. For simplification, instead of the Jacobi elliptic function cn, the trigonometric function cosine is introduced with the frequency Ω^* that satisfies the equality of the period of vibration

$$T = \frac{4K(k)}{\Omega_1} = \frac{2\pi}{\Omega^*} \tag{4.3.28}$$

The modified version of the approximate solution is

$$y = Y\left(\cos\Omega^* t - \cos\Omega t\right) \tag{4.3.29}$$

where Ω^* satisfies Equation (4.3.28). The solution (4.3.29) has two parts: one, which describes the oscillatory motion with the excitation frequency Ω and the second, which gives the correction to the frequency of the free oscillator caused by excitation. Usually, only the solution with excitation frequency Ω called the '*steady-state* solution at the excitation frequency' is considered. In this section both the terms in Equation (4.3.29) are taken into consideration.

Substituting Equation (4.3.29) and its second time derivative into Equation (4.3.25) and separating the terms with $\cos\Omega^* t$ and $\cos\Omega t$, two algebraic equations are obtained

$$\left(\alpha - \Omega^{*2}\right) + \frac{9}{4}\gamma Y^2 = 0 \tag{4.3.30}$$

$$Y\left(\Omega^2 - \alpha\right) - \frac{9}{4}\gamma Y^3 = F \tag{4.3.31}$$

Equation (4.3.31) is cubic in Y and, in general, can have one, two or three real solutions depending on the sign of the expression

$$D = \frac{16}{81\gamma^2}\left(\left(\frac{F}{2}\right)^2 - \frac{4}{9\gamma}\left(\frac{\Omega^2 - \alpha}{3}\right)^3\right) \tag{4.3.32}$$

For $D < 0$, three real solutions of Equation (4.3.25) exist

$$y_i = Y_i\left(\cos\left(t\sqrt{\alpha + \frac{9}{4}\gamma Y_i^2}\right) - \cos\Omega t\right) \tag{4.3.33}$$

where $i = 1, 2, 3$ and

$$\cos\phi = \frac{2F}{9\gamma}\left(\frac{4}{27\gamma}\left(\Omega^2 - \alpha\right)\right)^{-3/2} \tag{4.3.34}$$

with

$$
\begin{aligned}
Y_1 &= 2\sqrt{\frac{4(\Omega^2-\alpha)}{27\gamma}}\cos\frac{\phi}{3}, \\
Y_2 &= 2\sqrt{\frac{4(\Omega^2-\alpha)}{27\gamma}}\cos\frac{\phi+2\pi}{3}, \\
Y_3 &= 2\sqrt{\frac{4(\Omega^2-\alpha)}{27\gamma}}\cos\frac{\phi+4\pi}{3}
\end{aligned}
\tag{4.3.35a-c}
$$

For $D=0$, which holds when

$$
\left(\frac{F}{2}\right)^2-\frac{4}{9\gamma}\left(\frac{\Omega^2-\alpha}{3}\right)^3=0 \tag{4.3.36}
$$

two real solutions are

$$
\begin{aligned}
y_1 &= -2\left(\frac{2F}{9\gamma}\right)^{1/3}\left(\cos t\sqrt{\alpha+9\gamma\left(\frac{2F}{9\gamma}\right)^{2/3}}-\cos\Omega t\right), \\
y_{2,3} &= \left(\frac{2F}{9\gamma}\right)^{1/3}\left(\cos t\sqrt{\alpha+\frac{9\gamma}{4}\left(\frac{2F}{9\gamma}\right)^{2/3}}-\cos\Omega t\right)
\end{aligned}
\tag{4.3.37a,b}
$$

From Equation (4.3.36),

$$
F=\frac{4}{\sqrt{3\gamma}}\left(\Omega^2-\alpha\right)^{3/2} \tag{4.3.38}
$$

and the solutions are therefore

$$
\begin{aligned}
y_1 &= -\frac{4}{\sqrt[3]{3}}\sqrt{\frac{\Omega^2-\alpha}{3\gamma}}\left(\cos t\sqrt{\alpha+\frac{12}{\sqrt[3]{9}}(\Omega^2-\alpha)}-\cos\Omega t\right), \\
y_{2,3} &= \frac{2}{\sqrt[3]{3}}\sqrt{\frac{\Omega^2-\alpha}{3\gamma}}\left(\cos t\sqrt{\alpha+3\frac{\Omega^2-\alpha}{\sqrt[3]{9}}}-\cos\Omega t\right)
\end{aligned}
\tag{4.3.39a,b}
$$

If $D>0$, only one real solution exists, where

$$
Y=\sqrt[3]{-\frac{2F}{9\gamma}+\sqrt{D}}+\sqrt[3]{-\frac{2F}{9\gamma}-\sqrt{D}} \tag{4.3.40}
$$

and

$$\Omega^* = \sqrt{\alpha + \frac{9}{4}\gamma Y} \tag{4.3.41}$$

If $\Omega = \sqrt{\alpha}$, the amplitude of vibration is $Y = \sqrt[3]{-4F/(9\gamma)}$ and the oscillations are

$$\begin{aligned} y &= \sqrt[3]{-\frac{4F}{9\gamma}}\left(\cos t\sqrt{\alpha + \frac{9}{4}\sqrt[3]{\frac{16F^2}{81\gamma^2}}} - \cos t\sqrt{\alpha}\right) \\ &= 2\sqrt[3]{-\frac{4F}{9\gamma}}\sin t\frac{\sqrt{\alpha + \frac{9}{4}\sqrt[3]{\frac{16F^2}{81\gamma^2}}} + \sqrt{\alpha}}{2}\sin t\frac{\sqrt{\alpha + \frac{9}{4}\sqrt[3]{\frac{16F^2}{81\gamma^2}}} - \sqrt{\alpha}}{2} \end{aligned} \tag{4.3.42}$$

Equation (4.3.29) can conveniently be written as

$$y = -2Y\sin\left(\frac{\Omega^* + \Omega}{2}t\right)\sin\left(\frac{\Omega^* - \Omega}{2}t\right) \tag{4.3.43}$$

which means that the periods of oscillation are

$$T_1 = \frac{4\pi}{\Omega^* + \Omega}, \quad T_2 = \frac{4\pi}{\Omega^* - \Omega} \tag{4.3.44}$$

Introducing the *detuning parameter* σ, which quantitatively describes the proximity of Ω to $\sqrt{\alpha}$, and when $\Omega - \sqrt{\alpha} = \varepsilon\sigma$, where $\varepsilon \ll 1$ is a small parameter, Equation (4.3.31) transforms into

$$\left(\sigma - \frac{9}{8}\frac{\gamma}{\sqrt{\alpha}}Y^2\right)Y = \frac{F}{2\sqrt{\alpha}} \tag{4.3.45}$$

The relation (4.3.45) has form of the frequency–amplitude equation given by Nayfeh and Mook [3]. Nayfeh and Mook [3] considered the steady-state solution at the excitation frequency of Equation (4.3.25) when the nonlinearity and the amplitude of the excitation force are both small. Due to averaging, the coefficient of the second term in Equation (4.3.45) is 3/8 in [3]. This procedure is also given in Chapter 5 for the case when the damping exists in the system.

For the pure cubic oscillator, when $\alpha = 0$, the relationship in Equation (4.3.45) gives the amplitude

$$Y = -\sqrt[3]{\frac{2F}{9\gamma}} = -0.60571\sqrt[3]{\frac{F}{\gamma}} \tag{4.3.46}$$

The result in Equation (4.3.46) is equal to that calculated in Equation (4.3.37) for Y if $D = 0$ and when Equation (4.3.38) is satisfied, i.e., when

$$F = \frac{4\Omega^3}{9\sqrt{3\gamma}} = 0.2566\frac{\Omega^3}{\sqrt{\gamma}} \tag{4.3.47}$$

For the linear case, when $\gamma = 0$, and

$$\Omega^* = \sqrt{\alpha}, \quad Y = \frac{F}{\Omega^2 - \alpha} \tag{4.3.48a,b}$$

only one real solution exists

$$y = \frac{F}{\Omega^2 - \alpha}\left(\cos t\sqrt{\alpha} - \cos \Omega t\right) = -\frac{2F}{\Omega^2 - \alpha}\sin\frac{t(\sqrt{\alpha}+\Omega)}{2}\sin\frac{t(\sqrt{\alpha}-\Omega)}{2} \tag{4.3.49}$$

This solution is well known. The case when the excitation frequency is equal to the frequency of the linear oscillator $\Omega = \sqrt{\alpha}$, corresponds to the resonant case. For the linear oscillator in the resonant case the amplitude of oscillations tends to infinity (see Equation (4.3.48b)).

For $\alpha = \gamma = 0$, the harmonically excited Duffing equation (4.3.25) transforms into

$$\ddot{y} = F_0 \cos \Omega t \tag{4.3.50}$$

Solving Equations (4.3.30) and (4.3.31) gives

$$Y = \frac{F}{\Omega^2}, \quad \Omega^* = 0 \tag{4.3.51a,b}$$

and the corresponding solution (4.3.29) has the form

$$y = \frac{F}{\Omega^2}(1 - \cos \Omega t) = \frac{2F}{\Omega^2}\sin^2(2\Omega t) \tag{4.3.52}$$

The solution (4.3.52) is equal to the exact solution of Equation (4.3.50) for the initial conditions (4.3.26).

4.3.4 The harmonically excited pure cubic Duffing equation

For the pure cubic Duffing oscillator ($\alpha = 0$) the steady-state solution at the excitation frequency

$$y = Y \cos \Omega t \tag{4.3.53}$$

where Y is the solution of the modified relation (4.3.31)

$$\frac{9}{4}\gamma Y^3 - Y\Omega^2 + F = 0 \tag{4.3.54}$$

is investigated here. The displacement Y depends on F, γ and Ω and the influence of these parameters on the steady-state solution at the excitation frequency is considered.

For the case when $\Omega = 0.1$, $F = 0.1$ and $\gamma = 1$, the solution of Equation (4.3.54) is $Y = 0.3584$, so $y = 0.3584 \cos 0.1t$. If $F = 1$ the displacement y as a function of time

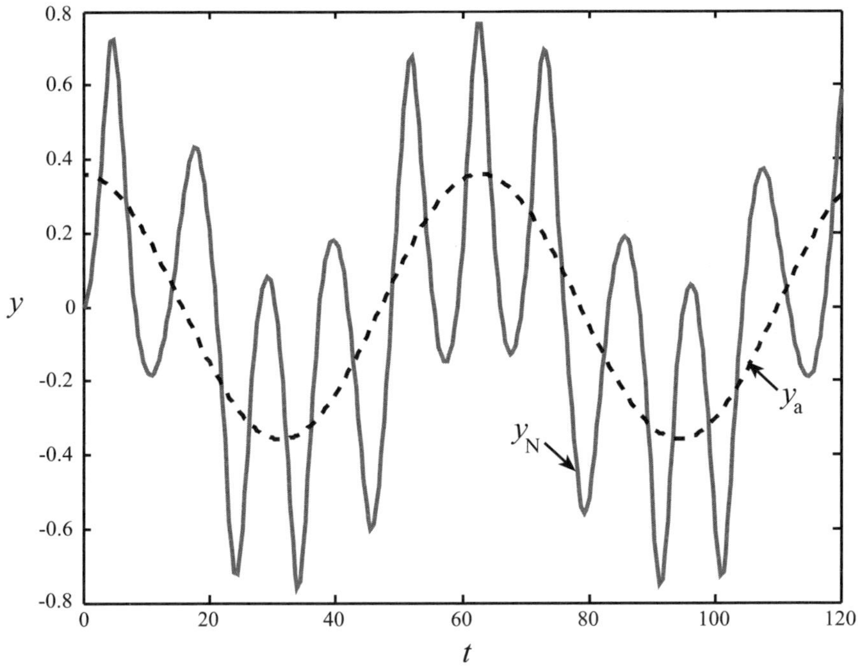

Figure 4.4 Time history obtained numerically (grey solid line) and the solution at the excitation frequency obtained analytically (black dashed line) for $F=0.1$, $\Omega=0.1$ and $\gamma=1$.

is $y=0.76363\cos\Omega t$. In Figure 4.4 and Figure 4.5 the numerically obtained solutions of the pure cubic differential equation

$$\ddot{y}+\gamma y^3 = F\cos\Omega \tag{4.3.55}$$

and the aforementioned analytical results are plotted. Examining the time histories, it can be seen that the *forced oscillations* are along a curve that correspond to the response at the excitation frequency. The higher the value of excitation amplitude F, the higher the amplitude Y and the frequency of vibration Ω.

For the parameter values $F=0.1$, $\Omega=0.1$ and $\gamma=0.1$, the solution obtained analytically at the excitation frequency is $y=0.78255\cos 0.1t$. On increasing the coefficient of nonlinearity to $\gamma=10$, the amplitude of the response at the excitation frequency decreases to $Y=0.16532$. In Figure 4.6. the steady-state solutions at the excitation frequency and numerically obtained solutions for $F=0.1$, $\Omega=0.1$ and $\gamma=0.1$ and $\gamma=10$ are plotted. Comparing the curves, it is seen that the amplitude of vibration is higher for smaller values of the coefficient of nonlinearity γ. The smaller the coefficient of nonlinearity, the higher the amplitude of vibration. For $\gamma=0$ the amplitude is $Y=F/\Omega^2$, i.e., for the parameters $F=0.1$, $\Omega=0.1$, it is $Y=10$ (see Equation (4.3.51)).

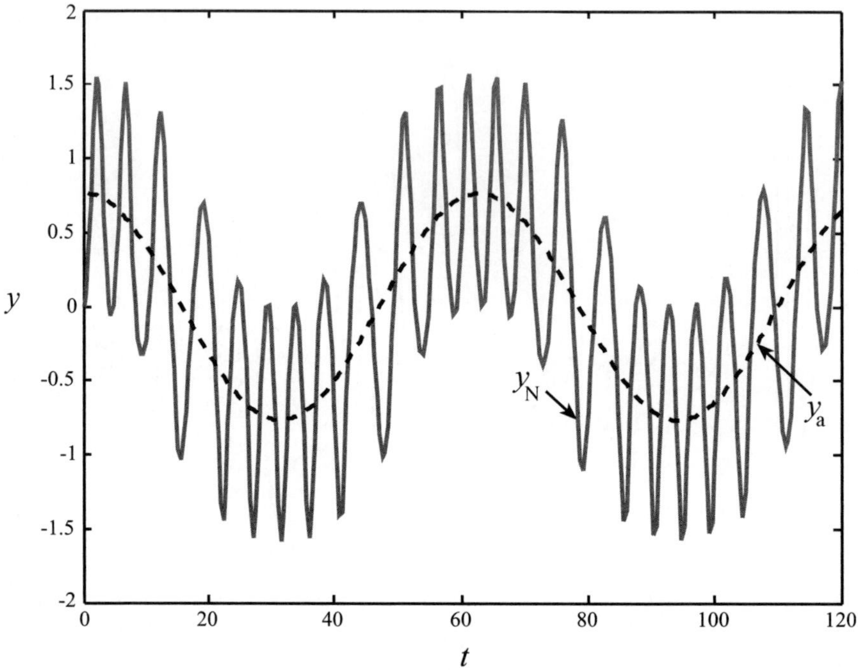

Figure 4.5 Time history obtained numerically (grey solid line) and the steady-state solution at the excitation frequency obtained analytically (black dashed line) for $F = 1$, $\Omega = 0.1$ and $\gamma = 1$.

In Figure 4.7 the numerical solution of Equation (4.3.55) and the steady-state solution at the excitation frequency, given by Equation (4.3.53) are plotted for $\Omega = 0.2$, $F = 0.1$ and $\gamma = 1$. Solving Equation (4.3.54) gives $Y = 0.37094$. Comparing the curves in Figure 4.4 and Figure 4.7, it can be seen that an increase in the frequency of excitation causes an increase of the amplitude of vibration Y.

4.4 The elliptic Galerkin method

The Galerkin method is one of the most frequently applied residual weighted methods [56] for solving the Duffing equation approximately. Usually, the trial solution is assumed to be a linear combination of trigonometric functions. Substituting the trial solution into the differential equation (4.1.2), a residual function is obtained. If the trial solution is the exact solution of the differential equation, then the residual function is zero. If the trial solution is not the exact solution then the residual function is not zero. The weighting function is then arbitrarily chosen to be harmonic and the averaging of the product of the residual and weighting function is carried out.

The period of the trigonometric function is taken as the time interval of the integration. The constants in the trial solution are chosen to make the average

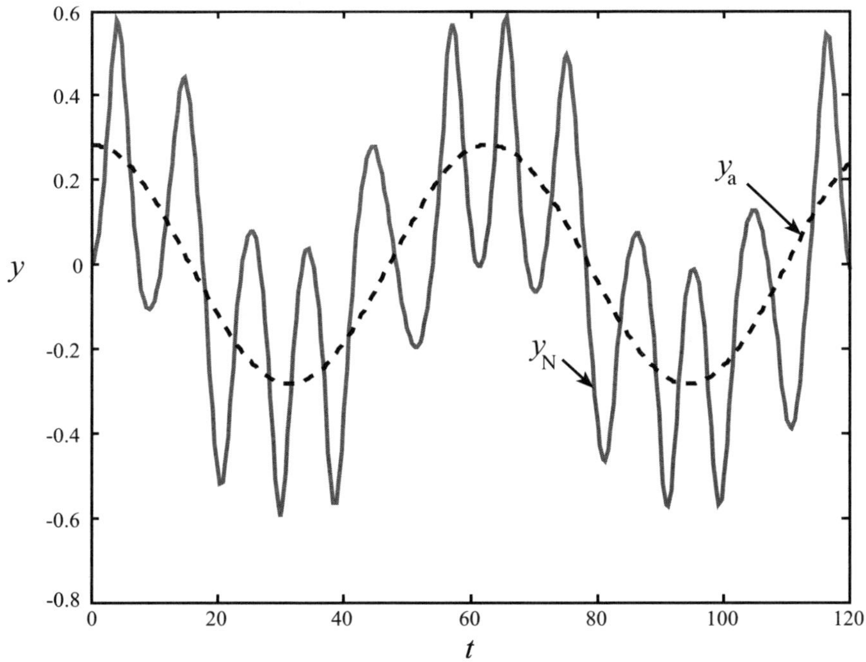

Figure 4.6 Time history obtained numerically (grey solid line) and the steady-state response at the excitation frequency obtained analytically (black dashed line) for: $F=0.1$, $\Omega=0.1$ and $\gamma=2$.

'residual' work over a certain time interval equal to zero [57–60]. The assumed trial solution is appropriate for the Duffing equation with small nonlinearities, including a small cubic term.

An improvement in the method can be made by assuming that the trial solution has the form of the Jacobi elliptic function or the linear combination of the Jacobi elliptic functions. There are two approaches of the Galerkin procedure for a strong nonlinear Duffing equation: one is based on the usual Galerkin method assuming that the functions are known a priori, and in the second it is assumed that the parameters of the Jacobi elliptic function are unknown. In this section only the first approach is considered.

For the Duffing equation of the form

$$\ddot{y} \pm \alpha y \pm \gamma y^3 = f(y, \dot{y}) + \sum_{i=1}^{n} F_i \, \mathrm{ep}_i(\omega_i t, k_{Fi}^2) \tag{4.4.1}$$

the trial solution is a Jacobi elliptic function or the linear combination of Jacobi elliptic functions

$$y^* = \sum_{i=1}^{n} Y_i \mathrm{ep}_i(\omega_i t, k_{Fi}^2) \equiv \sum_{i=1}^{n} Y_i \mathrm{ep}_i \tag{4.4.2}$$

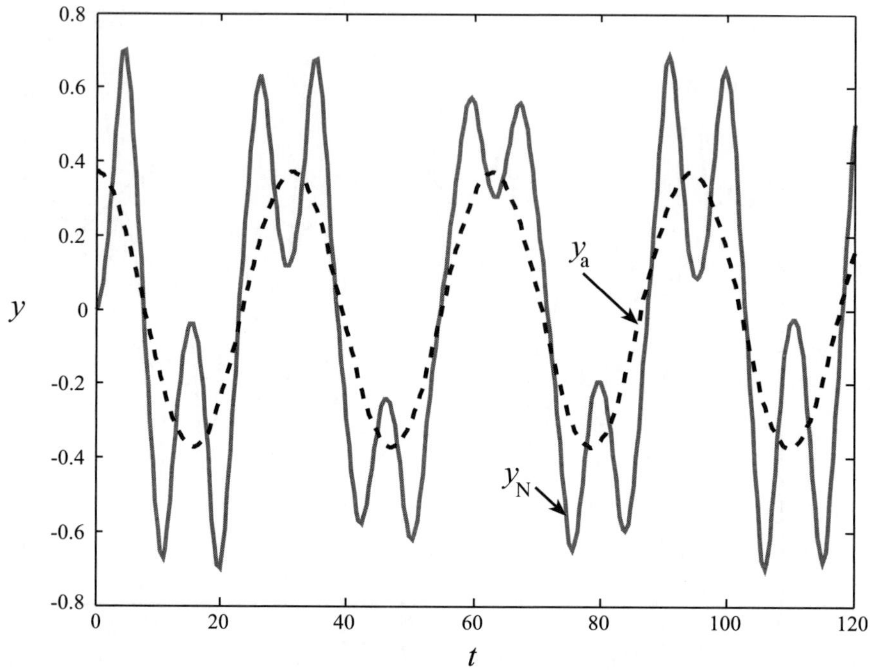

Figure 4.7 Time history obtained numerically (grey solid line) and the steady-state solution ate the excitation frequency obtained analytically (black dashed line) for: $F = 0.1$, $\Omega = 0.2$ *and* $\gamma = 1$.

where $F_i \mathrm{ep}_i(\omega_i t, k_{Fi}^2)$ is the periodical excitation force of elliptic type, ep_i is the general notation for an elliptic function, F_i are known constants and ω_i and k_{Fi} are the frequency and modulus of the *i*th elliptic function ep_i. The unknown constants Y_i need to be determined.

Substituting Equation (4.4.2) into Equation (4.4.1) gives the residual function

$$\begin{aligned} r(\psi) \equiv \sum_{i=1}^{n} Y_i \omega_i^2 (\mathrm{ep}_i)'' \pm \alpha \sum_{i=1}^{n} Y_i \mathrm{ep}_i \pm \gamma \left(\sum_{i=1}^{n} Y_i \mathrm{ep}_i \right)^3 \\ - f\left(\sum_{i=1}^{n} Y_i \mathrm{ep}_i, \sum_{i=1}^{n} Y_i \omega_i (\mathrm{ep}_i)' \right) - \sum_{i=1}^{n} F_i \mathrm{ep}_i \end{aligned} \tag{4.4.3}$$

where $(.)' = d(.)/d\psi$, $(.)'' = d^2(.)/d\psi^2$ and $\psi = \omega_i t$. As mentioned previously, the residual function (4.4.3) is not zero when the trial solution (4.4.2) is the approximate solution of Equation (4.4.1), but it is zero when it is the exact solution. In accordance with the Galerkin procedure, the residual function $r(\psi)$ is multiplied with the arbitrary weighting function $w_i(\psi)$, which is chosen as the derivative of the trial solution (4.4.2)

$$w_i(\psi) = \frac{\partial y^*}{\partial Y_i} \tag{4.4.4}$$

The product of the residual function (4.4.3) and the weighting function (4.4.4) is averaged in the interval $[0,4K(k_F)]$, where $4K(k_F)$ is the period of the elliptic function and $K(k_F)$ is the total elliptic integral of the first kind. The resulting expressions are set to zero, giving the algebraic equations

$$\int_0^{4K(k_F)} r(\psi)w_i(\psi)d\psi = 0 \tag{4.4.5}$$

The number of algebraic equations is equal to the number of unknown constants Y_i. By solving Equations (4.4.5), the values of Y_i are found. Equation (4.4.5) represents the ordinary Galerkin condition for finding the constants. The method is appropriate if ω_i and k_i are known.

4.4.1 Duffing oscillator with a strong excitation force of elliptic type

The Duffing equation with strong excitation of a Jacobi elliptic form is given by

$$\ddot{y} + \alpha y + \gamma y^3 = F\mathrm{cn}(\Omega t, k_F^2) \tag{4.4.6}$$

where F is the excitation amplitude and Ω and k_F are the frequency and the modulus of the Jacobi elliptic function. Due to the property of the cn function (4AI.6b) the excitation with a Jacobi elliptic function can be treated as a multifrequency excitation. Using the Fourier series expansion of the cn function (4AI.6b) and retaining only the first term on the right-hand side of Equation (4.4.6), which is a trigonometric function, the excitation transforms to a single frequency type.

It is assumed the solution in the form of a Jacobi elliptic function with the known parameters Ω and k_F

$$y^* = Y\,\mathrm{cn}(\Omega t, k_F^2) \equiv Y\,\mathrm{cn}(\psi, k_F^2) = Y\,\mathrm{cn} \tag{4.4.7}$$

where Y is an unknown parameter, which needs to be calculated. Substituting Equation (4.4.7) into Equation (4.4.6), the nonzero residual function follows

$$r(\psi) = -Y(1-2\Omega^2 + 2\Omega^2\mathrm{cn}^2)\mathrm{cn} + \alpha Y\,\mathrm{cn} + \gamma Y^3\,\mathrm{cn}^3 - F\,\mathrm{cn} \tag{4.4.8}$$

The function (4.4.8) is multiplied with the weighting function

$$w(\psi) = \mathrm{cn} \tag{4.4.9}$$

Applying the aforementioned procedure of averaging the product of the residual (4.4.8) and the weight function (4.4.9), the following cubic-order algebraic equation is obtained

$$Y^3 + \frac{\alpha C_2 - (1-2k_F^2)\Omega^2 C_2 - 2k_F^2\Omega^2 C_4}{\gamma C_4}Y - \frac{FC_2}{\gamma C_4} = 0 \tag{4.4.10}$$

where the values of the parameters C_2 and C_4 are given in Equations (4AI.8) and (4AI.9) of Appendix 4AI. Depending on the parameter values, Equation (4.4.10) can have one, two or three real solutions for Y. The number of real solutions depends on the sign of

$$D=\frac{(\alpha C_2-(1-2k_F^2)\Omega^2C_2-2\Omega^2k_F^2C_4)^3}{27\gamma^3C_4^3}+\frac{F^2C_2^2}{4\gamma^2C_4^2} \tag{4.4.11}$$

Three possible cases can be considered separately depending on the sign of the expression (4.4.11).

For $D<0$, the following three distinct real roots exist

$$Y_1=2\sqrt{-\frac{\alpha C_2-(1-2k_F^2)\Omega^2C_2-2k_F^2\Omega^2C_4}{3\gamma C_4}}\cos\frac{\phi}{3},$$

$$Y_2=2\sqrt{-\frac{\alpha C_2-(1-2k_F^2)\Omega^2C_2-2k_F^2\Omega^2C_4}{3\gamma C_4}}\cos\frac{\phi+2\pi}{3}, \tag{4.4.12a–c}$$

$$Y_3=2\sqrt{-\frac{\alpha C_2-(1-2k_F^2)\Omega^2C_2-2k_F^2\Omega^2C_4}{3\gamma C_4}}\cos\frac{\phi+4\pi}{3}$$

where

$$\phi=\cos^{-1}\left(\frac{3FC_2\sqrt{3\gamma C_4}}{2\left(-(\alpha C_2-(1-2k_F^2)\Omega^2C_2-2k_F^2\Omega^2C_4)\right)^{3/2}}\right) \tag{4.4.13}$$

For $D>0$, only one real root exists, which is given by

$$Y=\sqrt[3]{\frac{FC_2}{2\gamma C_4}+\sqrt{\frac{F^2C_2^3}{4\gamma^2C_4^2}+\frac{(\alpha C_2-(1-2k_F^2)\Omega^2C_2-2k_F^2\Omega^2C)^3}{27\gamma^3C_4^3}}}$$
$$+\sqrt[3]{\frac{FC_2}{2\gamma C_4}+\sqrt{\frac{F^2C_2^3}{4\gamma^2C_4^2}-\frac{(\alpha C_2-(1-2k_F^2)\Omega^2C_2-2k_F^2\Omega^2C)^3}{27\gamma^3C_4^3}}} \tag{4.4.14}$$

For $D=0$, two real roots exist. One is single and the other is a repeated root given by

$$Y_1=2\sqrt[3]{\frac{FC_2}{2\gamma C_4}},\quad Y_2=Y_3=\sqrt[3]{\frac{FC_2}{\gamma C_4}} \tag{4.4.15a,b}$$

Substituting the approximate values for the averaging parameters C_2 and C_4, as given in Equation (4AI.8) and Equation (4AI.9) of Appendix 4AI, respectively, gives

$$\frac{3}{4}\gamma Y^3+\alpha-\Omega^2-F+\frac{1}{2}k_F^2\Omega^2=0 \tag{4.4.16}$$

The solutions for the algebraic equation (4.4.16) are plotted in Figure 4.8. In Figure 4.8(a) the *frequency-response curves* are given for various values of the

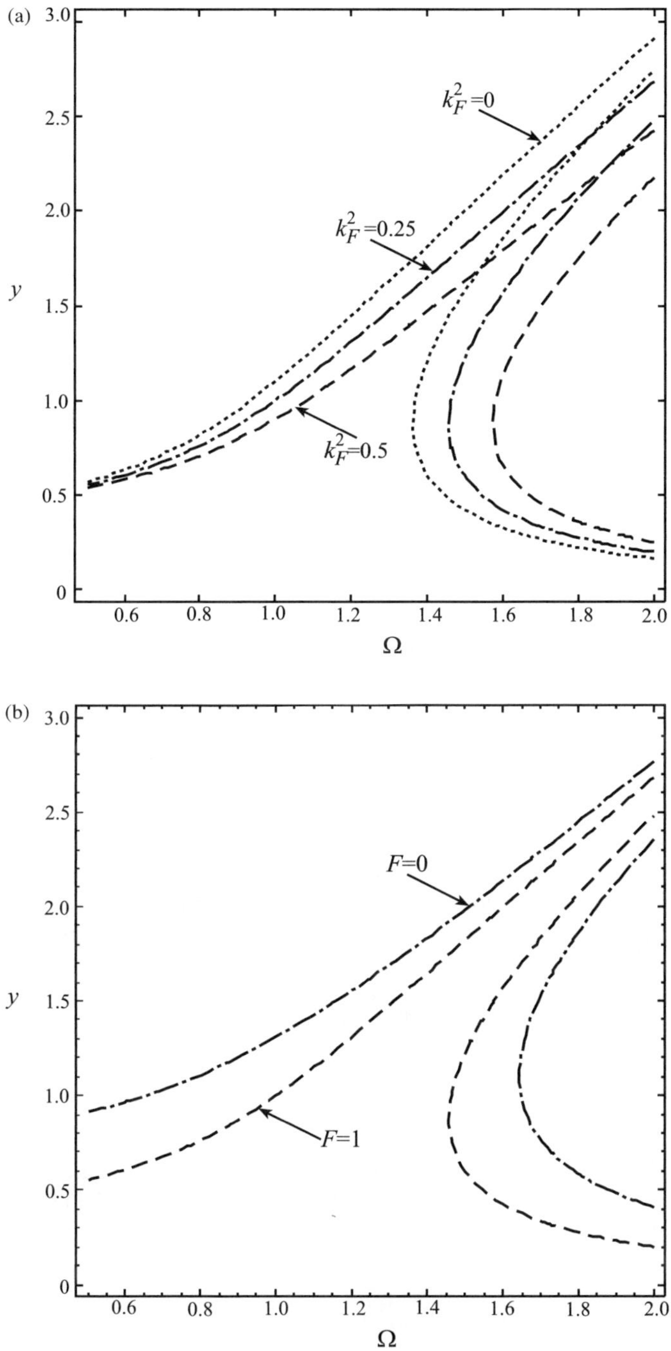

Figure 4.8 Frequency response curves: (a) for $F = 1$ and various k_F^2, (b) for $k_F^2 = 0.25$ and various F.

modulus of the Jacobi function. The parameter values in Equation (4.4.16) are $\alpha=\gamma=1$ and the excitation amplitude is assumed to be constant $F=1$. It is clear that the higher the modulus k_F, the lower the amplitude for certain frequencies. Moreover, the *jump* is shifted to higher frequencies for the higher values of the modulus.

In Figure 4.8(b) the frequency-response curves are shown for various values of F. The parameter values in Equation (4.4.16) are $\alpha=\gamma=1$ and $k_F^2=0.1$. The amplitude of vibration increases as the excitation amplitude increases.

4.4.1.1 Duffing oscillator with strong excitation force of a trigonometric type

For the case when the excitation function is a trigonometric cosine function, i.e., $k_F=0$, Equation (4.4.10) transforms into

$$Y^3+\frac{\alpha C_2-\Omega^2 C_2}{\gamma C_4}Y-\frac{FC_2}{\gamma C_4}=0 \tag{4.4.17}$$

After using Equation (4AI.8) and Equation (4AI.9), Equation (4.4.17) yields

$$\frac{3}{4}Y^3+\frac{\alpha-\Omega^2}{\gamma}Y-\frac{F}{\gamma}=0 \tag{4.4.18}$$

which is the well-known relationship given in many textbooks on nonlinear oscillations (see for example [3]). This equation is also discussed in Chapter 5, for the case when damping exists.

4.5 The straightforward expansion method

The straightforward expansion method is suitable for solving for *free vibration* of the Duffing equation that contains a small parameter ε. The following Duffing equation is considered

$$\ddot{y}+\alpha y+\gamma y^3=\varepsilon f(y,\dot{y}) \tag{4.5.1}$$

with the initial conditions

$$y(0)=y_0, \quad \dot{y}(0)=\dot{y}_0 \tag{4.5.2a,b}$$

where $\varepsilon \ll 1$ is a small parameter. The basic idea is to assume that the cubic nonlinearity is also small, and that the solution of Equation (4.5.1) has the form of a series in the small parameter

$$y=y_*+\sum_{i=1}^{\infty}\varepsilon^i y_i(y)\equiv y_*+\sum_{i=1}^{\infty}\varepsilon^i y_i \tag{4.5.3}$$

Substituting Equation (4.5.3) into Equation (4.5.1) and equating the terms with the same order of the small parameter ε, a system of linear differential equations is obtained. The solutions of the equations give the approximate analytical solution of Equation (4.5.1).

For the case when γ is not small, Equation (4.5.1) can be rewritten in the form

$$\ddot{y}+\omega^2(1-2k^2)y+\frac{2k^2\omega^2}{Y^2}y^3=\varepsilon f(y,\dot{y}) \tag{4.5.4}$$

where, in accordance with Equation (4.2.6)

$$\alpha=\omega^2(1-2k^2),\quad \gamma=\frac{2\omega^2k^2}{Y^2} \tag{4.5.5a,b}$$

This is a step where the unknown frequency ω_* and the modulus k_* of the Jacobi elliptic function are introduced, which represent the corrected values of the frequency ω and the modulus k of the function given by Equation (4.2.6)

$$\omega_*^2=\omega^2+\sum_{i=1}^{\infty}\varepsilon^i b_i,\quad k_*^2=k^2+\sum_{i=1}^{\infty}\varepsilon^i k_i \tag{4.5.6a,b}$$

where b_i is the frequency and k_i is the modulus correction factors which are to be calculated.

Introducing the series expansions (4.5.3) and (4.5.6) into Equation (4.5.4), and separating the terms with the same order of the small parameter ε, the following system of differential equations is obtained

$$\varepsilon^0:\ \ddot{y}_*+\omega_*^2(1-2k_*^2)y_*+\frac{2\omega_*^2k_*^2}{Y^2}y_*^3=0,$$

$$\varepsilon^1:\ \ddot{y}_1+\omega_*^2(1-2k_*^2)y_1+\frac{2\omega_*^2k_*^2}{Y^2}3y_*^2y_1=b_1y_*(1-2k_*^2)-2\omega_*^2k_1y_* + \frac{2}{Y^2}((b_1k_*^2+k_1\omega_*^2)y_*^3+f_0,$$

$$\varepsilon^2:\ \ddot{y}_2+\omega_*^2(1-2k_*^2)y_2+\frac{2\omega_*^2k_*^2}{Y^2}3y_*^2y_2=(b_2y_*+b_1y_1)(1-2k_*^2)+2k_1(b_1y_*-\omega_*^2y_1) + \frac{2}{Y^2}\left((b_2k_*^2-b_1k_1+k_2\omega_*^2)y_*^3+3(b_1k_*^2+k_1\omega_*^2)y_*^2y_1-3\omega_*^2k_*^2y_*y_1^2\right)+f_1 \tag{4.5.7a–c}$$

where $f_0, f_1, \ldots$, are the terms of the Taylor-series expansion [2] for the two variable function $f(y,\dot{y})$ around the values of $y_*, \dot{y}_*$

$$f_0=f(y_*,\dot{y}_*),\quad f_1=y_1\frac{\partial f(y_*,\dot{y}_*)}{\partial y}+\dot{y}_1\frac{\partial f(y_*,\dot{y}_*)}{\partial \dot{y}},\cdots \tag{4.5.8a,b}$$

The solution of Equation (4.5.7a) is

$$y_*=Y_0\,\mathrm{cn}(\omega_*t+\theta,k_*^2)\equiv Y_0\,\mathrm{cn} \tag{4.5.9}$$

where for $\varepsilon = 0$, the following holds

$$\omega_* = \omega, \quad k_*^2 = k^2 \tag{4.5.10a,b}$$

Substituting Equation (4.5.9) into Equation (4.5.7b), the following linear second-order differential equation with time variable parameter is obtained

$$\ddot{y}_1 + \omega_*^2(1-2k_*^2)y_1 + \frac{2\omega_*^2 k_*^2}{Y^2}(3Y^2 cn^2)y_1$$
$$= Y_0(b_1(1-2k^2)-2\omega^2 k_1)\text{cn} + 2(b_1k^2 + k_1\omega^2)Y_0\text{cn}^3 + f_0(Y_0\,\text{cn}, -\omega\dot{Y}_0\,\text{sn dn}) \tag{4.5.11}$$

Unfortunately, a closed form analytical solution of Equation (4.5.11) does not exist. The complementary function (solution of the homogeneous equation (4.5.11)) is $y_{1h} = (Y_0/3)\text{cn}$. Eliminating the *secular terms* on the right-hand side of Equation (4.5.11), i.e., equating to zero the terms with cn and cn^3, the frequency b_1 and the modulus k_1 corrections are obtained, which give the frequency and the modulus of the Jacobi elliptic function in the first approximation

$$\omega_*^2 = \omega^2 + \varepsilon b_1 = \alpha + \gamma Y_0^2 + \varepsilon b_1,$$
$$k_*^2 = k^2 + \varepsilon k_1 = \frac{\gamma Y_0^2}{2(\alpha + \gamma Y_0^2)} + \varepsilon k_1 \tag{4.5.12a,b}$$

Then, the solution in the first approximation is

$$y = \left(Y_0 + \frac{\varepsilon Y_0}{3}\right)\text{cn}\left(\omega_* t + \theta, k_*^2\right) + \varepsilon y_{1p} \tag{4.5.13}$$

where y_{1p} is the particular integral of Equation (4.5.11). In accordance with the initial conditions (4.5.2), the unknown values Y_0 and θ in Equation (4.5.13) are calculated. The same procedure should be repeated to obtain the next approximations.

The function $f_0(Y_0\text{cn}, -\omega Y_0\text{sn dn})$ is approximated with a function f^* in the form given in Appendix 4AII

$$f^* \approx f_1\,\text{cn} + f_3\,\text{cn}^3 \tag{4.5.14}$$

where f_1 and f_3 are the terms with cn and cn^3. Substituting Equation (4.5.14) into Equation (4.5.11), separating the terms with the same order of cn function, the following system of algebraic equations is obtained

$$Y_0(b_1(1-2k^2)-2\omega^2 k_1) = -f_1, \quad 2(b_1k^2 + k_1\omega^2)Y_0 = -f_3 \tag{4.5.15a,b}$$

The relations in Equation (4.5.15) give the frequency and the modulus correction factors

$$b_1 = -\frac{f_1 + f_3}{Y_0}, \quad k_1 = \frac{2k^2(f_1 + f_3) - f_3}{2\omega^2 Y_0} \tag{4.5.16a,b}$$

which yield the frequency and the modulus (4.5.12) in the first approximation

$$\omega_*^2 = \omega^2 + \varepsilon b_1 = (\alpha + \gamma Y_0^2) - \frac{\varepsilon(f_1 + f_3)}{Y_0},$$
$$k_*^2 = k^2 + \varepsilon k_1 = \frac{\gamma Y_0^2}{2(\alpha + \gamma Y_0^2)} + \frac{\varepsilon \gamma Y_0 (f_1 + f_3)}{2(\alpha + \gamma Y_0^2)^2} - \frac{\varepsilon f_3}{2(\alpha + \gamma Y_0^2) Y_0} \quad (4.5.17a,b)$$

For the initial conditions given in Equation (4.5.2), the solution in the first approximation is

$$y = y_0 \,\mathrm{cn}\left(t\sqrt{(\alpha + \gamma Y_0^2) - \frac{\varepsilon(f_1 + f_3)}{Y_0}}, \frac{\gamma Y_0^2}{2(\alpha + \gamma Y_0^2)} + \frac{\varepsilon \gamma Y_0 (f_1 + f_3)}{2(\alpha + \gamma Y_0^2)^2} - \frac{\varepsilon f_3}{2(\alpha + \gamma Y_0^2) Y_0} \right) \quad (4.5.18)$$

where $Y_0 = 3\, y_0/(3 + \varepsilon)$.

4.5.1 The Duffing equation with a small quadratic term

As an example, consider the Duffing equation with a weak quadratic term

$$\ddot{y} + \omega^2 (1 - 2k^2) y + \frac{2k^2 \omega^2}{Y^2} y^3 = -\varepsilon \beta y |y| \quad (4.5.19)$$

where $\varepsilon \ll 1$ is a small parameter. Substituting the solution of the strongly nonlinear cubic equation (4.5.9) into Equation (4.5.19) and using Equation (4.5.11), a parametrically excited linear second order differential equation is obtained. For $y_1 = (Y_0/3)\mathrm{cn}$, the right-hand side of this equation simplifies to

$$Y_0 \,\mathrm{cn}(b_1(1 - 2k^2) - 2\omega^2 k_1) + 2(b_1 k^2 + k_1 \omega^2) Y_0 \,\mathrm{cn}^3 - \beta Y_0^2 \,\mathrm{cn}|\mathrm{cn}| = 0 \quad (4.5.20)$$

Using approximation (4AII.15) given in Appendix 4AII, the last term in Equation (4.5.20) can be transformed into a sum: $\mathrm{cn}|\mathrm{cn}| \approx a_0 \,\mathrm{cn} + a_1 \,\mathrm{cn}^3 = \frac{16}{15\pi} \mathrm{cn} + \frac{32}{15\pi} \mathrm{cn}^3$. Using the harmonic balance method, the following correction parameters are obtained

$$b_1 = \beta |Y_0| (a_0 + a_1), \quad k_1 = \frac{\beta |Y_0|}{2\omega^2} \left(a_1 - 2k^2 (a_0 + a_1) \right) \quad (4.5.21a,b)$$

In accordance with Equation (4.5.12) and Equation (4.5.21), the frequency and the modulus of the Jacobi function in the first approximation are

$$\omega_*^2 = \alpha + \gamma Y_0^2 + \varepsilon \beta |Y_0|_0 (a_0 + a_1),$$

$$k_*^2 = \frac{\gamma Y_0^2}{2(\alpha + \gamma Y_0^2)} + \frac{\varepsilon \beta |Y_0|_0 a_1}{2(\alpha + \gamma Y_0^2)} - \frac{\varepsilon \beta \gamma Y_0^2 |Y_0| a_1}{2(\alpha + \gamma Y_0^2)^2} (a_0 + a_1) \quad (4.5.22a,b)$$

and the corresponding first-order solution of Equation (4.5.19) is

$$y = y_0 \operatorname{cn}\left(t\sqrt{\alpha + \gamma Y_0^2 + \varepsilon\beta|Y_0|(a_0 + a_1)}, \frac{\gamma Y_0^2 + \varepsilon\beta|Y_0|a_1}{2(\alpha + \gamma Y_0^2)} - \frac{\varepsilon\beta\gamma Y_0^2|Y_0|a_1(a_0 + a_1)}{2(\alpha + \gamma Y_0^2)^2}\right) \tag{4.5.23}$$

For the initial conditions given in Equation (4.5.2), the initial phase angle θ is zero, and the initial amplitude is $Y_0 = 3\ y_0/(3 + \varepsilon)$.

Three numerical simulations are carried out. For $\alpha = \beta = \gamma = 1$, $\varepsilon = 0.1$, and the initial conditions $y_0 = 0.1$ and $\dot{y}_0 = 0$, the solution in the first approximation is

$$y_a = 0.1 \operatorname{cn}(1.00960\, t, 0.00786) \tag{4.5.24}$$

In Figure 4.9(a), the approximate solution y_a given by Equation (4.5.24) is compared with the numerical solution y_N, obtained by using the Runge–Kutta method to solve Equation (4.5.19). It can be seen that the difference between the solutions is negligible.

For the same parameter values and the initial conditions, but for $\varepsilon = 0.8$, the approximate first-order solution is

$$y_a = 0.8 \operatorname{cn}(1.04320\, t, 0.02427) \tag{4.5.25}$$

The solution given by Equation (4.5.25) is compared with that calculated numerically in Figure 4.9(b). The difference is evident, especially after a longer period of time.

For the initial conditions $y_0 = 1$ and $\dot{y}_0 = 0$ and parameter values $\alpha = \beta = \gamma = 1$, $\varepsilon = 0.1$, the analytical approximate solution is

$$y_a = \operatorname{cn}(1.42660\, t, 0.25041) \tag{4.5.26}$$

In Figure 4.9(c), the analytical solution given in Equation (4.5.26) and the numerical solution of Equation (4.5.19) are plotted. It can be seen that there is a difference between the solutions, which increases as time passes.

Examining Figure 4.9, it can be seen that the analytical approximate solution is accurate if ε is small enough, and if the initial displacement is also small. For higher values of ε and y_0, the difference between the analytical approximate and the exact numerical solutions is evident, particularly for longer time intervals. To improve the solution, higher-order approximations should be included.

4.6 The elliptic Lindstedt–Poincaré method

The Lindstedt–Poincaré method can also be used to solve the Duffing equation (4.5.1), where $\varepsilon \ll 1$ is a small parameter. The method requires the series expansion of the solution (4.5.3) and the following new parameters are introduced

$$\alpha^* = \alpha + \varepsilon\alpha_1 + \ldots, \quad \gamma^* = \gamma + \varepsilon\gamma_1 + \ldots \tag{4.6.1a,b}$$

where α_1, ..., and γ_1, ..., are unknown parameters that are to be calculated. Substituting Equation (4.5.3) and $\alpha = \alpha^* - \varepsilon\alpha_1 - \ldots, \gamma = \gamma^* - \varepsilon\gamma_1 - \ldots$, into

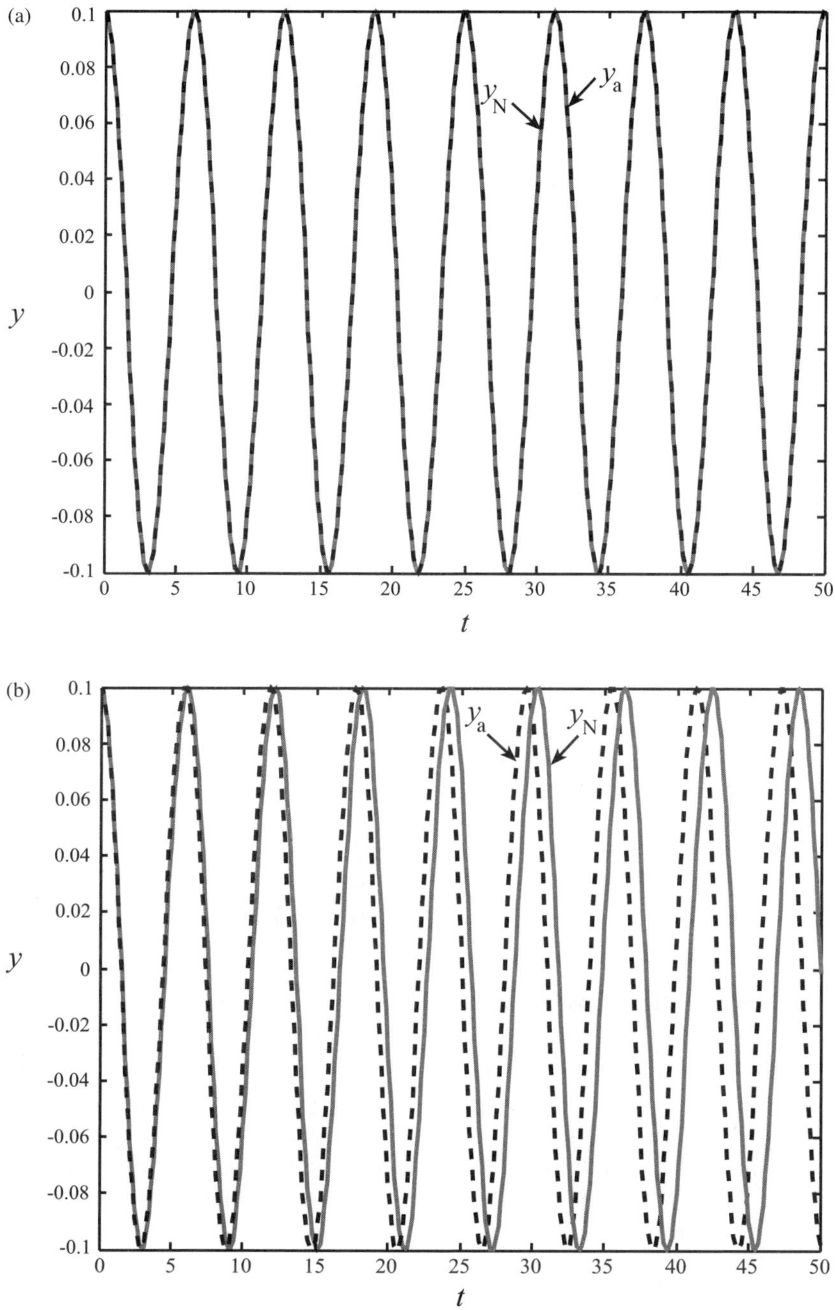

Figure 4.9 Time histories obtained analytically y_a (black dashed line) and numerically y_N (grey solid line) for: (a) $y_0 = 0.1$ and $\varepsilon = 0.1$, (b) $y_0 = 0.1$ and $\varepsilon = 0.8$, (c) $y_0 = 1$ and $\varepsilon = 0.1$.

(c)

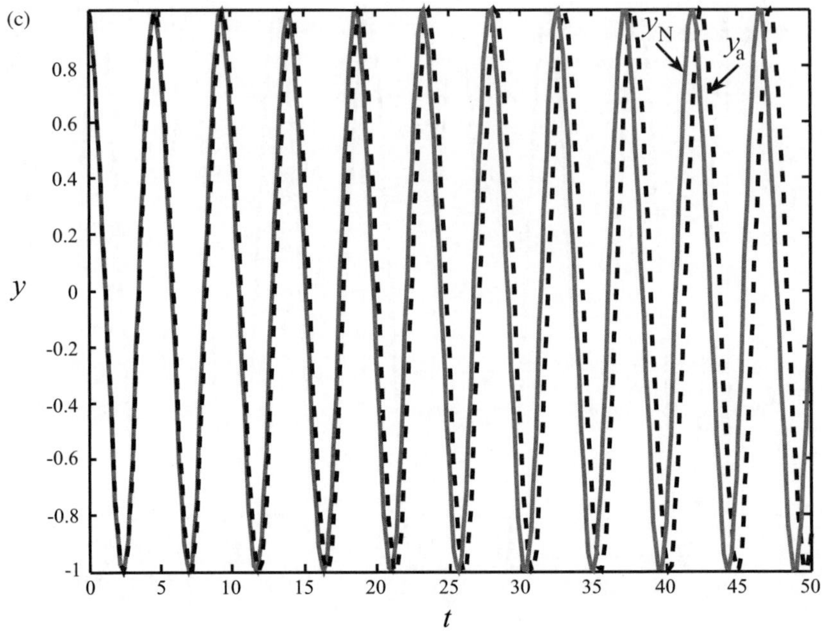

Figure 4.9 (Continued)

Equation (4.5.1), and separating the terms with the same order of the small parameter ε, the following system of differential equations is obtained

$$\begin{aligned}&\varepsilon^0 : \ddot{y}_* + \alpha^* y_* + \gamma^* y_*^3 = 0,\\ &\varepsilon^1 : \ddot{y}_1 + \alpha^* y_1 + 3\gamma^* y_*^2 y_1 = \alpha_1 y_* + \gamma_1 y_*^3 + f(y_*, \dot{y}_*)\\ &\ldots\end{aligned} \tag{4.6.2a,b}$$

The exact solution of Equation (4.6.2a) is

$$y_* = Y_0 \operatorname{cn}(\omega t + \theta, k^2) \equiv Y_0 \operatorname{cn} \tag{4.6.3}$$

where Y_0 and θ are arbitrary constants that are dependent on the initial conditions, and for $\varepsilon = 0$ the parameters are $\alpha^* = \alpha$ and $\gamma^* = \gamma$, which give the frequency and the modulus of the Jacobi elliptic function as in Equation (4.2.6). Substituting these relations and Equation (4.6.3) into Equation (4.6.2b), the following linear differential equation is obtained

$$\ddot{y}_1 + \alpha^* y_1 + 3\gamma^* Y_0^2 \operatorname{cn}^2 y_1 = \alpha_1 Y_0 \operatorname{cn} + \gamma_1 Y_0^3 \operatorname{cn}^3 + f(Y_0 \operatorname{cn}, -\omega Y_0 \operatorname{sn} \operatorname{dn}) \tag{4.6.4}$$

The complementary function (homogenous solution) of Equation (4.6.4) is

$$y_{1h} = \frac{Y_0}{3} \operatorname{cn} \tag{4.6.5}$$

In order to eliminate the secular terms on the left-hand side of Equation (4.6.4), $f(Y_0\text{cn}, -\omega Y_0\text{sn dn})$ is approximated with a function f^* (see Appendix 4AII) which has the form

$$f^* \approx f_1 \,\text{cn} + f_3 \,\text{cn}^3 \tag{4.6.6}$$

where f_1 and f_3 are the coefficients of cn and cn^3, respectively. Substituting Equation (4.6.6) into Equation (4.6.4) and equating the terms with the same order of the cn function, the unknown parameters are as follows

$$\alpha_1 = -\frac{f_1(Y_0)}{Y_0}, \quad \gamma_1 = -\frac{f_3(Y_0)}{Y_0^3} \tag{4.6.7a,b}$$

Substituting Equation (4.6.7) into Equation (4.6.1), the approximate first-order solution is obtained

$$y = \left(Y_0 + \frac{\varepsilon Y_0}{3}\right)\text{cn}(\omega t + \theta, k^2) \tag{4.6.8}$$

with the frequency and the modulus of the Jacobi elliptic function given in the form

$$\omega^2 = (\alpha + \varepsilon\alpha_1) + (\gamma + \varepsilon\gamma_1)Y_0^2, \quad k^2 = \frac{(\gamma + \varepsilon\gamma_1)Y_0^2}{2(\alpha + \varepsilon\alpha_1) + 2(\gamma + \varepsilon\gamma_1)Y_0^2} \tag{4.6.9a,b}$$

By using Equation (4.6.7), these parameters are

$$\omega^2 = (\alpha + \gamma Y_0^2) - \frac{\varepsilon}{Y_0}(f_1 + f_3), \quad k^2 = \frac{\gamma Y_0^2 - \dfrac{\varepsilon f_3}{Y_0}}{2(\alpha + \gamma Y_0^2) - \dfrac{2\varepsilon}{Y_0}(f_1 + f_3)} \tag{4.6.10a,b}$$

The arbitrary constants Y_0 and θ in Equation (4.6.8) satisfy the initial conditions given in Equation (4.2.1).

Comparing the frequency and the modulus of the Jacobi elliptic function in the first approximation obtained by using the straightforward expansion (4.5.17) with that obtained by the elliptic Lindstedt–Poincaré method (4.6.10), it is evident that the frequency is the same but the module are different. Using the series expansion of the modulus of the Jacobi function (4.6.10b) in the small parameter ε, gives

$$k^2 = \frac{\gamma Y_0^2}{2(\alpha + \gamma Y_0^2)}\left(1 + \frac{\varepsilon}{Y_0(\alpha + \gamma Y_0^2)}(f_1 + f_3)\right) - \frac{\varepsilon f_3}{2(\alpha + \gamma Y_0^2)Y_0} + O(\varepsilon) \tag{4.6.11}$$

Comparing the first three terms of Equation (4.6.11) with Equation (4.5.17b), it is evident that for small values of the parameter ε, the straightforward expansion method and the elliptic Lindstedt–Poincaré method give the same values for the modulus of the Jacobi function.

4.6.1 The Duffing equation with a small quadratic term

To compare the accuracy of the straightforward expansion method and the elliptic Lindstedt–Poincaré method, the example in the previous section is repeated.

Consider the Duffing equation with an additional small quadratic term

$$\ddot{y} + \alpha y + \gamma y^3 = -\varepsilon\beta y|y| \tag{4.6.12}$$

Using the elliptic Lindstedt–Poincaré procedure and the relations obtained in the previous section, the differential equation modified version of Equation (4.6.4) for Equation (4.6.12) is rewritten as

$$\ddot{y}_1 + \alpha^* y_1 + 3\gamma^* Y_0^2 \operatorname{cn}^2 y_1 = \alpha_1 Y_0 \operatorname{cn} + \gamma_1 Y_0^3 \operatorname{cn}^3 - \beta Y_0|Y_0|\operatorname{cn}|\operatorname{cn}| \tag{4.6.13}$$

After transforming the product cn|cn|, as shown in Appendix 4AII (see Equation (4AII.15)):

$$\operatorname{cn}|\operatorname{cn}| \approx a_0 \operatorname{cn} + a_1 \operatorname{cn}^3 = \frac{16}{15\pi}\operatorname{cn} + \frac{32}{15\pi}\operatorname{cn}^3 \tag{4.6.14}$$

and substituting Equation (4.6.5) into Equation (4.6.13), the secular terms are grouped and equated to zero

$$\alpha_1 Y_0 \operatorname{cn} + \gamma_1 Y_0^3 \operatorname{cn}^3 - \beta Y_0|Y_0|(a_0 \operatorname{cn} + a_1 \operatorname{cn}^3) = 0 \tag{4.6.15}$$

It is clear from Equation (4.6.15) that the frequency and the modulus correction coefficients are

$$\alpha_1 = \beta|Y_0|a_0, \quad \gamma_1 = \frac{\beta a_1}{Y_0} \tag{4.6.16a,b}$$

which give the frequency and the modulus in the first approximation

$$\omega^2 = (\alpha + \gamma Y_0^2) + \varepsilon\beta|Y_0|(a_0 + a_1), \quad k^2 = \frac{1}{2}\frac{\gamma Y_0^2 + \varepsilon\beta a_1|Y_0|}{(\alpha + \gamma Y_0^2) + \varepsilon\beta|Y_0|(a_0 + a_1)} \tag{4.6.17a,b}$$

Using Equation (4AII.14), results in

$$\omega^2 = (\alpha + \gamma Y_0^2) + \frac{16}{5\pi}\varepsilon\beta|Y_0|, \ k^2 = \frac{15\pi\gamma Y_0^2 + 32\varepsilon\beta|Y_0|}{30\pi(\alpha + \gamma Y_0^2) + 96\varepsilon\beta|Y_0|} \tag{4.6.18a,b}$$

Applying the expressions given in Equation (4.6.16), the solution in the first approximation is

$$y = \left(Y_0 + \frac{\varepsilon Y_0}{3}\right)\operatorname{cn}\left(t\sqrt{(\alpha + \gamma Y_0^2) + \varepsilon\beta\frac{16}{5\pi}|Y_0|} + \theta, \frac{15\pi\gamma Y_0^2 + 32\varepsilon\beta|Y_0|}{30\pi(\alpha + \gamma Y_0^2) + 96\varepsilon\beta|Y_0|}\right) \tag{4.6.19}$$

For the initial values given in Equation (4.2.9), the initial phase angle is $\theta = 0$ and the initial amplitude $Y_0 = 3y_0/(3+\varepsilon)$. The solution in the first approximation is then,

$$y = y_0 \,\mathrm{cn}\left(t\sqrt{(\alpha+\gamma Y_0^2)+\varepsilon\beta\frac{16}{5\pi}|Y_0|}, \frac{15\pi\gamma Y_0^2+32\varepsilon\beta|Y_0|}{30\pi(\alpha+\gamma Y_0^2)+96\varepsilon\beta|Y_0|}\right) \tag{4.6.20}$$

Three numerical examples, considered in the previous section, are presented here again by using this method.

For $\alpha=\beta=\gamma=1$, $\varepsilon=0.1$, and the initial conditions $y_0 = 0.1$, $\dot{y}_0 = 0$, the approximate analytic solution is

$$y_a = 0.1\,\mathrm{cn}(1.00960t, 0.00465) \tag{4.6.21}$$

In Figure 4.10(a), the approximate solution y_a given in Equation (4.6.21) is compared with the numerical result y_N obtained by applying the Runge–Kutta to Equation (4.6.12). The difference between solutions is negligible.

For the same values of the parameters and the initial conditions, but for $\varepsilon = 0.8$, the approximate first-order solution is

$$y_a = 0.1\,\mathrm{cn}(1.04320t, 0.00454) \tag{4.6.22}$$

The solution (4.6.22) is compared with the numerically obtained one in Figure 4.10(b). The difference is quite small and it is apparent only after some time.

For $y_0 = 1$ and $\dot{y}_0 = 0$, and $\alpha=\beta=\gamma=1$, $\varepsilon=0.1$, the analytical approximate solution is

$$y_a = \mathrm{cn}(1.42660t, 0.23025) \tag{4.6.23}$$

In Figure 4.10(c), the analytical solution (4.6.23) and the numerical solution of Equation (4.6.12) are plotted. The time histories shown start to differ slightly after a long time.

Comparing the solution of the Duffing equation with a quadratic term obtained by the straightforward expansion and the elliptic Lindstedt–Poincaré method, it can be concluded that the second method gives a better solution even for higher values of the initial displacement and the small parameter.

The analytical solution in the first approximation obtained by the elliptic Lindstedt–Poincaré method (4.6.20) is the special case of the solution obtained by using the harmonic balance method (4.3.11), when the quadratic term is small and $\varepsilon \neq 1$.

4.7 Averaging methods

The averaging methods originally proposed by Krylov and Bogolubov have been applied widely in the solution of the Duffing equation. Besides the generalized method of averaging, the Krylov–Bogolubov method, the Krylov-Bogolubov–Mitropolski method, averaging using Lagrangians etc. are the most often utilised procedures. For

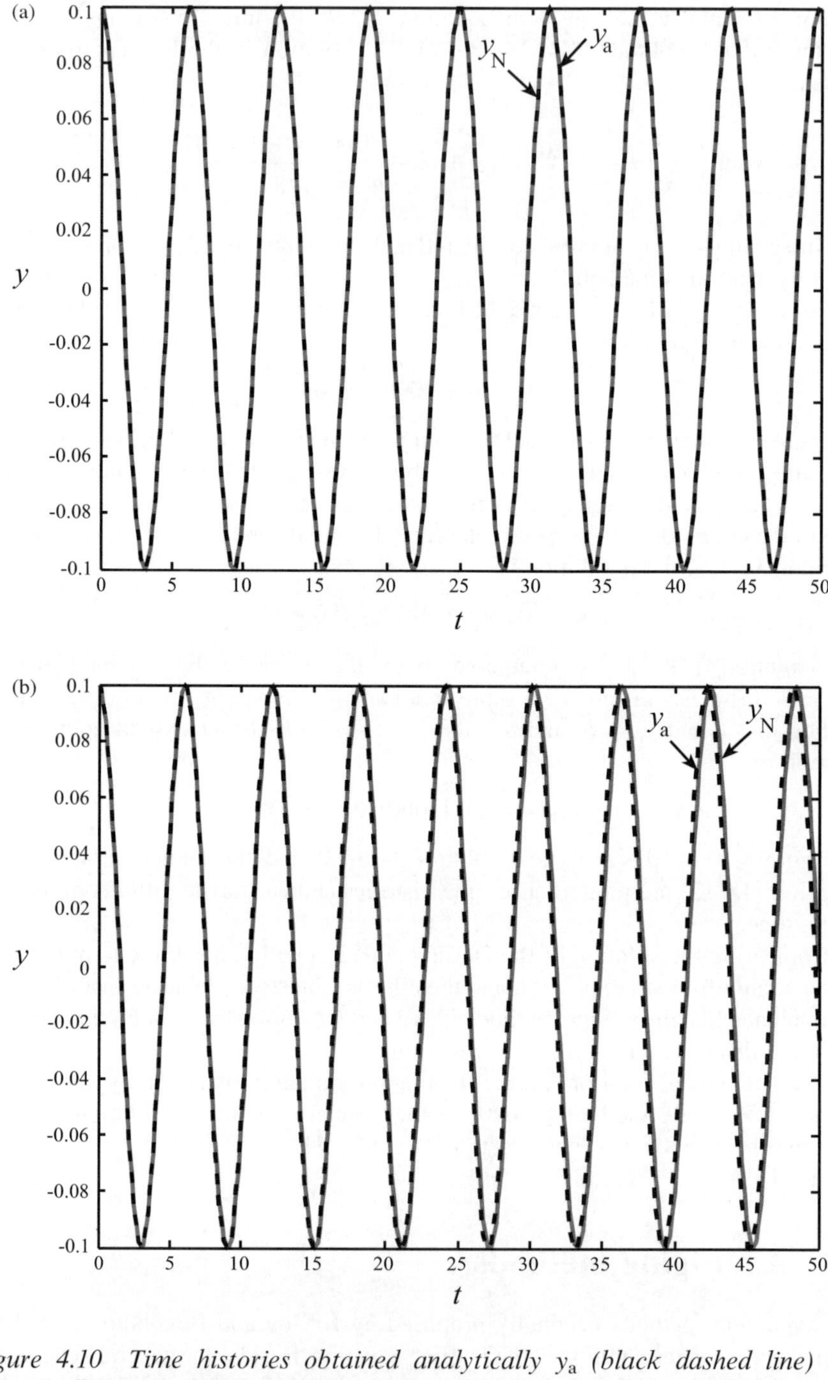

Figure 4.10 Time histories obtained analytically y_a (black dashed line) and numerically (grey solid line) y_N for: (a) $y_0 = 0.1$ and $\varepsilon = 0.1$, (b) $y_0 = 0.1$ and $\varepsilon = 0.8$, (c) $y_0 = 1$ and $\varepsilon = 0.1$.

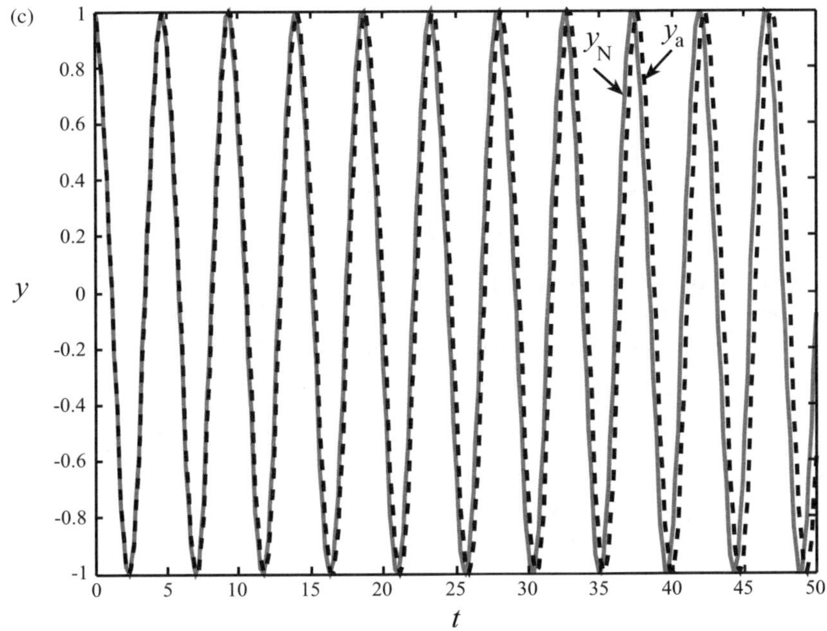

Figure 4.10 (Continued)

all the methods, the variation of the parameters of the assumed solution is used to transform the second order Duffing equation into two first-order differential equations. This is the point at which the methods differ. Very often the analytical assumed solution is a trigonometric function, where the amplitude and the phase are varied. Unfortunately, the accuracy of this approximate solution for a strongly nonlinear Duffing equation is poor. This is because the periodic solution of the nonlinear Duffing equation contains many harmonic components and their influence can be significant. To improve the accuracy without diminishing the advantages of the method, the Jacobi elliptic function is introduced as the more appropriate generating solution for solving the Duffing equation. In this case the variation of the amplitude and the phase, and the modulus of the Jacobi elliptic function, are determined. This is because the modulus of the elliptic function is the most basic and important parameter, having a significant influence on the accuracy of approximate solutions. The elliptic averaging method enables this parameter to be obtained as well.

4.7.1 The generalised elliptic averaging method

In the averaging method, the amplitude and phase of vibration (4.2.5) which is the solution of the Duffing equation (4.2.4) with a strong cubic term are assumed to be time dependent and these functions are obtained using an averaging procedure.

The solution of the perturbed equation (4.5.1) is assumed to be of the form

$$y = Y(t)\text{cn}(\psi, k^2) \equiv Y\,\text{cn} \tag{4.7.1}$$

where the amplitude Y and the argument ψ of the elliptic function cn are time dependent. The Coppola-Rand elliptic averaging procedure [17] is presented in this section. By introducing the new variable φ, the argument ψ is rewritten in the form [17]

$$\psi = 4K(k)\varphi(t) \tag{4.7.2}$$

The period of the elliptic function $4K(k)$ depends on the amplitude Y. Here, K is a function of the modulus k that means that it is a function of the amplitude Y, i.e., $k \equiv k(Y)$. Then, the first time derivative of Equation (4.7.1) is

$$\dot{y} = \dot{Y}(\text{cn} + 4Y\varphi K_k k_Y\,\text{cn}_\psi + Yk_Y\,\text{cn}_k) + 4KY\dot{\varphi}\,\text{cn}_\psi \tag{4.7.3}$$

where the subscripts denote differentiations with respect to the corresponding variables. Equation (4.7.3) is to be compared with the first time derivative of Equation (4.7.1), which is given by

$$\dot{y} = Y\omega\,\text{cn}_\psi(\psi, k^2) \equiv Y\omega\,\text{cn}_\psi \tag{4.7.4}$$

where $\text{cn}_\psi \equiv \frac{\partial\,\text{cn}(\psi,k^2)}{\partial\psi}$ is given in Appendix 4AI. Equation (4.7.3) is forced to have the form defined by Equation (4.7.4):

$$\dot{Y}(\text{cn} + 4Y\varphi K_k k_Y\,\text{cn}_\psi + Yk_Y\,\text{cn}_k) + 4KY\dot{\varphi}\,\text{cn}_\psi = Y\omega\,\text{cn}_\psi \tag{4.7.5}$$

Differentiating Equation (4.7.4) gives

$$\ddot{y} = \dot{Y}((\omega + \omega_Y Y)\,\text{cn}_\psi + 4K_k k_Y \varphi Y\omega\,\text{cn}_{\psi\psi} + Y\omega k_Y\,\text{cn}_{\psi k}) + 4KY\omega\,\text{cn}_{\psi\psi}\dot{\varphi} \tag{4.7.6}$$

Substituting Equation (4.7.1) and Equation (4.7.6) into Equation (4.5.1) and solving for $\dot{Y}$ and $\dot{\varphi}$, gives the following two first-order differential equations

$$\begin{aligned} \dot{Y} &= \frac{\varepsilon f}{\omega}\text{cn}_\psi, \\ 4K\dot{\varphi} &= \omega - \frac{\varepsilon f}{Y\omega}(\text{cn} + 4Y\varphi K_k k_Y\,\text{cn}_\psi + Yk_Y\,\text{cn}_k) \end{aligned} \tag{4.7.7a,b}$$

Introducing the corresponding time derivatives (see Appendix 4AI), the system of Equation (4.7.7) is transformed into

$$\begin{aligned} \dot{Y} &= \frac{\varepsilon f}{\omega}\text{cn}_\psi, \\ 4K\dot{\varphi} &= \omega - \frac{\varepsilon f}{Y\omega}\left(\text{cn} - \frac{1-2k^2}{1-k^2}\left(Z\,cn' + k^2\,\text{cn}(1-\text{cn}^2)\right)\right) \end{aligned} \tag{4.7.8a,b}$$

with the Jacobi zeta function $Z(4K\varphi, k)$

$$Z(4K\varphi, k) = E(4K\varphi, k) - 4\varphi E \tag{4.7.9}$$

where $E(4K\varphi, k)$ is the incomplete elliptic integral of the second kind and $E \equiv E(k)$ is the complete elliptic integral of the second kind. The terms on the right-hand side of Equation (4.7.8) are functions of the variables Y and φ. The system of Equations (4.7.8) is rewritten in the following abbreviated form

$$\dot{Y} = \varepsilon F_1(Y, \varphi), \qquad \dot{\varphi} = \Omega(Y) + \varepsilon F_2(Y, \varphi) \tag{4.7.10a,b}$$

where

$$\Omega(Y) = \frac{\omega}{4K} = \frac{\sqrt{\alpha + \gamma Y^2}}{4K} \tag{4.7.11}$$

and F_1 and F_2 are periodic functions of φ. Equations (4.7.10) are periodic in φ and as such, they are in the correct form for averaging.

Averaging over the period of the elliptic functions $4K(k)$ is introduced resulting in

$$\dot{Y} = \frac{\varepsilon}{4K}\int_0^{4K} F_1(Y, u)du, \qquad \dot{\varphi} = \Omega(Y) + \frac{\varepsilon}{4K}\int_0^{4K} F_2(Y, u)du \tag{4.7.12a,b}$$

where $u = 4K\varphi$.

The averaging method is based on posing a near-identity transformation from (Y,φ) to $(\bar{Y}, \bar{\varphi})$ (see [17])

$$\dot{\bar{Y}} = \varepsilon\left(F_1(\bar{Y}, \varphi) - \Omega(\bar{Y})\frac{\partial w_1}{\partial \varphi}\right) + \varepsilon^2\left(H_1(\bar{Y}, \varphi) - \Omega(\bar{Y})\frac{\partial v_1}{\partial \varphi}\right), \qquad \dot{\bar{\varphi}} = \Omega(\bar{Y}) + \varepsilon\left(F_2(\bar{Y}, \varphi) + \frac{d\Omega(\bar{Y})}{d\bar{Y}} - \Omega(\bar{Y})\frac{\partial w_2}{\partial \varphi}\right) \tag{4.7.13a,b}$$

The generating functions w_1, v_1 and w_2 are chosen so that Equations (4.7.13) are in the averaged form, i.e.,

$$\dot{\bar{Y}} = \varepsilon \bar{F}_1(\bar{Y}) + \varepsilon^2 \bar{H}_1(\bar{Y}), \qquad \dot{\bar{\varphi}} = \Omega(\bar{Y}) + \varepsilon \bar{F}_2(\bar{Y}) \tag{4.7.14a,b}$$

where $\bar{F}_1$, $\bar{F}_2$ and $\bar{H}_1$ are the mean values of F_1, F_2 and H_1 taken over one period in the periodic variable φ.

In the method of averaging, Equations (4.7.10) are replaced with the more useful Equations (4.7.14), which are, however, approximate and valid in the small limit of ε. Solving the averaged differential equations (4.7.14), the approximate amplitude–time and phase–time variations are obtained, which give the approximate solution of Equation (4.5.1). In [61], the computer algebra program MACSYMA for averaging was developed.

4.7.2 Elliptic Krylov–Bogolubov (EKB) method for the pure cubic Duffing oscillator

For the case of a pure cubic Duffing Equation (4.2.23), the modulus of the elliptic function is constant $k^2 = 1/2$, so is the corresponding complete elliptic integral of the first kind $K = K(1/2) = 1.85407$. For this oscillator, the time derivative of the phase angle (4.7.2) simplifies to

$$\dot{\psi} = 4K(1/2)\dot{\varphi} = 7.41628\ \dot{\varphi} \tag{4.7.15}$$

Since ω depends on the amplitude Y (see Equation (4.2.6a)), the phase angle and its corresponding first derivative are

$$\psi = \theta(t) + \int_t \omega(Y)dt, \quad \dot{\psi} = \dot{\theta} + \omega \tag{4.7.16a,b}$$

From Equations (4.7.15) and (4.7.16), it can be seen that

$$4K\dot{\varphi} = \dot{\theta} + \omega \tag{4.7.17}$$

Introducing a new variable $\dot{\theta}$ into Equation (4.7.7b) and by noting that the modulus of the elliptic function is constant, according to (4.7.7) the following two coupled first-order differential equations are obtained

$$\begin{aligned} \dot{Y} &= \frac{\varepsilon f}{\omega}\mathrm{cn}_{\psi}, \\ \dot{\theta} &= -\frac{\varepsilon f}{Y\omega}\mathrm{cn} \end{aligned} \tag{4.7.18a,b}$$

For $\omega = Y\sqrt{\gamma}$ and $\mathrm{cn}_{\psi} = -\mathrm{sn\ dn}$, it follows

$$\begin{aligned} \dot{Y} &= -\frac{\varepsilon f(Y\,\mathrm{cn}, -Y\omega\,\mathrm{sn\ dn})}{Y\sqrt{\gamma}}\mathrm{sn\ dn}, \\ Y\dot{\theta} &= -\frac{\varepsilon f(Y\,\mathrm{cn}, -Y\omega\,\mathrm{sn\ dn})}{Y\sqrt{\gamma}}\mathrm{cn} \end{aligned} \tag{4.7.19a,b}$$

At this point, the averaging procedure is introduced:

$$\begin{aligned} \dot{Y} &= -\frac{\varepsilon}{\omega}\frac{1}{4K}\int_0^{4K} \varepsilon f(Y\,\mathrm{cn}, -Y\omega\,\mathrm{sn\ dn})\mathrm{sn\ dn}\, d\psi, \\ Y\dot{\theta} &= -\frac{\varepsilon}{\omega}\frac{1}{4K}\int_0^{4K} \varepsilon f(Y\,\mathrm{cn}, -Y\omega\,\mathrm{sn\ dn})\mathrm{cn}\, d\psi \end{aligned} \tag{4.7.20a,b}$$

where $K = K(1/2) = 1.85407$ and $\mathrm{cn} = \mathrm{cn}(\psi,1/2)$, $\mathrm{sn} = \mathrm{sn}(\psi,1/2)$, $\mathrm{dn} = \mathrm{dn}(\psi,1/2)$.

Two different cases are analysed with respect to the form of the nonlinear function f:

a. If the nonlinear function depends only on the displacement, i.e., $f \equiv f(y)$, Equations (4.7.20) simplify to

$$\dot{Y} = 0, \quad \dot{\theta} = -\frac{\varepsilon}{Y\omega}\frac{1}{4K}\int_0^{4K} f(Y\,\mathrm{cn})\mathrm{cn}\,d\psi \tag{4.7.21a,b}$$

For $Y=$const. and $\omega = Y\sqrt{\gamma}$ we have

$$\theta = \theta_0 - \frac{\varepsilon t}{4K\sqrt{\gamma}}\int_0^{4K} f(Y\,\mathrm{cn})\mathrm{cn}\,d\psi \tag{4.7.22}$$

Then, the EKB approximate solution is

$$y = Y cn\left(\left(\omega - \frac{\varepsilon}{7.41728\sqrt{\gamma}}\int_0^{4K} f(Y\mathrm{cn})\mathrm{cn}d\psi\right)t + \theta_0, 1/2\right) \tag{4.7.23}$$

with the values for Y and θ_0 defined by the initial conditions.
b. For $f \equiv f(\dot{y})$, Equations (4.7.20) have the form

$$\dot{Y} = -\frac{\varepsilon}{\omega}\frac{1}{4K}\int_0^{4K} f(-Y\omega\,\mathrm{sn}\,\mathrm{dn})\mathrm{sn}\,\mathrm{dn}\,d\psi, \quad \dot{\theta} = 0 \tag{4.7.24a,b}$$

The EKB solution yields

$$y = Y(t)\mathrm{cn}\left((\sqrt{\gamma}\int_0^t Y(t)dt) + \theta_0, 1/2\right) \tag{4.7.25}$$

where $Y(t)$ is the solution of Equation (4.7.24a).

4.7.2.1 Pure cubic oscillator with a weak linear term

For the differential equation with a strong nonlinear cubic term and a weak linear term

$$\ddot{y} + \gamma y^3 = -\varepsilon y \tag{4.7.26}$$

Equation (4.7.22) gives the phase angle

$$\theta = \theta_0 + \frac{\varepsilon t}{4K\omega}\int_0^{4K} \mathrm{cn}^2\,d\psi = \theta_0 + \frac{\varepsilon t}{\omega}\left(\frac{2E}{K} - 1\right) = \theta_0 + \frac{0.4569\varepsilon t}{\omega} \tag{4.7.27}$$

In accordance with Equations (4.7.27) and (4.7.23), the approximate solution of Equation (4.7.26) is

$$y = Y\,\mathrm{cn}\left(\left(\omega + \frac{0.4569\varepsilon}{\omega}\right)t + \theta_0, 1/2\right) \tag{4.7.28}$$

where $E = E(1/2) = 1.35064$ is the complete elliptic integral of the second kind for the modulus $k^2 = 1/2$ and $\omega = Y\sqrt{\gamma}$.

It should be noted that Equation (4.7.26) represents a special case of Equation (4.2.4), where the coefficient of the linear term has a small value ($\varepsilon \ll 1$). Using Equations (4.2.5) and (4.2.6), the exact solution of Equation (4.7.26) is as follows

$$y = Y\,\mathrm{cn}\left(t\sqrt{\varepsilon+\gamma Y^2}+\theta_0, \frac{\gamma Y^2}{2(\varepsilon+\gamma Y^2)}\right) \tag{4.7.29}$$

Using the series expansion of the functions in Equation (4.7.29) and the relation $\omega = Y\sqrt{\gamma}$, the approximate solution is obtained

$$y = Y\,\mathrm{cn}\left(t\left(\omega+\frac{\varepsilon}{2\omega}\right)+\theta_0, 1/2\right) \tag{4.7.30}$$

Comparing the solutions given by Equations (4.7.28) and (4.7.30), it is evident that the difference is negligible.

4.7.2.2 Pure cubic oscillator with linear damping

For the differential equation with a small linear damping term

$$\ddot{y}+2\zeta\dot{y}+\gamma y^3 = 0 \tag{4.7.31}$$

Equation (4.7.24a) yields the amplitude of vibration

$$\dot{Y} = -\frac{\varepsilon(2\zeta)Y}{4K}\int_0^{4K}\mathrm{sn}^2\,\mathrm{dn}^2\,d\psi = -\frac{\varepsilon(2\zeta)Y}{4K}Q \tag{4.7.32}$$

i.e.,

$$Y = Y_0 e^{-\frac{\varepsilon(2\zeta)t}{4K}Q} \tag{4.7.33}$$

where

$$Q = \int_0^{4K}\mathrm{sn}^2\,\mathrm{dn}^2\,d\psi = 4K(1-k^2)-(1-2k^2)C_2-k^2C_4 \tag{4.7.34}$$

and the averaged values C_2 and C_4 are given in Appendix 4AI.

Using Equations (4.7.25) and (4.7.33) gives

$$y = Y_0 e^{-\frac{\varepsilon(2\zeta)t}{4K}Q}\mathrm{cn}\left(Y_0\sqrt{\gamma}\frac{4K}{\varepsilon(2\zeta)Q}\left(1-\exp\left(-\frac{\varepsilon t(2\zeta)Q}{4K}\right)\right)+\theta_0, 1/2\right) \tag{4.7.35}$$

After some simplification, the approximate solution is found to be

$$y = Y_0 e^{-\frac{\varepsilon(2\zeta)t}{4K}Q}\mathrm{cn}(Y_0 t\sqrt{\gamma}+\theta_0, 1/2) \tag{4.7.36}$$

Equation (4.7.36) implies that the amplitude of vibration decreases exponentially. The period of vibration increases, but very slowly. This means that it can be assumed that the frequency of vibration is constant.

Note that the EKB method is usually known as a method in which the amplitude and phase vary as a function of time, as it is assumed that the perturbed amplitude and phase of the solution differs for a small value with respect to the generating solution.

4.8 Elliptic homotopy methods

The so-called homotopy methods are based on homotopy, which is a fundamental concept in topology and differential geometry, which can be traced back to Poincaré [62]. By means of the homotopy method, a continuous mapping of an initial guess of the exact solution of the Duffing equation is constructed. In this section, two homotopy methods are presented: the homotopy perturbation method (HPM) and the homotopy analysis method (HAP).

4.8.1 The elliptic homotopy perturbation method

The homotopy perturbation method is applied to solve Equation (4.2.31a) with the initial conditions given in Equation (4.2.9). If $f=0$, Equation (4.2.4) has the exact solution described by Equations (4.2.5) and (4.2.6). Based on this result, the initial guess for y is assumed to have the form

$$Y_0(t) \equiv Y_0 = y_0 \,\mathrm{cn}(\omega_1 t, k_1^2) = y_0 \,\mathrm{cn}_1 \tag{4.8.1}$$

where ω_1 and k_1^2 transform into ω and k^2 when $f=0$. As suggested by He [63], a definition of the homotopy $X{:}\Omega x[0,1] \to R$ means that two continuous functions from one topological space can be 'continuously deformed' one into another by introducing an artificial embedding parameter p, often called a 'homotopy parameter', the values of which are in the interval [0,1]. Hence, a homotopy can be constructed

$$H(y,p) = (1-p)((\ddot{X}+\alpha X+\gamma X^3)-(\ddot{Y}_0+\alpha Y_0+\gamma Y_0^3))+p(\ddot{X}+\alpha X+\gamma X^3-f(X,\dot{X})) \tag{4.8.2}$$

As p increases from 0 to 1, $H(y,p)$ varies continuously i.e., 'deforms' as it is usually called in topology. Enforcing homotopy to zero, $H(y,p)=0$, i.e.,

$$(1-p)((\ddot{X}+\alpha X+\gamma X^3)-(\ddot{Y}_0+\alpha Y_0+\gamma Y_0^3))+p(\ddot{X}+\alpha X+\gamma X^3-f(X,\dot{X}))=0 \tag{4.8.3}$$

gives the zeroth-order deformation equation with the initial conditions

$$X(0,p) = y_0, \quad \dot{X}(0,p) = 0 \tag{4.8.4a,b}$$

For $p=0$, Equation (4.8.3) simplifies to

$$\ddot{X}+\alpha X+\gamma X^3 = 0 \tag{4.8.5}$$

The exact solution of Equation (4.8.5) is

$$X(t,0) = y^0(t) = Y\,\mathrm{cn}(\omega t + \theta, k^2) \tag{4.8.6}$$

For $p = 1$, the equation has the same form as the original equation

$$\ddot{X} + \alpha X + \gamma X^3 = f(X, \dot{X}) \tag{4.8.7}$$

and the solution is

$$X(t,1) = y(t) \tag{4.8.8}$$

It can be concluded that for the change of p from zero to unity, the solution continually changes from the one given in Equation (4.8.6) to the one described by Equation (4.8.8).

As $X(t,p)$ is a function of the homotopy parameter $p \in [0\ 1]$ and it is smooth enough to have the kth-order partial derivatives with respect to p at $p = 0$, it can be expanded into the Maclaurin series

$$X(t,p) = Y_0(t) + \sum_{k=1}^{\infty} (y_k(y))p^k \tag{4.8.9}$$

where $X(t,0) = Y_0$ is employed, and the kth-order homotopy derivative of $X(t,p)$ is

$$y_k(t,p) = \frac{1}{k!}\frac{\partial^k X(t,p)}{\partial p^k} \tag{4.8.10}$$

It is at this point where a crucial assumption and restriction to the homotopy is introduced. It is supposed that the homotopy series (4.8.9) is convergent at $p = 1$. This assumption is valid because of the hypothesis given in [62]: 'If the Duffing equation has at least one solution, then there exists at least one zeroth-order deformation equation such that its homotopy series solution converges to the solution of the original nonlinear equation'. Now, using the relation $X(t,1) = y$, gives the homotopy series solution

$$y = Y_0 + \sum_{k=1}^{n} y_k \tag{4.8.11}$$

In accordance with the fundamental theorem of calculus about the Taylor series, the coefficient y_k of the homotopy series (4.8.10) is unique. Therefore, the governing equation of y_k is unique, too, and can be deduced directly from the zeroth-order deformation equation (4.8.3).

Substituting Equation (4.8.11) into Equation (4.8.3) and separating the terms with the same order of the parameter p, a system of linear differential equations is obtained. The terms next to p^1 form the following first-order deformation equation

$$\ddot{y}_1 + \alpha y_1 + 3\gamma Y_0^2 y_1 = -(\ddot{Y}_0 + \alpha Y_0 + \gamma Y_0^3 - f(Y_0, \dot{Y}_0)) \tag{4.8.12}$$

with the initial conditions

$$y_1(0) = 0, \quad \dot{y}_1(0) = 0 \tag{4.8.13a,b}$$

Substituting Equation (4.8.1) into Equation (4.8.12), gives

$$\begin{aligned}&\ddot{y}_1 + \alpha y_1 + 3\gamma y_0^2 \operatorname{cn}_1^2 y_1 \\ &\quad = -(-y_0\omega_1^2 \operatorname{cn}_1(1-2k_1^2 + 2k_1^2 \operatorname{cn}_1^2) + \alpha y_0 \operatorname{cn}_1 + \gamma y_0^3 \operatorname{cn}_1^3 - f(y_0 \operatorname{cn}_1, -y_0\omega_1 \operatorname{sn}_1 \operatorname{dn}_1))\end{aligned} \tag{4.8.14}$$

where sn_1 and dn_1 denote the following Jacobi elliptic functions $\operatorname{sn}_1 \equiv \operatorname{sn}(\omega_1 t, k_1)$ and $\operatorname{dn}_1 \equiv \operatorname{dn}(\omega_1 t, k_1)$, respectively.

For

$$\alpha = 0, \quad f(y, \dot{y}) = -\beta y^2, \quad \gamma > \beta \tag{4.8.15a–c}$$

Equation (4.8.14) is

$$\ddot{y}_1 + 3\gamma y_0^2 \operatorname{cn}_1^2 y_1 = -(-y_0\omega_1^2 \operatorname{cn}_1(1-2k_1^2 + 2k_1^2 \operatorname{cn}_1^2) + \gamma y_0^3 \operatorname{cn}_1^3 + \beta y_0^2 \operatorname{cn}_1^2)) \tag{4.8.16}$$

The solution of Equation (4.8.16) is assumed to be the sum of a constant and a linear term in the elliptic function cn_1

$$y_1 = K_0 + K_1 \operatorname{cn}_1 \tag{4.8.17}$$

Substituting Equation (4.8.17) into Equation (4.8.16) and separating the terms with the same order of the elliptic function cn_1, the following system of algebraic equations is obtained

$$\begin{aligned}&(K_1 + y_0)\omega_1^2(1-2k_1^2) = 0, \\ &3\gamma y_0^2 K_0 + \beta K_0^2 = 0, \\ &3\beta y_0^2 K_1 + \gamma y_0^3 - 2(K_1 + y_0)k_1^2\omega_1^2 = 0\end{aligned} \tag{4.8.18a–c}$$

The initial conditions (4.8.13) impose the following relation for K_0 and K_1

$$K_0 + K_1 = 0 \tag{4.8.19}$$

Solving Equations (4.8.18) with the condition (4.8.19), it follows

$$\omega_1^2 = 3\gamma y_0^2 \frac{\gamma y_0 + \beta}{3\gamma y_0 + \beta}, \quad k_1^2 = \frac{1}{2}, \quad K_0 = -K_1 = -\frac{\beta}{3\gamma} \tag{4.8.20a–c}$$

Using Equations (4.8.1), (4.8.11) and (4.8.20), gives the first-order homotopy series approximation

$$y(t) = -\frac{\beta}{3\gamma} + \left(y_0 + \frac{\beta}{3\gamma}\right) \operatorname{cn}\left(y_0 t\sqrt{3\gamma \frac{\gamma y_0 + \beta}{3\gamma y_0 + \beta}}, \frac{1}{2}\right) \tag{4.8.21}$$

Examining Equation (4.8.20b), it is clear that the coefficient β has no influence on the modulus of the Jacobi function. The frequency and the argument of the Jacobi function and also the accuracy of the approximate solution (4.8.21) depend on the coefficient ratio β/γ. For smaller values of the ratio $(\beta/\gamma) \ll 1$, the difference between the exact solution and the approximate solution is negligible. For higher values of the ratio β/γ, the difference between these solutions is significant and the solution in the first approximation is not acceptable.

4.8.2 The elliptic homotopy analysis method

To overcome the restriction given in the HPM about the convergence of the homotopy series at $p=1$, the homotopy analysis method is developed, based on the appropriate choice of an auxiliary parameter that is used to ensure the convergence of the solution series. The method provides greater freedom to choose the auxiliary parameter $\hbar \neq 0$, usually called the 'convergence-control parameter'.

Introducing the convergence-control parameter $\hbar \neq 0$, in accordance with Equation (4.8.3), the following zeroth-order deformation equation can be constructed

$$(1-p)((\ddot{X}+\alpha X+\gamma X^3)-(\ddot{Y}_0+\alpha Y_0+\gamma Y_0^3)) = p\hbar(\ddot{X}+\alpha X+\gamma X^3-f(X,\dot{X})) \quad (4.8.22)$$

Since $\hbar \neq 0$, the above equation at $p=1$ becomes

$$p\hbar(\ddot{X}+\alpha X+\gamma X^3-f(X,\dot{X})) = 0 \quad (4.8.23)$$

which is equivalent to the original equation (4.2.31a), provided that $y = X(t,1)$. Similarly, taking the first-order homotopy derivative on both sides of Equation (4.8.22), the first-order deformation equation is

$$\ddot{y}_1+\alpha y_1+3\gamma Y_0^2 y_1-\hbar(\ddot{Y}_0+\alpha Y_0+\gamma Y_0^3-f(Y_0,\dot{Y}_0)) = 0 \quad (4.8.24)$$

Introducing Equation (4.8.1) into Equation (4.8.24), yields

$$\begin{aligned}&\ddot{y}_1+\alpha y_1+3\gamma y_0^2 \,\mathrm{cn}_1^2 y_1 \\ &\quad = \hbar(-y_0\omega_1^2 \,\mathrm{cn}_1(1-2k_1^2+2k_1^2 \,\mathrm{cn}_1^2)+\alpha y_0 \,\mathrm{cn}_1+\gamma y_0^3 \,\mathrm{cn}_1^3-f(y_0 \,\mathrm{cn}_1, -y_0\omega_1 \,\mathrm{sn}_1 \,\mathrm{dn}_1))\end{aligned} \quad (4.8.25)$$

where sn_1 and dn_1 denote the Jacobi elliptic functions, i.e., $\mathrm{sn}_1 \equiv sn(\omega_1 t, k_1)$ and $\mathrm{dn}_1 \equiv \mathrm{d}n(\omega_1 t, k_1)$. For

$$\alpha = 0, \quad f(y,\dot{y}) = -\beta y^2, \quad \gamma > \beta \quad (4.8.26a–c)$$

Equation (4.8.25) is

$$\ddot{y}_1+3\gamma y_0^2 \,\mathrm{cn}_1^2 y_1 = \hbar(-y_0\omega_1^2 \,\mathrm{cn}_1(1-2k_1^2+2k_1^2 \,\mathrm{cn}_1^2)+\gamma y_0^3 \,\mathrm{cn}_1^3+\beta y_0^2 \,\mathrm{cn}_1^2)) \quad (4.8.27)$$

To obtain the solution of Equation (4.8.27), the convergence-control parameter $\hbar$ has to be calculated. The convergence-control parameter $\hbar$ in Equation (4.8.27) can be regarded as an iteration factor that is widely used in numerical computation. If a proper value of $\hbar$ is chosen, the convergence of the homotopy series solutions is obtained. The calculation of the solution of Equation (4.8.27) and the values of $\hbar$ has to be done numerically.

The HAM is a method for the era of computers: without a high-performance computer and symbolic computation software such as MATHEMATICA or MAPLE (see [64]), it is impossible to solve high-order deformation equations quickly and to choose an appropriate value of the convergence-control parameter $\hbar$ by means of analysing the high-order approximations. It is true that expressions given by the HAM are often lengthy. However, by means of the computer, it often takes only a few seconds to solve these lengthy expressions.

Numerical calculations of Equations (4.8.27) show that for $\alpha = f = 0$, the homotopy series solution is convergent for any value of the physical parameter $0 \leq \gamma < \infty$ by using $\hbar = -(1+\gamma)^{-1}$ [64].

It should be noted that for the case when $\hbar = -1$, the homotopy analysis method transforms into the homotopy perturbation method.

4.9 Summary

After considering and comparing some analytic approximate solution procedures for the Duffing equation in this chapter, the following conclusions can be drawn:

1. The analytical procedures can be divided into two groups. In the first group are the methods that require the existence of a small physical parameter of the system. The second group comprises the methods that do not depend on a small physical parameter. In this chapter, the straightforward expansion method, the general elliptic averaging method, the elliptic Krylov–Bogolubov method, which are usually called perturbation techniques have been considered. The harmonic balance method, the Galerkin method, the homotopy perturbation and the homotopy analysis method have also been considered. The methods from the first group are applicable to the Duffing equation that has a strong cubic term and all the other nonlinearities associated with stiffness and damping are small. The methods from the second group can be used even when all the nonlinearities in the Duffing equation are large.

2. Most of the perturbation methods require the existence of a small parameter and are therefore not valid, in principle, for the Duffing equation in which the nonlinearity is large. The following problems appear regarding the assumption of a small parameter:
   - Many Duffing oscillators do not have small physical parameters.
   - There are no criteria for determining a small parameter and its choice is up to the individual.
   - The assumption of a small parameter greatly restricts the application of the perturbation techniques.

3. The advantage of the perturbation techniques is their simplicity and an easy calculation procedure, for which additional numerical software is not needed.

4. The appropriate choice of a small parameter may give ideal results that are valid for a restricted time interval from the initial motion. An inappropriate

choice of the small parameter affects an outcome crucially in a negative way. If the term with a small parameter becomes sufficiently large, the results are unsatisfactory.

5. The advantage of nonperturbation techniques is that they do not require the existence of a small parameter and are applicable to all Duffing oscillators. Unfortunately, the harmonic balance method and the Galerkin method often give results that have some deficiencies or are not sufficiently accurate.

6. Homotopy methods are suitable for solving the Duffing equation that has strong nonlinearity. Unlike most of the methods, homotopy methods are independent of the small parameters: no matter whether the Duffing equation contains small physical parameters or not, a homotopy parameter $p \in [0\ 1]$ can always be introduced to obtain a series solution. While the homotopy perturbation method has a restriction (it is assumed that the homotopy series solution is convergent), the homotopy analysis method gives the possibility of proving convergence of the series expansion of the solution, which is not possible by applying other approximate analytic solutions. An auxiliary parameter, the so-called 'convergence-control parameter' is introduced and its appropriate value assures the convergence of the series solution. Unfortunately, the parameter cannot be calculated without the use of a computer and symbolic computation software. The homotopy perturbation technique gives good results in the first approximation.

7. The suggested elliptic analytical methods based on the exact solution of the strong nonlinear Duffing equation yield much more appropriate solutions than those which use trigonometric functions. The methods give more accurate results for a wide class of problems and for longer time intervals. The main disadvantage in the application of the elliptic methods to the Duffing equations in comparison to the standard methods is the complexity of the elliptic functions.

References

[1] G. Duffing. *Erzwungene Schwingungen bei veränderlicher Eigenfrequenz und ihre technische Bedeutung*, Vieweg & Sohn, Braunschweig, 1918.

[2] I.S. Gradstein, I.M. Rjizhik, *Table of Integrals*, Series and Products, Moscow, Nauka, 1971 (in Russian).

[3] A.H. Nayfeh, D.T. Mook, *Nonlinear Oscillations*, Wiley, New York, 1979.

[4] N.N. Bogolubov, Y.A. Mitropolskij, *Asymptotic Methods in the Theory of Non-linear Vibrations*, Nauka, Moscow, 1968 (in Russian).

[5] N.N. Moiseev, *Asymptotic Methods of the Non-Linear Science*, Nauka, Moscow, 1981.

[6] L. Cveticanin, H. Yamakawa, O. Matsushita, An asymptotic method applied to nonlinear systems with coupled deflection. *Journal of the Franklin Institute*, 328, 71–83, 1991.

[7] L. Cveticanin, An asymptotic solution for weak nonlinear vibrations of the rotor. *Mechanism and Machine Theory*, 28, 495–506, 1993.

[8] L. Cveticanin, The vibrations of a textile machine rotor with nonlinear characteristics. *Mechanism and Machine Theory*, 21, 29–32, 1986.

[9] L. Cveticanin, Vibrations of a textile machine rotor. *Journal of Sound and Vibration*, 97, 181–187, 1987.

[10] L. Cveticanin, The oscillations of a textile machine rotor on which the textile is wound up. *Mechanism and Machine Theory*, 26, 253–260, 1991.

[11] L. Cveticanin, Dynamic behaviour of a rotor with time-dependent parameter. *Japanese Society of Mechanical Engineers, Series C*, 37, 41–48, 1994.

[12] L. Cveticanin, Approximate solution of a time-dependent differential equation. *Meccanica*, 30, 665–671, 1995.

[13] P.F. Byrd, M.D. Friedman, *Handbook of Elliptic Integrals for Engineers and Physicists*, Springer-Verlag, Berlin, 1953.

[14] P.G.D. Barkham, A.C. Soudack, An extension to the method of Krylov and Bogolubov. *International Journal of Control*, 10, 377–392, 1969.

[15] P.A. Christopher, An approximate solution to a strongly nonlinear, second order, differential equation. *International Journal of Control*, 17, 597–608, 1973.

[16] P.A. Christopher, A. Brocklehurst, A generalized form of an approximate solution to a strongly nonlinear, second order, differential equation. *International Journal of Control*, 19, 831–839, 1974.

[17] V.T. Coppola, R.H. Rand, Averaging using elliptic functions: Approximation of limit cycles. *Acta Mechanica*, 81, 125–142, 1990.

[18] S.B. Yuste, J.D. Bejarano, Construction of approximate analytical solutions to a new class of nonlinear oscillator equations. *Journal of Sound and Vibration*, 110, 347–350, 1986.

[19] S.B. Yuste, J.D. Bejarano, Extension and improvement to the Krylov–Bogolubov methods using elliptic functions. *International Journal of Control*, 49, 1127–1141, 1989.

[20] S.B. Yuste, J.D. Bejarano, Improvement of a Krylov–Bogolubov method that uses Jacobi elliptic functions. *Journal of Sound and Vibration*, 139, 151–163, 1990.

[21] S.B. Yuste, On Duffing oscillators with slowly varying parameters. *International Journal of Non-Linear Mechanics*, 26, 671–677, 1991.

[22] S.B. Yuste, Quasi-pure-cubic oscillators studied using a Krylov–Bogolubov method. *Journal of Sound and Vibration*, 158, 267–275, 1992.

[23] R.E. Mickens, Comments on the method of harmonic balance. *Journal of Sound and Vibration*, 94, 456–460, 1984.

[24] R.E. Mickens, A generalization of the method of harmonic balance. *Journal of Sound and Vibration*, 111, 515–518, 1986.

[25] S.B. Yuste, Comments on the method of harmonic balance in which Jacobi elliptic functions are used. *Journal of Sound and Vibration*, 145, 381–390, 1991.

[26] S.B. Yuste, Cubication of non-linear oscillators using the principle of harmonic balance. *International Journal of Non-Linear Mechanics*, 27, 347–356, 1992.

[27] S. Hiamang, R.E. Mickens, Harmonic balance: Comparison of equation of motion and energy methods. *Journal of Sound and Vibration*, 164, 179–181, 1993.

[28] C.R. Handy, Combining the methods of harmonic balance and Kryloff–Bogoliuboff. *Journal of Sound and Vibration*, 102, 243–246, 1985.

[29] S.H. Chen, Y.K. Cheung, An elliptic perturbation method for certain strongly non-linear oscillators. *Journal of Sound and Vibration*, 192, 453–464, 1996.

[30] S.H. Chen, Y.K. Cheung, An elliptic Lindstedt–Poincaré method for analysis of certain strongly non-linear oscillators. *Nonlinear Dynamics*, 12, 199–213, 1997.

[31] M. Belhaq, F. Lakrad, The elliptic multiple scales method for a class of autonomous strongly non-linear oscillators, *Journal of Sound and Vibration*, 234, 547–553, 2000.

[32] S.B. Yuste, A generalized Galerkin method for cubic oscillators, *Journal of Sound and Vibration*, 130, 332–336, 1989.

[33] L. Cveticanin, Homotopy-perturbation method for pure non-linear differential equation, Chaos, *Solitons and Fractals*, 30, 1221–1230, 2006.

[34] L. Cveticanin, Analytical methods for solving strongly non-linear differential equations, *Journal of Sound and Vibration*, 214, 325–338, 1998.

[35] L. Cveticanin, Vibrations of the non-linear oscillator with quadratic non-linearity, *Physica A*, 341, 123–135, 2004.

[36] L. Cveticanin, Vibrations of the system with quadratic non-linearity and a constant excitation force, *Journal of Sound and Vibration*, 261, 169–176, 2003.

[37] L. Cveticanin, Oscillator with fraction order restoring force, *Journal of Sound and Vibration*, 320, 1064–1077, 2009.

[38] L. Cveticanin, The motion of a two-mass system with non-linear connection. *Journal of Sound and Vibration*, 252, 361–369, 2002.

[39] L. Cveticanin, Forced non-linear vibrations of a symmetrical two-mass-system, *Journal of Sound and Vibration*, 265, 451–458, 2003.

[40] L. Cveticanin, Vibrations of a free two-mass-system with quadratic non-linearity and a constant excitation force. *Journal of Sound and Vibration*, 270, 441–449, 2004.

[41] L. Cveticanin, Approximate analytical solutions to a class of nonlinear equations with complex functions. *Journal of Sound and Vibration*, 157, 289–302, 1992.

[42] L. Cveticanin, An approximate solution for a system of two coupled differential equations. *Journal of Sound and Vibration*, 153, 375–380, 1992.

[43] L. Cveticanin, Some particular solutions which describe the motion of the rotor. *Journal of Sound and Vibration*, 212, 173–178, 1998.

[44] L. Cveticanin, The approximate solving methods for the cubic Duffing equation based on the Jacobi elliptic functions. *International Journal of Nonlinear Sciences and Numerical Simulation*, 10, 1491–1516, 2009.

[45] L. Cveticanin, Vibrations of a coupled two-degree-of-freedom system. *Journal of Sound and Vibration*, 247, 279–292, 2001.

[46] L. Cveticanin, Analytic approach for the solution of the complex-valued strong non-linear differential equation of Duffing type. *Physica A*, 297, 348–360, 2001.

[47] L. Cveticanin, Analytical solutions of the system of two coupled pure cubic nonlinear oscillators equations. *Journal of Sound and Vibration*, 245, 571–580, 2001.

[48] L. Cveticanin, Free vibration of a strong non-linear system described with complex function. *Journal of Sound and Vibration*, 277, 815–824, 2004.

[49] L. Cveticanin, A new approach for solving of a complex-valued differential equation. *Journal of Sound and Vibration*, 278, 1181–1195, 2004.

[50] L. Cveticanin, Approximate solution of a strongly non-linear complex differential equation. *Journal of Sound and Vibration*, 284, 503–512, 2005.

[51] L. Cveticanin, The homotopy-perturbation method applied for solving complex-valued differential equations with strong cubic nonlinearity. *Journal of Sound and Vibration*, 285, 1171–1179, 2005.

[52] L. Cveticanin, Free vibration on a Jeffcott rotor with pure cubic non-linear elastic property of the shaft. *Mechanism and Machine Theory*, 49, 1330–1344, 2005.

[53] L. Cveticanin, Vibration of strongly nonlinear rotors with time variable parameters. *Machine Vibration*, 4, 40–45, 1995.

[54] M. Abramowitz, I.A. Stegun, *Handbook of Mathematical Functions*, Dover, New York, 1972.

[55] A.C. Soudack, P.G. Barkham, On the transient solution of the unforced Duffing equation with large damping. *International Journal of Control*, 13, 767–769, 1971.

[56] H. Grandin, *Fundamentals of the Finite Element Method*, Macmillan Press, New York, 1986.

[57] A.B. Finlayson, *The Method of Weighted Residuals and Variational Principles*, Academic Press, New York, 1972.

[58] J. Petrolito, Approximate solutions of differential equations using Galerkin's method and weighted residuals. *Journal of Mechanical Engineering Education*, 28, 14–25, 1998.

[59] G. Chen, Applications of a generalized Galerkin's method to non-linear oscillations of two-degree-of-freedom systems. *Journal of Sound and Vibration*, 119, 225–242, 1987.

[60] G. Chen, H. Du, The Galerkin method for initial value problems based on the principle of total virtual action. *Journal of Sound and Vibration*, 203, 457–472, 1997.

[61] V.T. Coppola, R.H. Rand, Computer algebra, elliptic functions and chaos, *Computers in Engineering*, Editors: G.I. Kinzel, S.M. Rohde, D.W. Bennett, C. Born, A.A. Businaina, G. Gabriele, V. Ghazarian, J. T. ShoupBook No. G0515A- 1990, American Society of Mechanical Engineers, 193–200, 1990.

[62] S. Liao, Notes on the homotopy analysis method: Some definitions and theorems. *Communications in Nonlinear Science and Numerical Simulations*, 14, 983–99, 2008.

[63] J.H. He, An approximate solution technique depending upon an artificial parameter, *Communications in Nonlinear Science and Numerical Simulations*, 3, 92–97, 1998.

[64] S.J. Liao, Y. Tan, A general approach to obtain series solutions of nonlinear differential equations. *Studies in Applied Mathematics*, 119, 297–355, 2007.

Appendix 4AI: Jacobi elliptic functions and elliptic integrals

For the convenience of the readers, some facts about Jacobi elliptic functions are given here. Jacobi elliptic functions are periodic functions defined over a complex plane. Three fundamental elliptic functions are the Jacobi elliptic sine $\mathrm{sn}(\psi,k^2)\equiv\mathrm{sn}$, cosine $\mathrm{cn}(\psi,k^2))\equiv\mathrm{cn}$ and delta $\mathrm{dn}(\psi,k^2))\equiv\mathrm{dn}$ functions with the argument ψ and the modulus k^2. Note that the standard treatments use two versions for modulus of Jacobi elliptic functions: k [13] and k^2 [54]. The elliptic functions sn and cn can be thought of as the generalisations of the sine and cosine functions, respectively. For $k^2=0$ and $k^2=1$, the following identities hold

$$\begin{aligned}&\mathrm{sn}(\psi,0)=\sin\psi, \quad \mathrm{cn}(\psi,0)=\cos\psi, \quad \mathrm{dn}(\psi,0)=1,\\&\mathrm{sn}(\psi,1)=\tan h\psi, \quad \mathrm{cn}(\psi,1)=\mathrm{sech}\,\psi, \quad \mathrm{dn}(\psi,1)=\mathrm{sech}\,\psi\end{aligned} \tag{4AI.1a–f}$$

where $\sin\psi$ and $\cos\psi$ are trigonometric functions and $\tanh\psi$ and $\mathrm{sech}\psi$ are hyperbolic trigonometric functions.

The elliptic functions satisfy the following expressions:

$$\mathrm{sn}^2+\mathrm{cn}^2=1, \quad \mathrm{dn}^2+k^2\mathrm{sn}^2=1, \quad 1-k^2+k^2\mathrm{cn}^2=\mathrm{dn}^2 \tag{4AI.2a–c}$$

where the first is reminiscent of the comparable trigonometric identity. Only two of these three relations are independent.

In Table 4AI.1, additional properties of the Jacobi elliptic functions are summarised. There, some additional abbreviations are introduced. Thus, $l \equiv \sqrt{1-k^2}$ is the complementary modulus, $Z(\psi,k)$ is the Jacobi zeta function, $K(k) \equiv K$ is the complete elliptic integral of the first kind and $E(k) \equiv E$ is the complete elliptic integral of the second kind. According to [13], the development of the complete elliptic

Table 4AI.1 Some properties of the Jacobi elliptic functions.

Property	$\text{sn}(\psi,k^2) \equiv \text{sn}$	$\text{cn}(\psi,k^2) \equiv \text{cn}$	$\text{dn}(\psi,k^2) \equiv \text{dn}$
$\frac{\partial(\cdot)}{\partial\psi} = (\cdot)_\psi \equiv (\cdot)'$	cn dn	−sn dn	$-k^2$ sn cn
$\frac{\partial^2(\cdot)}{\partial\psi^2} = (\cdot)_{\psi\psi} \equiv (\cdot)''$	$2\rho_1\text{cn}^3 + \rho_2\text{cn}$	$2\rho_1\text{sn}^3 + \rho_2\text{sn}$	$2\rho_1\text{dn}^3 + \rho_2\text{dn}$
$\frac{\partial(\cdot)}{\partial k^2} \equiv (\cdot)_k$	$\frac{1}{kl^2}\left(\left(\left(l^2-\frac{E}{K}\right)\psi - Z(\psi,k)\right)\text{sn}' + k^2(1-\text{sn}^2)\text{sn}\right)$	$\frac{1}{kl^2}\left(\left(\left(l^2-\frac{E}{K}\right)\psi - Z(\psi,k)\right)\text{cn}' - k^2(1-\text{cn}^2)\text{cn}\right)$	$\frac{1}{kl^2}\left(\left(\left(l^2-\frac{E}{K}\right)\psi - Z(\psi,k)\right)\text{dn}' - (1-\text{dn}^2)\text{dn}\right)$
$\frac{\partial^2(\cdot)}{\partial\psi\partial k^2} \equiv (\cdot)_{\psi,k}$	$\frac{1}{kl^2}\left(\left(\left(l^2-\frac{E}{K}\right)\psi - Z(\psi,k)\right)(2\rho_1\,\text{sn}^3 + \rho_2\,\text{sn}) - (2\rho_1\,\text{sn}^2+\rho_3)\text{sn}'\right)$	$\frac{1}{kl^2}\left(\left(\left(l^2-\frac{E}{K}\right)\psi - Z(\psi,k)\right)(2\rho_1\,\text{cn}^3 + \rho_2\,\text{cn}) - (2\rho_1\,\text{sn}^2+\rho_3)\text{cn}'\right)$	$\frac{1}{kl^2}\left(\left(\left(l^2-\frac{E}{K}\right)\psi - Z(\psi,k)\right)(2\rho_1\,\text{dn}^3 + \rho_2\,\text{dn}) - (2\rho_1\,\text{dn}^2+\rho_3)\text{dn}'\right)$
ρ_1	K^2	$-k^2$	-1
ρ_2	$-(k^2+1)$	$2k^2-1$	$-(k^2-2)$
ρ_3	0	k^2	k^2
Period	$4K(k)$	$4K(k)$	$2K(k)$
Odd/Even	Odd	Even	Even
Max and min	1 and −1	1 and −1	1 and l

integrals of the first and second kind are

$$K(k) = \frac{\pi}{2}\left(1 + \frac{1}{4}k^2 + \frac{9}{64}k^4 + \frac{25}{256}k^6 + \ldots\right) \tag{4AI.3}$$

$$E(k) = \frac{\pi}{2}\left(1 - \frac{1}{4}k^2 - \frac{3}{64}k^4 - \frac{5}{256}k^6 - \ldots\right) \tag{4AI.4}$$

As k^2 goes from zero to unity, K monotonically increases from $\pi/2$ to infinity as

$$K(0) = \pi/2, \quad K(1) = \infty \tag{4AI.5}$$

and E decreases monotonically from $\pi/2$ to 1 when k^2 increases from 0 to 1.

The sn, cn and dn Jacobi elliptic functions are periodical. The period of sn and cn in their argument ψ is $4K$ and of dn is $2K$. In Figure 4.11, the three Jacobi elliptic functions for $k^2 = 1/2$ and the sine and cosine trigonometric functions, i.e., sn and cn functions for $k^2 = 0$, are plotted. It is evident that the periods of the Jacobi elliptic functions differ for a different modulus: the period of sn and cn functions is longer for $k^2 = 1/2$ than for $k^2 = 0$.

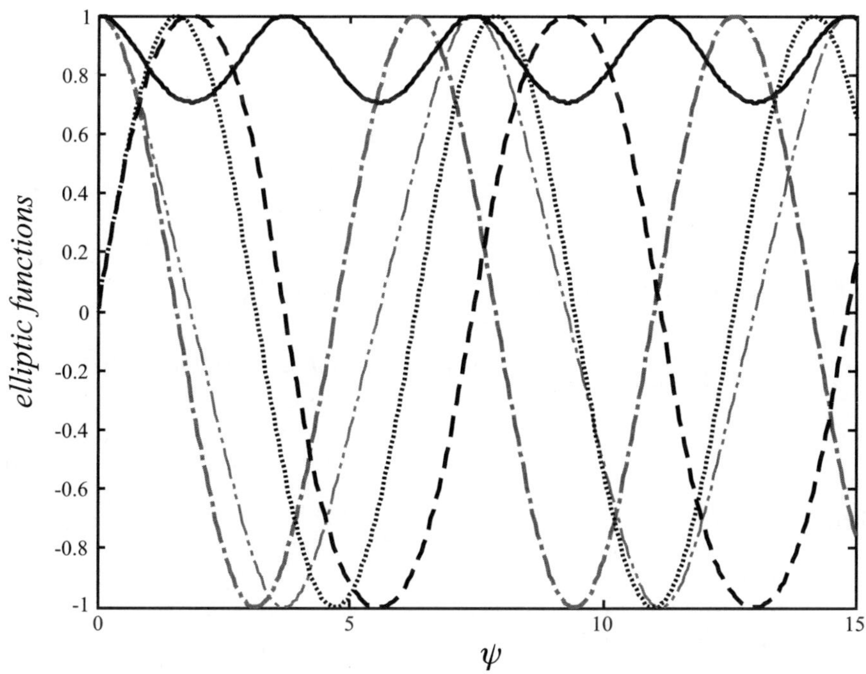

Figure 4.11 The elliptic functions: $\sin\psi = \mathrm{sn}(\psi,0)$ (black dotted line), $\mathrm{sn}(\psi,1/2)$ (black dashed line), $\cos\psi = \mathrm{cn}(\psi,0)$ (grey dashed-dotted line), $\mathrm{cn}(\psi,1/2)$ (grey dashed-dotted-dotted line) and $\mathrm{dn}(\psi,1/2)$ (black solid line).

Using the Fourier expansion, the sn and cn Jacobi elliptic functions are presented as a series of corresponding trigonometric sine and cosine functions, respectively:

$$\mathrm{sn}(\psi, k^2) = \frac{2\pi}{kK}\sum_{m=0}^{\infty}\frac{q^{m+\frac{1}{2}}}{1-q^{2m+1}}\sin\left((2m+1)\frac{\pi\psi}{2K}\right),$$
$$\mathrm{cn}(\psi, k^2) = \frac{2\pi}{kK}\sum_{m=0}^{\infty}\frac{q^{m+\frac{1}{2}}}{1+q^{2m+1}}\cos\left((2m+1)\frac{\pi\psi}{2K}\right) \tag{4AI.6a,b}$$

where $q = \exp(-\pi K'/K)$ and $K' = K(l)$ is the associated complete elliptic integral of the first kind.

Some useful averaged values of the cn function are:

$$C_0 = \int_0^{4K(k)} d\psi = 4K \tag{4AI.7}$$

$$C_2 = \int_0^{4K(k)} \mathrm{cn}^2 d\psi = 4\frac{E-(1-k^2)K}{k^2} \tag{4AI.8}$$

and for $m = 1, 2, 3,$

$$C_{2m+2} = \int_0^{4K(k)} \mathrm{cn}^{2m+2} d\psi = \frac{2m(2k^2-1)C_{2m} + (2m-1)(1-k^2)C_{2m-2}}{(2m+1)k^2} \tag{4AI.9}$$

For an odd-order cn function,

$$C_{2m+3} = \int_0^{4K(k)} cn^{2m+3} d\psi = 4\frac{(2m+1)(2k^2-1)C_{2m+1} + 2m(1-k^2)C_{2m-1}}{(2m+1)k^2} \tag{4AI.10}$$

where

$$C_1 = \int_0^{4K(k)} \mathrm{cn}\, d\psi = 4\int_0^{K(k)} cn d\psi = \frac{4\sin^{-1}(k)}{k},$$
$$C_3 = \int_0^{4K(k)} \mathrm{cn}^3 d\psi = 4\int_0^{K(k)} cn^3 d\psi = 2\frac{(2k^2-1)\sin^{-1}(k) + k\sqrt{1-k^2}}{k^3} \tag{4AI.11a,b}$$

Appendix 4AII: The best L_2 norm approximation

Here, an approximate function $f^*(\psi)$ for the function $f(\psi)$ is calculated as a sum of the known functions $\phi_i(\psi)$

$$f^*(\psi) = \sum_{i=0}^{n} a_i \phi_i(\psi) \tag{4AII.1}$$

where ψ is the independent variable and a_i are the coefficients that are to be determined. The L_2 norm of the function $f(\psi) \equiv f$ is formed

$$F(a) = \left\| \sum_{i=0}^{n} a_i \phi_i - f \right\|^2 = \int_0^T \omega(\psi) \left(\sum_{i=0}^{n} a_i \phi_i - f \right)^2 d\psi \tag{4AII.2}$$

The norm L_2 has the minimum

$$F(a^*) = \min F(a), a \in R^{n+1} \tag{4AII.3}$$

for those values of the coefficients a_i that satisfy the relation

$$\text{grad}\, F(a)_{a*} = 0 \tag{4AII.4}$$

Using the suggested procedure and assuming only two terms of the series expansion

$$f^*(\psi) = a_1 \phi_1(\psi) + a_2 \phi_2(\psi) \tag{4AII.5}$$

Equation (4AII.2) transforms into

$$F(a) = \int_0^T (a_0 \phi_0 + a_1 \phi_1 - f)^2 d\psi \tag{4AII.6}$$

where $\omega(\psi) = 1$. For the Jacobi elliptic function ep(ψ,k^2), the period of integration is $4K(k)$, where $K(k)$ is the complete first-order elliptic integral. Using the condition for a minimum (4AII.4), the following two algebraic equations are obtained

$$\begin{aligned} (\phi_0, \phi_0)a_0^* + (\phi_0, \phi_1)a_1^* &= (\phi_0, f), \\ (\phi_1, \phi_0)a_0^* + (\phi_1, \phi_1)a_1^* &= (\phi_1, f) \end{aligned} \tag{4AII.7a,b}$$

where

$$(\phi_0,\phi_0)=\int_0^{4K(k)}\phi_0(\psi)\phi_0(\psi)d\psi,\quad (\phi_0,\phi_1)=(\phi_1,\phi_0)=\int_0^{4K(k)}\phi_0(\psi)\phi_1(\psi)d\psi,$$

$$(\phi_1,\phi_1)=\int_0^{4K(k)}\phi_1(\psi)\phi_1(\psi)d\psi,\quad (\phi_0,f)=\int_0^{4K(k)}\phi_0(\psi)f(\psi)d\psi,$$

$$(\phi_1,f)=\int_0^{4K(k)}\phi_1(\psi)f(\psi)d\psi \tag{4AII.8a–e}$$

Solving Equations (4AII.7a,b), the parameters $a_0^*=a_0, a_1^*=a_1$ are obtained.

Examples

1. An approximate form for the function

$$f(\psi)=\mathrm{cn}(\psi,k^2)\left|\mathrm{cn}(\psi,k^2)\right| \tag{4AII.9}$$

is assumed as

$$f^*(\psi)=a_0\,\mathrm{cn}(\psi,k^2)+a_1\,\mathrm{cn}^3(\psi,k^2) \tag{4AII.10}$$

where

$$\phi_0=\mathrm{cn}(\psi,k^2),\quad \phi_1=\mathrm{cn}^3(\psi,k^2) \tag{4AII.11a,b}$$

and the coefficients a_0 and a_1 need to be calculated. Using the suggested procedure and Equations (4AII.8) gives

$$(\phi_0,\phi_0)=\int_0^{4K(k)}\mathrm{cn}^2(\psi,k^2)d\psi=C_2,\quad (\phi_0,\phi_1)=\int_0^{4K(k)}\mathrm{cn}^4(\psi,k^2)d\psi=C_4,$$

$$(\phi_1,\phi_1)=\int_0^{4K(k)}\mathrm{cn}^6(\psi,k^2)d\psi=C_6,$$

$$(\phi_0,f)=\int_0^{4K(k)}\mathrm{cn}^3(\psi,k^2)d\psi=C_3,\quad (\phi_1,f)=\int_0^{4K(k)}\mathrm{cn}^5(\psi,k^2)d\psi=C_5$$

(4AII.12a–e)

where the averaged elliptic functions C_2, C_4 and C_6 are given in Equations (4AI.8) and (4AI.9), while C_3 and C_5 are defined by Equations (4AI.11b) and (4AI.10). For $k^2 \ll 1$, the series expansion for the constants C_2 - C_6 is (see [13])

$$C_2 = \pi + \frac{\pi}{8}k^2 + O(k^4), \quad C_3 = \frac{8}{3} + \frac{4}{15}k^2 + O(k^4), \quad C_4 = \frac{3\pi}{4} + \frac{\pi}{16}k^2 + O(k^4),$$
$$C_5 = \frac{32}{15} + \frac{16}{105}k^2 + O(k^4), \quad C_6 = \frac{5\pi}{8} + \frac{5\pi}{128}k^2 + O(k^4) \tag{4AII.13a–e}$$

Substituting Equations (4AII.12) with Equations (4AII.13) into Equation (4AII.7), the parameters a_0 and a_1 are obtained:

$$a_0 = \frac{16}{15\pi}, \quad a_1 = \frac{32}{15\pi} \tag{4AII.14a,b}$$

so Equation (A.II.10) can be approximated to

$$f^*(\psi) = \frac{16}{15\pi}\mathrm{cn}(\psi, k^2) + \frac{32}{15\pi}\mathrm{cn}^3(\psi, k^2) \tag{4AII.15}$$

2. Using the suggested procedure, the following approximations to the products $\mathrm{sn}(\psi,k^2)\mathrm{dn}(\psi,k^2)$, $\mathrm{cn}(\psi,k^2)\mathrm{dn}(\psi,k^2)$ and $\mathrm{sn}(\psi,k^2)\mathrm{cn}^2(\psi,k^2)$, are found

$$\mathrm{sn}(\psi, k^2)\mathrm{dn}(\psi, k^2) \approx \frac{\pi k^2}{4(K-E)}\mathrm{sn}(\psi, k^2),$$
$$\mathrm{cn}(\psi, k^2)\mathrm{dn}(\psi, k^2) \approx \frac{\pi k^2}{4(E-(1-k^2)K)}\mathrm{cn}(\psi, k^2), \tag{4AII.16a–c}$$
$$\mathrm{sn}(\psi, k^2)\mathrm{cn}^2(\psi, k^2) \approx \frac{(2-k^2)E-2(1-k^2)K}{3k^2(E-K)}\mathrm{sn}(\psi, k^2)$$

Using the series expansion for $K(k)$ and $E(k)$ given in [13], the relations given in Equations (4AII.16) simplify to

$$\mathrm{sn}(\psi, k^2)\mathrm{dn}(\psi, k^2) \approx \left(1-\frac{3}{8}k^2\right)\mathrm{sn}(\psi, k^2),$$
$$\mathrm{cn}(\psi, k^2)\mathrm{dn}(\psi, k^2) \approx \left(1-\frac{k^2}{8}\right)\mathrm{cn}(\psi, k^2), \tag{4AII.17a–c}$$
$$\mathrm{sn}(\psi, k^2)\mathrm{cn}^2(\psi, k^2) \approx \frac{1}{4} - \frac{k^2}{32}\mathrm{sn}(\psi, k^2)$$

5

Forced harmonic vibration of a Duffing oscillator with linear viscous damping

Tamás Kalmár-Nagy[1] and Balakumar Balachandran[2]
[1]*Department of Aerospace Engineering, Texas A&M University, USA*
[2]*Department of Mechanical Engineering, University of Maryland, USA*

5.1 Introduction

The Duffing oscillator has become a classical paradigm for illustrating the remarkable *jump phenomenon* and other nonlinear behaviour [1,2]. The understanding gained on the basis of this low-order nonlinear system has helped in the development of reduced-order models of complex mechanical systems ranging from microscales to macroscales [3,4].

The nondimensional Duffing equation with damping and external forcing studied in this chapter has the form

$$\ddot{y} + 2\zeta\,\dot{y} + y + \gamma\, y^3 = F\cos\Omega\, t \tag{5.1.1}$$

where y, t, ζ, F, Ω and γ are the displacement, time, damping ratio, excitation amplitude, excitation frequency and cubic stiffness parameter, respectively. With $\gamma = 0$, Equation (5.1.1) reduces to the forced linear oscillator. On the other hand, with

$\gamma \neq 0$, Equation (5.1.1) describes a forced anharmonic nonlinear oscillator, which is characterised by a nonlinear force–displacement relationship. A positive (negative) cubic stiffness parameter corresponds to a hardening (softening) spring. In the literature, the system described by Equation (5.1.1) is also referred to as an externally excited Duffing oscillator to distinguish it from the system referred to as a parametrically excited Duffing oscillator, which is given by

$$\ddot{y} + 2\zeta\,\dot{y} + (1 + F\cos\Omega\,t)y + \gamma\,y^3 = 0 \tag{5.1.2}$$

In the parametrically excited system (5.1.2), the excitation modulates the linear stiffness parameter, while in the externally excited system (5.1.1), the excitation appears as an nonhomogeneous term that is external to the system. Furthermore, Equation (5.1.2) has the trivial solution $y = 0$, while Equation (5.1.1) does not.

Due to the time-dependent forcing, the Duffing equation (5.1.1) is a second-order nonautonomous system, and this forcing effectively adds another dimension to the system. To better understand this, Equation (5.1.1) is recast in the extended state-space form:

$$\begin{aligned} \dot{y}_1 &= y_2, \\ \dot{y}_2 &= -2\zeta y_2 - y_1 - \gamma\,y_1^3 + F\cos\Omega t, \\ \dot{t} &= 1 \end{aligned} \tag{5.1.3a}$$

The system (5.1.3a), in which the time t is treated as the third state-variable in addition to y_1 and y_2, is a three-dimensional *autonomous* system. Equation (5.1.1) can also be written as the following three-dimensional autonomous system:

$$\begin{aligned} \dot{y}_1 &= y_2, \\ \dot{y}_2 &= -2\zeta y_2 - y_1 - \gamma\,y_1^3 + F\cos\theta, \\ \dot{\theta} &= \Omega \end{aligned} \tag{5.1.3b}$$

The divergence of the vector field of the autonomous system (5.1.3a) is given by

$$\frac{\partial \dot{y}_1}{\partial y_1} + \frac{\partial \dot{y}_2}{\partial y_2} + \frac{\partial \dot{t}}{\partial t} = -2\zeta \tag{5.1.4}$$

which is negative for positive damping ratio ζ; that is, local volumes in the $(y_1,\ y_2,\ t)$ space are contracted. This means that there is dissipation in the system due to linear viscous damping, and this dissipation means that Equation (5.1.1) can have the solutions to which trajectories are attracted as $t \to \infty$.

The cubic nonlinearity and the time-dependent forcing are responsible for the rich dynamics exhibited by the Duffing equation. Nonlinear behaviour occurs such as the jump phenomenon where the steady-state behaviour changes dramatically due to a transition from one stable solution to another stable solution as a control parameter such as the excitation frequency or the excitation amplitude is quasistatically varied. *Local bifurcations* can also occur leading to chaotic responses (for example, the period-doubling *route to chaos*), and *global bifurcations* due to breakdown of *homoclinic and heteroclinic orbits* of the unforced system. These have been extensively documented for a variety of systems (see, for example, [3,5–7]).

In this chapter, the authors intend to provide a glimpse of the fascinating behaviour exhibited by the Duffing oscillator described by Equation (5.1.1). The rest of the chapter is organised as follows. In the next section, the behaviour of a linear forced oscillator is reviewed to introduce the notions of *resonance*, amplitude response, and phase response. Following this, the amplitude and phase response of the Duffing oscillator is studied in Section 5.3. Both the *primary* and *secondary resonance* are considered. In Section 5.4, periodic solutions of this system and their qualitative behaviour are discussed as a control parameter is varied. Some aspects of the global system dynamics are examined in Section 5.5. Finally, concluding remarks are presented in Section 5.6.

5.2 Free and forced responses of the linear oscillator

Setting the nonlinear term to zero in Equation (5.1.1), results in

$$\ddot{y} + 2\zeta\,\dot{y} + y = F\cos\Omega\,t \tag{5.2.1}$$

The initial conditions are assumed to be

$$y(0) = y_0, \quad \dot{y}(0) = \dot{y}_0 \tag{5.2.2a,b}$$

In order to examine *free oscillations* of this system, it is first assumed that $F = 0$.

5.2.1 Free oscillations and timescales

Here, the underdamped case ($|\zeta| < 1$) is considered to illustrate the presence of different *timescales* in the system. Equation (5.2.1) along with Equation (5.2.2a,b) can be solved by assuming solutions of the form $y(t) = ce^{\lambda t}$, which leads to

$$y(t) = e^{-\zeta t}\left(y_0\cos\left(\sqrt{1-\zeta^2}t\right) + \frac{\dot{y}_0 + \zeta y_0}{\sqrt{1-\zeta^2}}\sin\left(\sqrt{1-\zeta^2}t\right)\right) \tag{5.2.3}$$

Equation (5.2.3) can also be written as

$$y(t) = Ce^{-\zeta t}\sin(\omega_{\mathrm{d}}t + \varphi) \tag{5.2.4}$$

where the damped natural frequency ω_{d}, the constant C, and the phase φ are, respectively, given by

$$\begin{aligned}\omega_{\mathrm{d}} &= \sqrt{1-\zeta^2},\\ C &= \sqrt{y_0^2 + \left(\frac{\dot{y}_0 + \zeta y_0}{\omega_{\mathrm{d}}}\right)^2},\\ \varphi &= \tan^{-1}\left(\frac{y_0\sqrt{1-\zeta^2}}{\dot{y}_0 + \zeta y_0}\right)\end{aligned} \tag{5.2.5a–c}$$

Equation (5.2.3) describes the oscillations with an exponentially decaying envelope. The timescale corresponding to this decay is characterised by ζt, which is referred as to the *slow timescale*, and the oscillation corresponding to the timescale $\omega_d t$ is referred to as the *fast timescale* (compared to the timescale of the decay). The existence of different timescales is an important characteristic of both linear and nonlinear oscillators, and the existence of these timescales is taken advantage of in developing approximate solutions of weakly nonlinear systems, which is illustrated in Section 5.3.

5.2.2 Forced oscillations

In the presence of an external forcing, the linear oscillator is of the form

$$\begin{gathered} \ddot{y} + 2\zeta\dot{y} + y = F\cos\Omega t, \\ y(0) = y_0, \quad \dot{y}(0) = \dot{y}_0 \end{gathered} \tag{5.2.6a–c}$$

The solution of this nonhomogeneous system is the sum of the solution of the homogeneous equation and a particular solution of the nonhomogeneous equation. In physical terms, the motion of the system is the linear superposition of the free oscillation and a forced oscillation due to the external force. For an underdamped system ($|\zeta| < 1$) and nonresonance excitation ($\Omega \neq 1$; that is, an excitation whose frequency is not equal to the natural frequency), the solution has the form

$$y(t) = \underbrace{Ce^{-\zeta t}\cos(\omega_d t + \varphi)}_{\text{homogeneous part}} + \underbrace{a\cos(\Omega t + \phi)}_{\text{nonhomogeneous part}} \tag{5.2.7}$$

where ω_d is defined in Equation (5.2.5a), and C and φ are determined by the initial conditions. The constant ϕ is the phase shift relative to the phase of the external force, and it satisfies the relation

$$\phi(\Omega) = \begin{cases} -\tan^{-1}\left(\frac{2\zeta\Omega}{1-\Omega^2}\right) & \Omega \leq 1, \\ -\left(\pi - \tan^{-1}\left(\frac{2\zeta\Omega}{\Omega^2-1}\right)\right) & \Omega > 1 \end{cases} \tag{5.2.8a,b}$$

The nonhomogeneous part describes the steady-state response; that is, the response reached as $t \to \infty$. This response which occurs at the forcing frequency has the amplitude

$$a = M|F|$$

where

$$M(\Omega) = \frac{a}{|F|} = \frac{1}{\sqrt{(1-\Omega^2)^2 + 4\zeta^2\Omega^2}} \tag{5.2.9}$$

is called the magnification factor, which is a measure of the magnification of the displacement at each frequency compared to the static displacement.

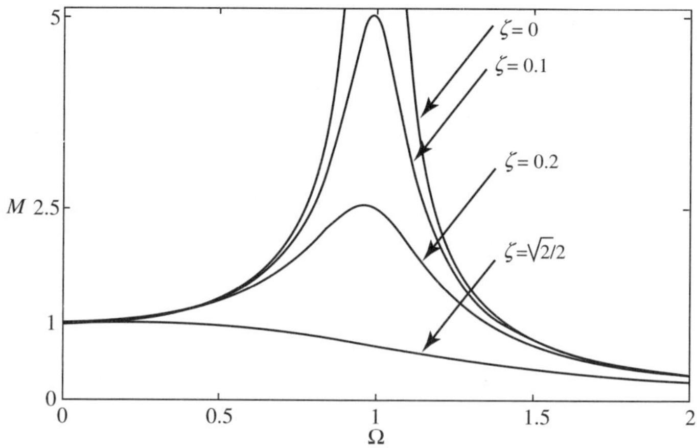

Figure 5.1 Amplitude-response curves: variation of the magnification factor with respect to the excitation frequency for different values of the damping ratio.

In Figure 5.1, the magnification factor is plotted with respect to Ω for various values of ζ. Each graph in this plot is called an amplitude-response curve, and each of these curves is representative of how the steady-state amplitude of the system changes as the excitation frequency is varied. For a given damping ratio, the excitation frequency at which the highest value of the magnification factor $M(\Omega)$ occurs can be determined from

$$\frac{dM(\Omega)}{d\Omega} = 0, \quad \frac{d^2M(\Omega)}{d\Omega^2} < 0 \tag{5.2.10a,b}$$

For $0 < \zeta < \sqrt{2}/2$, there is a maximum at $\Omega = \sqrt{1-2\zeta^2}$. Hence, the excitation frequency corresponding to the maximum steady-state response amplitude is a little lower than the natural frequency of the system but close to it for low damping. The variation of phase ϕ with respect to the forcing frequency is shown in Figure 5.2. In the case of an undamped system (i.e., $\zeta = 0$) the phase shift ϕ is zero for $\Omega < 1$ (the forced oscillation is in phase with the external forcing) and $\phi = -\pi$ for $\Omega > 1$ (the forced oscillation and the forcing are out-of-phase).

The case when $\Omega = 1$, that is when the excitation frequency is equal to the natural frequency, is of particular interest. The special relationship between the excitation frequency and the natural frequency is called a resonance. For the undamped system (i.e., $\zeta = 0$), the particular solution for $\Omega = 1$ takes the form

$$y_{\text{nonhom}}(t) = C_1 \cos(t + \phi) + C_2 t \sin t \tag{5.2.11}$$

where the constants C_1 and C_2 are appropriately determined constants. The second term in Equation (5.2.11) grows with time t and the solution becomes unbounded as $t \to \infty$. Expressions of the form $t^n \cos t$ or $t^n \sin t$ are called *secular terms*.

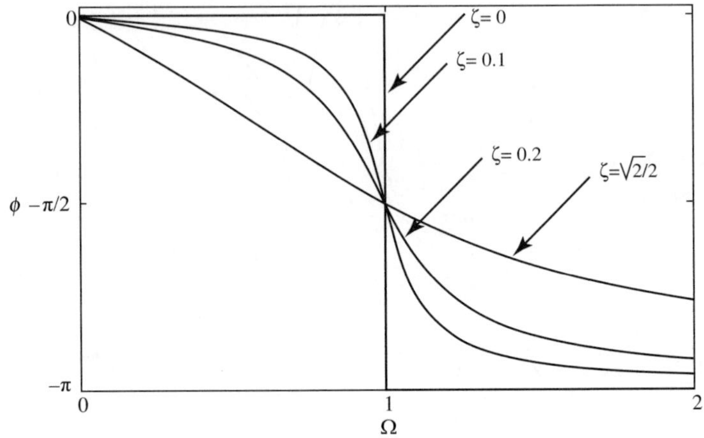

Figure 5.2 Phase-response curves: variation of the steady-state response phase with respect to the excitation frequency for different values of the damping ratio.

5.3 Amplitude and phase responses of the Duffing oscillator

In the previous section, the notions of slow and fast timescales, resonance, and secular terms were introduced. The addition of the nonlinear term y^3 to the simple *harmonic oscillator* – leading to the Duffing oscillator dramatically changes the picture. The principle of linear superposition can no longer be used to obtain the forced response of the linear system. In addition, the steady-state response depends on the initial conditions unlike that of the linear system where the steady-state response is independent of the chosen initial conditions. The maximum response also does not occur close to the system natural frequency as in the linear system. Due to the cubic nonlinearity, the system can experience resonances even when the excitation frequency is away from the natural frequency of the system. As the damped, forced nonlinear oscillator does not permit a closed-form solution, analytical approximations are sought for the forced response through perturbation analysis, as discussed in Chapter 4. The complexity of the response of the forced nonlinear oscillator is explored in the rest of this chapter.

Recognising the importance of the amplitude and phase responses of the linear oscillator, the same information is sought for the forced Duffing oscillator

$$\ddot{y} + 2\zeta\,\dot{y} + y + \gamma\,y^3 = F\cos\Omega\,t \qquad (5.3.1)$$

As there is no closed form solution for Equation (5.3.1), perturbation analysis is used to determine an analytical approximation for the forced response, assuming that the system has weak nonlinearity and weak damping. The goal of this exercise is to understand the influence of nonlinearity and compare the behaviour of the forced nonlinear system with that of the forced linear system, whose behaviour was discussed in the previous section. To facilitate the nonlinear analysis, a small parameter $\varepsilon \ll 1$, is introduced as an asymptotic ordering parameter and the damping and nonlinear terms are written, respectively, as $\zeta = \varepsilon\bar{\zeta}$ and $\gamma = \varepsilon\bar{\gamma}$, where $\bar{\zeta}$ and $\bar{\gamma}$ are O(1) quantities. With this rescaling, the unforced oscillator takes the following form:

$$\ddot{y} + y + \varepsilon\left(2\bar{\zeta}\,\dot{y} + \bar{\gamma}\,y^3\right) = 0 \tag{5.3.2}$$

Examining Equation (5.3.2), it is clear that it is a perturbation of the corresponding undamped and unforced linear oscillator. In order to focus on the system response during a resonance excitation, a weak or soft forcing $F = \varepsilon\bar{F}$, where $\bar{F}$ is O(1) is also assumed.

With the assumptions of weak damping, weak nonlinearity, and weak forcing, Equation (5.3.1) can be rewritten as

$$\ddot{y} + y + \varepsilon(2\bar{\zeta}\,\dot{y} + \bar{\gamma}\,y^3) = \varepsilon\bar{F}\cos\Omega t \tag{5.3.3}$$

To find the different resonances possible in the system, a straightforward expansion of the following form is carried out:

$$y(t) = \varepsilon y_1(t) + \varepsilon^2 y_2(t) + \ \ldots \tag{5.3.4}$$

The expansion (5.3.4) is an example of a Poincaré asymptotic series. On substituting Equation (5.3.4) into Equation (5.3.3), collecting terms of the same order, and solving the differential systems that correspond to the orders $O(\varepsilon)$ and $O(\varepsilon^2)$, it is found that small divisor terms occur in the particular response at $O(\varepsilon)$ when

$$\Omega \approx 1 \tag{5.3.5a}$$

and at $O(\varepsilon^2)$ when

$$\Omega \approx 1/3 \quad \text{or} \quad \Omega \approx 3 \tag{5.3.5b}$$

Based on the order at which the small divisor terms occur, Equation (5.3.5a) is said to describe a primary resonance while conditions (5.3.5b) are said to describe secondary resonances. While the primary resonance is identical to the resonance relation observed in the corresponding linear system, the secondary resonances are particular to the nonlinear system. These resonances are also referred to as nonlinear resonances. The resonance associated with the case, where the system is forced close to 1/3 of the system natural frequency, is called a *superharmonic resonance*, while the resonance associated with the case, where the system is forced close to 3 times the system natural frequency, is called a *subharmonic resonance*.

5.3.1 Primary resonance

Next, the system response during the resonance excitation, $\Omega \approx 1$, is considered. The proximity of the excitation frequency to the system natural frequency is expressed as

$$\Omega = 1 + \varepsilon\sigma \tag{5.3.6}$$

where σ is called the *detuning parameter*, which is a measure of how close the excitation frequency is to the natural frequency. With the assumptions of weak damping, weak nonlinearity, and weak forcing close to the system natural frequency, Equation (5.3.1) is rewritten as

$$\ddot{y} + y + \varepsilon(2\bar{\zeta}\,\dot{y} + \bar{\gamma}y^3) = \varepsilon\bar{F}\cos((1 + \varepsilon\sigma)t) \tag{5.3.7}$$

Noting that the steady-state solution for the forced linear oscillator is $a\cos(\Omega t + \phi)$ (see Equation (5.2.7), for small ε, an analytical approximation for Equation (5.3.7) is assumed to have the form

$$y(t) = a(t)\cos(\Omega t + \phi(t)) + O(\varepsilon) \tag{5.3.8}$$

where the amplitude a and phase ϕ are slowly varying quantities. This analytical approximation is an example of a generalised asymptotic series, as the coefficients are also functions of the asymptotic ordering parameter, which is ε in this case. Such analytical approximations can be constructed by using the method of multiple scales or the method of averaging [8,9]. This construction is illustrated here by using the method of multiple scales. Let

$$y(t;\ \varepsilon) = y_0(T_0,\ T_1) + \varepsilon y_1(T_0,\ T_1) + \ldots \tag{5.3.9}$$

where the fast *timescale* T_0 and slow timescale T_1 are given by

$$T_0 = t,\ \ T_1 = \varepsilon t \tag{5.3.10a,b}$$

With the introduction of the timescales, the time derivative with respect to time t is transformed as

$$\frac{d}{dt} = \frac{\partial}{\partial T_0} + \frac{\partial}{\partial T_1} = D_0 + D_1 \tag{5.3.11}$$

After substituting Equation (5.3.9) into Equation (5.3.7) and noting Equations (5.3.10a,b) and (5.3.11), the following hierarchy of equations can be obtained for O(1) and O(ε), respectively.

$$\begin{aligned} D_0^2 y_0 + y_0 &= 0, \\ D_0^2 y_1 + y_1 &= -2D_0D_1y_0 - 2\bar{\zeta}D_0y_0 - \bar{\gamma}y_0^3 + \bar{F}\cos(T_0 + \sigma T_1) \end{aligned} \tag{5.3.12a,b}$$

Then, the solution for the first component of the series (5.3.9) can be written as

$$y_0(T_0,\ T_1) = A(T_1)e^{jT_0} + A^*(T_1)e^{-jT_0} \tag{5.3.13}$$

where $j = \sqrt{-1}$, $A(T_1)$ is a complex valued amplitude function, and * indicates a complex conjugate of that quantity. On substituting Equation (5.3.13) into Equation (5.3.12b), the result is

$$D_0^2 y_1 + y_1 = -j(2A' + 2\bar{\zeta}A)e^{jT_0} - 3\bar{\gamma}A^2A^*e^{jT_0} - \bar{\gamma}A^3e^{j3T_0} + \frac{\bar{F}}{2}e^{jT_0}e^{j\sigma T_1} + \text{c.c.} \tag{5.3.14}$$

where the prime indicates a time derivative with respect to the slow time T_1 and c.c. indicates the complex conjugate of the preceding terms. On setting the source of the secular terms to zero in Equation (5.3.14), the result is

$$-j(2A'+2\bar{\zeta}A)-3\bar{\gamma}A^2A^* + \frac{\bar{F}}{2}e^{j\sigma T_1} = 0 \tag{5.3.15}$$

Introducing the polar form of the complex amplitude

$$A(T_1) = \frac{1}{2}a(T_1)e^{j\beta(T_1)} \tag{5.3.16}$$

where the amplitude $a(T_1)$ and the angle $\beta(T_1)$ are real-valued quantities, into Equation (5.3.15), separating the real and imaginary parts, and introducing the phase $\phi(T_1) = -(\sigma T_1 - \beta)$ leads to

$$\begin{aligned} a' &= -\bar{\zeta}a - \frac{\bar{F}}{2}\sin\phi, \\ a\phi' &= -\left(\sigma a - \frac{3}{8}\bar{\gamma}a^3 + \frac{\bar{F}}{2}\cos\phi\right) \end{aligned} \tag{5.3.17a,b}$$

These equations, which describe the slow time evolutions of the amplitude and phase, are referred to as the modulation equations, slow-flow equations or averaged equations. The *fixed points* of Equation (5.3.17a,b) correspond to solutions with constant amplitude and phase. These solutions satisfy

$$\begin{aligned} &\bar{\zeta}a + \frac{\bar{F}}{2}\sin\phi = 0, \\ &\sigma a - \frac{3}{8}\bar{\gamma}a^3 + \frac{\bar{F}}{2}\cos\phi = 0 \end{aligned} \tag{5.3.18a,b}$$

or equivalently,

$$\begin{aligned} \frac{\bar{F}}{2}\sin\phi &= -\bar{\zeta}a, \\ \frac{\bar{F}}{2}\cos\phi &= -a\left(\sigma - \frac{3}{8}\bar{\gamma}a^2\right) \end{aligned} \tag{5.3.19a,b}$$

Squaring and adding the equations in (5.3.19a,b) yields the frequency-response equation (amplitude–frequency equation)

$$\bar{F}^2 = 4a^2\left(\bar{\zeta}^2 + \left(\sigma - \frac{3}{8}\bar{\gamma}a^2\right)^2\right) \tag{5.3.20}$$

With this, the amplitude response (magnification factor) can be obtained as

$$M = \frac{a}{|\bar{F}|} = \frac{1}{2\sqrt{\bar{\zeta}^2 + \left(\sigma - \frac{3}{8}\bar{\gamma}a^2\right)^2}} \tag{5.3.21}$$

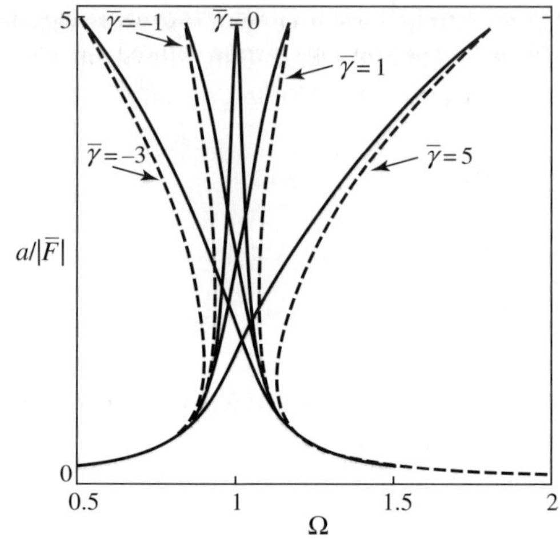

Figure 5.3 Amplitude-response curves for a varying strength of nonlinearities $\bar{\gamma}$ (from left to right). The parameter values used to construct these response curves are $\bar{F}=0.3$, $\bar{\zeta}=0.1$ and $\varepsilon=0.2$.

In Figure 5.3 the amplitude-response curves for nonlinearities of different strengths $\bar{\gamma}$ are shown. Unlike the amplitude response in the linear case, the amplitude response in the nonlinear case can be multivalued. For negative values of $\bar{\gamma}$, the response curves lean toward the lower frequencies, resulting in a softening response. The more positive the nonlinearity, the higher is the shift of the peak value of the magnification factor away from $\Omega = 1$ towards higher frequencies.

This is the hallmark of a hardening response. Increasing $\bar{F}$ results in a harder (for positive $\bar{\gamma}$) or a softer (for negative $\bar{\gamma}$) characteristic. As discussed later in this section, the amplitude of the peak response is given by $\bar{F}/(2\bar{\zeta})$, and as the excitation amplitude is increased, this peak amplitude increases, and the corresponding response curve leans further to the right (left) of $\Omega = 1$ for positive (negative) $\bar{\gamma}$.

The influence of damping on the magnification factor is illustrated in Figure 5.4. The phase response of the Duffing oscillator is obtained from Equation (5.3.19) as

$$\tan\phi = \frac{\bar{\zeta}}{\sigma - \frac{3}{8}\bar{\gamma}a^2} \tag{5.3.22}$$

Phase-response curves are plotted in Figure 5.5 for the damping coefficients used in Figure 5.4. As is evident from Equation (5.3.22), the nonlinearity affects the phase response, which is now a function of the response amplitude; this dependence on the response amplitude distorts the shape of the phase-response curve.

A profound difference between the responses of the linear oscillator and that of the Duffing oscillator is that the response of the latter is multivalued; that is, for a fixed value of the driving frequency there can be as many as three different response amplitudes, as seen in Figures 5.3–5.5. This is a consequence of the fact that

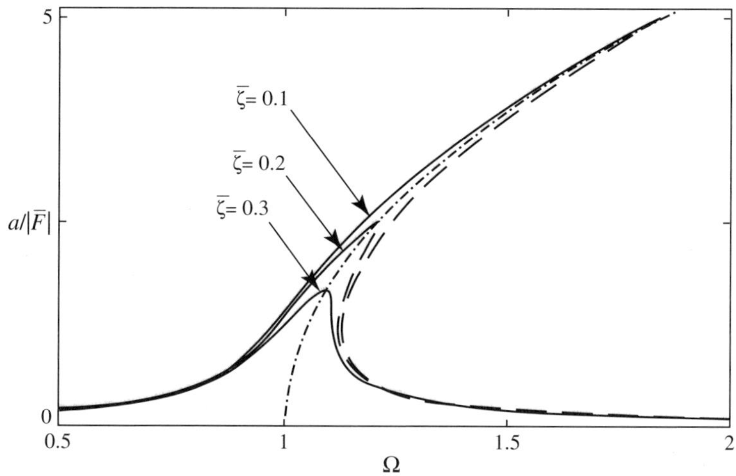

Figure 5.4 Amplitude-response curves: variation of the amplitude response with respect to the excitation frequency for different values of the damping ratio $\bar{\zeta}$. The parameter values used to construct these response curves are $\bar{F} = 0.3$, $\varepsilon = 0.2$ and $\bar{\gamma} = 5$. The backbone curve is shown as a dotted line.

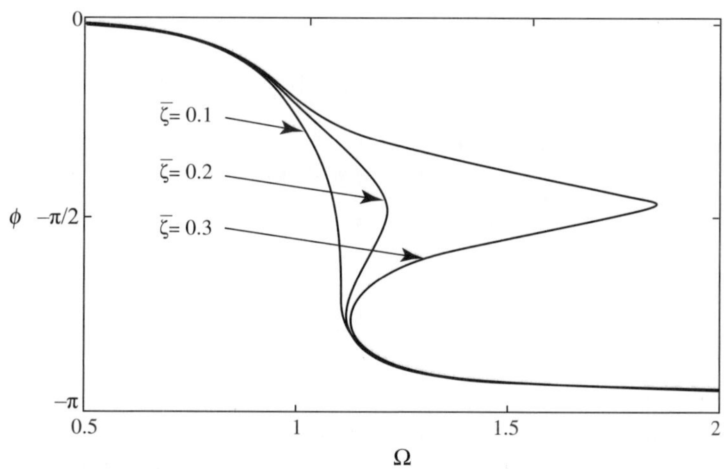

Figure 5.5 Phase-response curves: variation of the phase response with respect to the excitation frequency for different values of the damping ratio $\bar{\zeta}$. The parameter values used to construct these response curves are $\bar{F} = 0.3$, $\varepsilon = 0.2$ and $\bar{\gamma} = 5$.

Equation (5.3.20) is a cubic equation in a^2. Similar to the case of the linear oscillator, the maximum value of the magnification factor can be found from

$$\frac{dM}{d\Omega} = 0 \quad \text{and} \quad \frac{d^2M}{d\Omega^2} < 0 \tag{5.3.23a,b}$$

Differentiating Equation (5.3.20) with respect to Ω yields

$$\frac{1}{32\varepsilon^2} a\left(3\bar{\gamma}\varepsilon a^2 - 8\Omega + 8\right)\left(3a\bar{\gamma}\varepsilon\frac{da}{d\Omega} - 4\right) + \left(\bar{\zeta}^2 + \left(\frac{\Omega-1}{\varepsilon} - \frac{3a^2\bar{\gamma}}{8}\right)^2\right)\frac{da}{d\Omega} = 0 \tag{5.3.24}$$

which can be solved for $da/d\Omega$ as

$$\frac{da}{d\Omega} = \frac{8a(3\bar{\gamma}\varepsilon a^2 - 8\Omega + 8)}{27\bar{\gamma}^2\varepsilon^2 a^4 - 96\bar{\gamma}\varepsilon(\Omega-1)a^2 + 64\left(\varepsilon^2\bar{\zeta}^2 + (\Omega-1)^2\right)} \tag{5.3.25}$$

This derivative vanishes (and so does $dM/d\Omega$) when

$$3\bar{\gamma}\varepsilon a^2 - 8\Omega + 8 = 0 \Rightarrow a_p = \sqrt{\frac{8(\Omega-1)}{3\varepsilon\bar{\gamma}}} = \sqrt{\frac{8(\Omega-1)}{3\gamma}} \tag{5.3.26a}$$

which on the basis of Equations (5.3.6) and (5.3.20) can be rewritten as

$$a_p = \sqrt{\frac{8\varepsilon\sigma}{3\varepsilon\bar{\gamma}}} = \sqrt{\frac{8\sigma}{3\bar{\gamma}}} = \frac{\bar{F}}{2\bar{\zeta}} \tag{5.3.26b}$$

From Equations (5.3.26a) and (5.3.26b), it follows that

$$M_p = \frac{a_p}{|\bar{F}|} = \frac{\sqrt{\frac{8(\Omega-1)}{3\varepsilon\bar{\gamma}}}}{|\bar{F}|} = \frac{\sqrt{\frac{8(\Omega-1)}{3\gamma}}}{|\bar{F}|} \tag{5.3.27a}$$

and

$$M_p = \frac{a_p}{|\bar{F}|} = \frac{1}{2\bar{\zeta}} \tag{5.3.27b}$$

respectively. Equation (5.3.26a) describes the so-called *backbone curve*, which is also plotted in Figure 5.4. From Equations (5.3.26b) and (5.3.27b), it follows that the peak amplitude and the associated magnification factor are independent of the strength of the nonlinearity $\bar{\gamma}$ (however, this is only true for weak nonlinearity. The peak amplitude actually decreases for a hardening nonlinearity and increases for a softening nonlinearity. This is discussed in more detail in [10]). This is evident in Figure 5.3, where all of the peaks have the same magnitude for a fixed forcing amplitude and constant damping ratio. However, the peak amplitude location that can be determined from Equation (5.3.26b) as

$$\sigma_{\mathrm{p}} = \frac{3\bar{\gamma}}{8}\left(\frac{\bar{F}}{2\bar{\zeta}}\right)^2 \tag{5.3.27c}$$

depends on the strength of the nonlinearity.

In Figure 5.6, a representative amplitude-response curve is shown to illustrate the jump phenomenon. There is exactly one solution branch for $\Omega < \Omega_1, \Omega_2 < \Omega$ and three coexisting solutions for $\Omega_1 < \Omega < \Omega_2$, the so-called interval of bistability. At the frequency location $\Omega = \Omega_1$ ($\Omega = \Omega_2$), there are only two solutions S_1 and S_2 (S_3 and S_4), since at this frequency location the two solution branches merge. At $\Omega = \Omega_1$ and Ω_2, the periodic response of the forced Duffing oscillator loses *stability*, leading to a jump in the response, as discussed later.

In Figure 5.7, a representative solution of the Duffing equation is shown for specific initial conditions. As the harmonically forced oscillator is a second-order nonautonomous system, there are trajectory crossings in the $(y, \dot{y})$ plane. Although there is a unique solution associated with any initial condition, for some system and excitation parameter values, more than one solution satisfying Equation (5.3.3) can exist; that is, trajectories initiated from different initial conditions can be attracted to different solutions. Coexisting solutions are illustrated in Figure 5.8. These solutions are marked as P and Q in Figure 5.6. To elucidate the importance of several coexisting solutions, Figure 5.6 is revisited. Assume that an experiment is conducted. As the

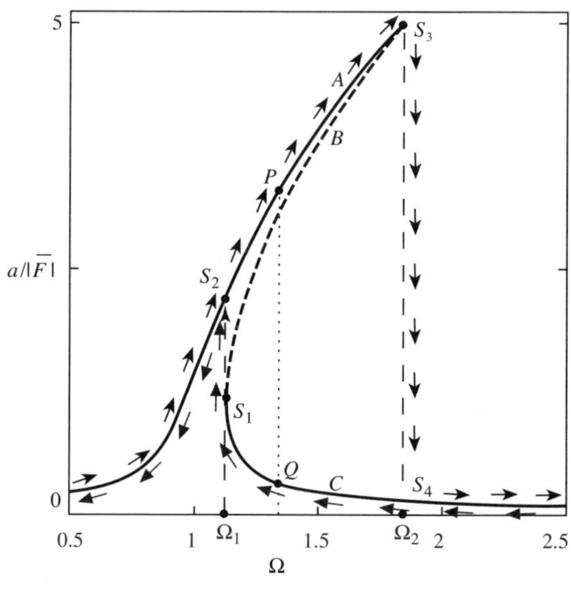

Figure 5.6 Illustration of the jump phenomenon or hysteresis in the response of the Duffing oscillator (5.3.3). The parameter values used to construct these response curves are $\bar{F} = 0.3$, $\bar{\zeta} = 0.1$, $\varepsilon = 0.2$ and $\bar{\gamma} = 5$. Phase plots corresponding to points P and Q are shown in Figure 5.8.

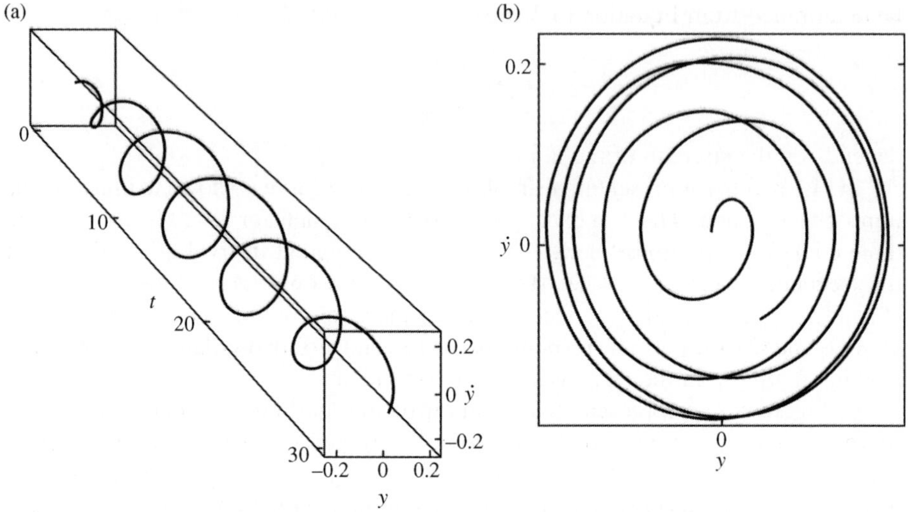

Figure 5.7 *(a) System response in the $(y, \dot{y}, t)$ space and (b) the phase plot in the $(y, \dot{y})$ plane for $\bar{F} = 0.3$, $\bar{\zeta} = 0.1$, $\varepsilon = 0.2$, $\bar{\gamma} = 5$ and $\Omega = 1.2$ over the time interval of $0 \leq t \leq 30$. The initial conditions are $y_0 = 0.01$, $\dot{y}_0 = 0$.*

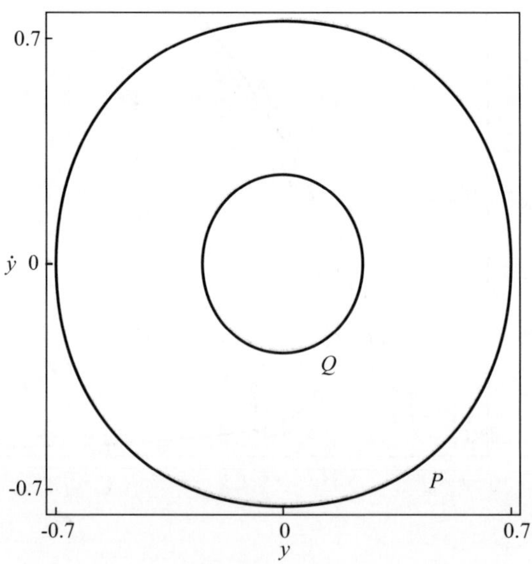

Figure 5.8 *Phase plot of coexisting stable steady-state solutions in the $(y, \dot{y})$ plane for $\bar{F} = 0.3$, $\bar{\zeta} = 0.1$, $\varepsilon = 0.2$, $\bar{\gamma} = 5$ and $\Omega = 1.34$.*

driving frequency – a natural control parameter – is gradually increased from $\Omega < \Omega_1$ in a quasistatic manner, the response amplitude will follow the upper branch or the large amplitude solution branch of the response diagram. Once Ω_2 is reached, this large amplitude *forced vibration* ceases to exist, and a fast (compared to the dominant timescale of the system) transition takes place to the lower branch consisting of small amplitude solutions; that is, a jump occurs from a large amplitude solution branch to a small amplitude solution branch. As the driving frequency Ω is further increased, the response follows the small amplitude solution branch. If the driving frequency is now slowly decreased from a frequency $\Omega > \Omega_2$, the amplitude of the steady-state forced response increases in accordance with the lower branch or the small-amplitude solution branch. At $\Omega = \Omega_1$, a transition occurs to a solution on the upper branch.

The transitions from the upper branch to the lower branch and vice versa occur at different values of the driving frequency, and as a consequence, depending on how the specific driving frequency is reached in the range of $\Omega_1 < \Omega < \Omega_2$, the response is different since it depends on the initial conditions; this phenomenon is called *hysteresis*.

In a physical experiment, the middle solution branch (the branch that joins the upper or large-amplitude solution branch and the lower or small amplitude solution branch) is not observed. The solutions on this middle solution branch are unstable, which means that if solutions on this middle branch are perturbed they will not return to that solution, but will be attracted to another solution. In the following, the stability of the solutions is examined.

To find the values of the critical points Ω_1 and Ω_2, the authors utilise the fact that these points correspond to vertical tangencies of the response curve; that is, where $d\Omega/dM = 0$. This condition can be found by equating the denominator of Equation (5.3.25) to zero, which translates to

$$27\bar{\gamma}^2\varepsilon^2a^4 - 96\bar{\gamma}\varepsilon(\Omega-1)a^2 + 64\left(\varepsilon^2\bar{\zeta}^2 + (\Omega-1)^2\right) = 0 \tag{5.3.28}$$

whose roots provide

$$\Omega_{1,2} = \frac{1}{8}\left(8 + 6\bar{\gamma}\varepsilon a^2 - \varepsilon\sqrt{9a^4\bar{\gamma}^2 - 64\bar{\zeta}^2}\right) \tag{5.3.29}$$

The condition for the existence of real solutions is

$$a \geq \sqrt{\frac{8\bar{\zeta}}{3\bar{\gamma}}} \tag{5.3.30}$$

The onset of bistability is characterised by the limiting case $a = \sqrt{8\bar{\zeta}/(3\bar{\gamma})}$. This corresponds to $\Omega_1 = \Omega_2 = 1 + 2\varepsilon\bar{\zeta}$, and the critical forcing amplitude $\bar{F} = 8\sqrt{\bar{\zeta}\left(\bar{\zeta}^2 + 2\bar{\gamma}\bar{\zeta} + 2\bar{\gamma}^2\right)/3\bar{\gamma}}$.

To characterise the stability of the solution branches depicted in Figure 5.6, the stability properties of the fixed points (a, ϕ) of Equations (5.3.17) need to be understood. The *Jacobian matrix* of this flow is

$$\mathbf{J} = \begin{bmatrix} -\bar{\zeta} & a\left(\sigma - \dfrac{3}{8}\bar{\gamma}a^2\right) \\ \dfrac{-\left(\sigma - \dfrac{9}{8}\bar{\gamma}a^2\right)}{a} & -\bar{\zeta} \end{bmatrix} \tag{5.3.31}$$

whose trace $\operatorname{tr}\mathbf{J}$ and determinant Δ are given by

$$\begin{aligned} \operatorname{tr}\mathbf{J} &= -2\bar{\zeta}, \\ \Delta &= \bar{\zeta}^2 + \left(\sigma - \frac{9}{8}\bar{\gamma}a^2\right)\left(\sigma - \frac{3}{8}\bar{\gamma}a^2\right) \end{aligned} \tag{5.3.32a,b}$$

respectively. The trace is equal to the sum of the eigenvalues of the Jacobian matrix $\mathbf{J}$, while the determinant Δ is equal to the product of its eigenvalues.

For the damped system, the trace, and thus, the sum of the eigenvalues of the Jacobian is negative, and therefore at least one of the eigenvalues has a negative real part. If the other eigenvalue has a negative (positive) real part, then, the fixed point (a, ϕ) is a stable node (saddle point). Branches of stable and unstable fixed points are shown as solid and dashed lines, respectively, in Figures 5.3, 5.4 and 5.6. If the other eigenvalue becomes zero, the system undergoes a static *bifurcation* (i.e., *saddle-node* or *pitchfork bifurcation*), but dynamic bifurcations such as *Hopf bifurcations* are not possible, as also discussed in Chapter 3. The condition for having a zero eigenvalue can be derived from Equations (5.3.32) and (5.3.20) (condition for the existence of a fixed point and one of the eigenvalues of the Jacobian matrix is zero)

$$\begin{aligned} \Delta &= \bar{\zeta}^2 + \left(\sigma - \frac{9}{8}\bar{\gamma}a^2\right)\left(\sigma - \frac{3}{8}\bar{\gamma}a^2\right) = 0, \\ \bar{F}^2 &= 4a^2\left(\bar{\zeta}^2 + \left(\sigma - \frac{3}{8}\bar{\gamma}a^2\right)^2\right) \end{aligned} \tag{5.3.33a,b}$$

This provides the following simple relationship between the system parameters at the static bifurcation point

$$\bar{F}^2 = 3\bar{\gamma}a^4\left(\sigma - \frac{3}{8}\bar{\gamma}a^2\right) \tag{5.3.34}$$

Having explored the stability of the solutions, the domains of attraction for the stable solutions are now discussed for the excitation parameter values corresponding to which three solutions exist. Let the stable fixed points A and C of Equation (5.3.17a,b) correspond to the upper branch and lower branch of solutions of Figure 5.6 for a certain set of parameter values, and the unstable fixed point B correspond to the middle branch of solutions for these parameter values. The unstable fixed point is a saddle point, and there is a one-dimensional *stable manifold* W^S associated with the eigenvalue with the negative real part and a one-dimensional *unstable manifold* W^U associated with the eigenvalue with the positive real part. It is noted that a stable manifold is tangent to the eigenvector associated with the eigenvalue with the negative real part, and any trajectory initiated on the stable manifold is attracted to B as $t \to \infty$. An unstable manifold is tangent to the eigenvector associated with the eigenvalue with the positive real part, and any trajectory initiated on the unstable manifold is attracted to B as $t \to -\infty$. A representative illustration of these manifolds is given in Figure 5.9 in the (a, ϕ) plane.

The stable manifold of B partitions this plane into two regions, which are the basins of attraction of the stable fixed points A and C. Depending on the initial conditions, the trajectories are attracted to either point A or C, as $t \to \infty$.

So far, the driving frequency has been considered as the control (bifurcation) parameter. An alternative way to capture the dynamics of the system is to find the amplitude of the response as the function of the amplitude of the driving force. The so-called force-response curve is depicted in Figure 5.10. Here, again, multiple and up to three coexisting solutions (the solid and dashed lines correspond to stable and unstable branches, respectively) can be observed. As F is increased quasi-statically and then decreased, a hysteresis phenomenon is seen as earlier noted in the context of Figure 5.6. In particular, there are three coexisting solutions for $F_1 < F < F_2$, and exactly one solution branch outside this region of bistability. The stable (thick solid line) and unstable (thick dashed line) branches merge at $F = F_1$ and $F = F_2$. At these points there is a *jump*, labelled by a thin dashed line, in the response.

The qualitative change or bifurcation associated with the jump phenomenon is an example of a catastrophic bifurcation (see, for example, [7]) since the states of the system vary discontinuously as the control parameter is varied gradually through its critical value. In the present case, the postbifurcation response is a bounded *attractor*, to be specific, a periodic attractor. However, this may not be true in all situations. It is noted that the jump phenomenon is related to the cusp catastrophe (see, for example, [2]) which is one of many elementary catastrophes proposed nearly four

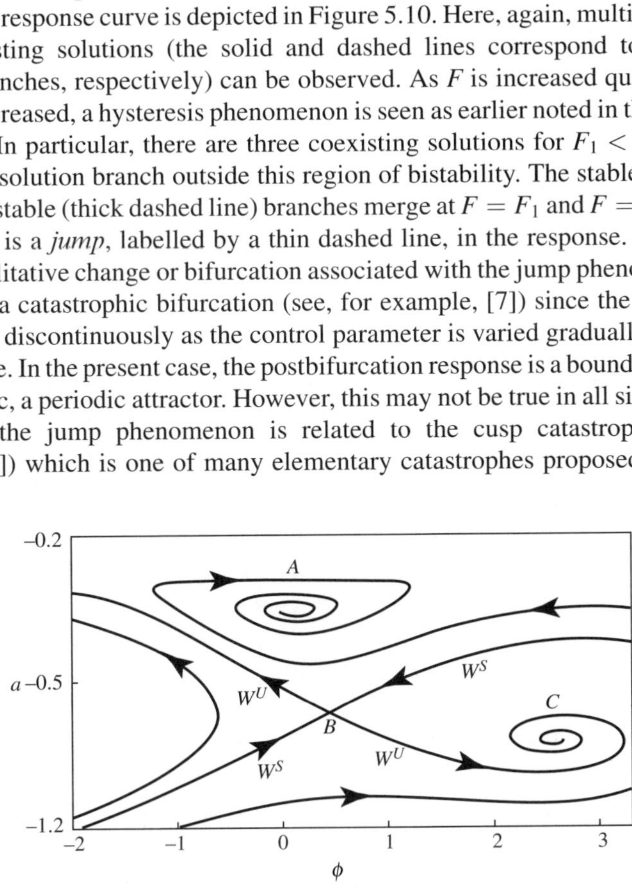

Figure 5.9 Phase plane for Equation (5.3.17a,b) with three coexisting equilibrium solutions. The stable manifold W^S of the saddle point B separates the domains of attraction of A and C. $\bar{F} = 0.3$, $\bar{\zeta} = 0.1$, $\varepsilon = 0.2$, $\bar{\gamma} = 5$ and $\Omega = 1.2$.

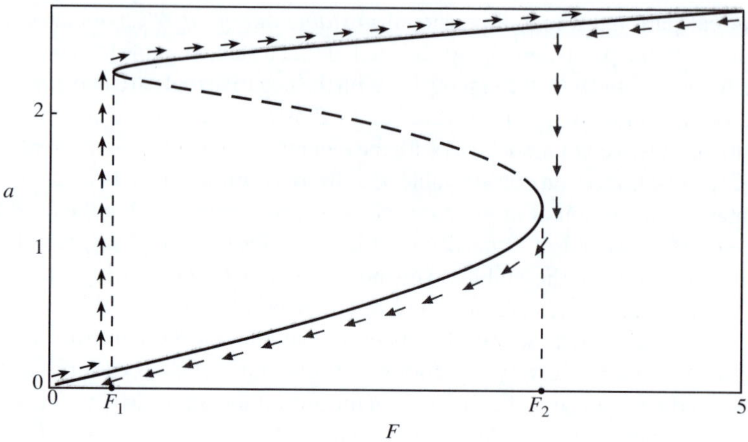

Figure 5.10 *Force-response curve for* $\bar{\zeta}=0.1$, $\varepsilon=0.1$, $\bar{\gamma}=1$ *and* $\Omega=1.2$.

decades ago [11]. Holmes and Rand [12] were the first to apply catastrophe theory to the Duffing oscillator.

Catastrophe theory is based on the behaviour of canonical functions of the form

$$f(y,\mu)=0 \tag{5.3.35}$$

close to the singular or critical point $y=0$ at the control parameter value $\mu=0$ of the system. The fixed-point equations given by Equation (5.3.18a,b) are in the form of Equation (5.3.35), and the jump location corresponds to a singular or critical point of this system. The theory of singularities, which encompasses catastrophe theory, can be used to understand the structural stability of bifurcations; that is, to understand whether a certain bifurcation would be stable to a perturbation to the system. For instance, this theory can be used to answer the question of whether the jump phenomenon seen in Figure 5.10 would still persist if a perturbation in the form of nonlinear viscous damping is added to the system described by Equation (5.3.7).

The analysis and numerical results discussed thus far are valid for a weakly nonlinear system subjected to a soft forcing at a primary resonance (i.e., $\Omega=1$) of the system. Subsequently, the cases of secondary resonances are analysed.

5.3.2 Secondary resonances

Due to the cubic nonlinearity in the system there are also resonances at other frequencies as mentioned earlier. These secondary resonances, occur at $\Omega=1/3$ and $\Omega=3$, which are different cases of the resonance relation $\Omega=(1-m)/n$, where m and n are integers such that $|m|+|n|=3$. Weakly nonlinear analyses can also be carried out to determine the system response as discussed in [2,7–9]. To illustrate this, the following system is considered

$$\ddot{y}+\varepsilon 2\bar{\zeta}\,\dot{y}+y+\varepsilon\gamma\, y^3=F\cos\Omega t \tag{5.3.36}$$

where the damping is weak and the nonlinearity is weak. However, the forcing is no longer weak. Compared to the excitation used in Section 5.3.1, here, since the excitation amplitude is $O(1)$, it is also referred to as a hard excitation. To determine an analytical approximation for the response of the oscillator (5.3.36), the method of multiple scales is used and an approximate solution in the form of Equation (5.3.9). After substituting Equation (5.3.9) into Equation (5.3.36) and making use of Equations (5.3.10a,b) and (5.3.11), the following hierarchy of equations is obtained for $O(1)$ and $O(\varepsilon)$, respectively:

$$\begin{aligned} D_0^2 y_0 + y_0 &= F\cos\Omega T_0, \\ D_0^2 y_1 + y_1 &= -2D_0 D_1 y_0 - 2\bar{\zeta} D_0 y_0 - \bar{\gamma} y_0^3 \end{aligned} \tag{5.3.37a,b}$$

Then, the solution of Equation (5.3.37a) can be written as

$$y_0(T_0,\, T_1) = A(T_1)e^{jT_0} + A^*(T_1)e^{-jT_0} + \Lambda(e^{j\Omega T_0} + e^{-j\Omega T_0}) \tag{5.3.38}$$

where $j = \sqrt{-1}$, $A(T_1)$ is a complex-valued amplitude function, A^* indicates a complex conjugate of that quantity, and

$$\Lambda = \frac{1}{2}\frac{F}{1-\Omega^2} \tag{5.3.39}$$

On substituting Equation (5.3.38) into Equation (5.3.37b), the result is

$$\begin{aligned} D_0^2 y_1 + y_1 = &-\left(2j(A' + \bar{\zeta}A) + 3\bar{\gamma}A^2A^* + 6\bar{\gamma}A\Lambda^2\right)e^{jT_0} - \bar{\gamma}\left(A^3 e^{j3T_0} + \Lambda^3 e^{j3\Omega T_0}\right) \\ &-3\bar{\gamma}\Lambda\left(A^2 e^{j(\Omega+2)T_0} + A^{*2} e^{j(\Omega-2)T_0}\right) - 3\bar{\gamma}A\Lambda^2\left(e^{j(1+2\Omega)T_0} + e^{j(1-2\Omega)T_0}\right) \\ &-\Lambda\left(2j\bar{\zeta}\Omega + 3\bar{\gamma}\Lambda^2 + 6\bar{\gamma}AA^*\right)e^{j\Omega T_0} + \text{c.c.} \end{aligned} \tag{5.3.40}$$

Examining the right-hand side of Equation (5.3.40), it is clear that the terms with $e^{\pm jT_0}$ lead to secular terms. In addition, secular terms arise when $\Omega = 1/3$, and $\Omega = 3$. When the excitation frequency is away from these frequencies, the excitation is said to be a nonresonant excitation. Next, three different cases are considered.

5.3.2.1 Case 1, nonresonant excitation: Ω away from 1/3 and 3

In this case, it follows from Equation (5.3.40) that setting the source of secular terms to zero results in

$$2j(A' + \bar{\zeta}A) + 3\bar{\gamma}A^2A^* + 6\bar{\gamma}A\Lambda^2 = 0 \tag{5.3.41}$$

On introducing the polar form given by Equation (5.3.16) into Equation (5.3.41) and separating real and imaginary parts, the modulation equations are obtained as

$$\begin{aligned} a' &= -\bar{\zeta}a, \\ a\beta' &= 3\bar{\gamma}\left(\frac{1}{8}\sigma a^3 + \Lambda^2 a\right) \end{aligned} \tag{5.3.42a,b}$$

Then, to the first approximation, the response of the oscillator (5.3.36) driven by a hard, nonresonant excitation can be expressed as

$$y(t) = a(t)\cos(t+\beta(t)) + \frac{F}{1-\Omega^2}\cos\,\Omega t + O(\varepsilon) \tag{5.3.43}$$

where the time evolutions of $a(t)$ and $\beta(t)$ are given by Equation (5.3.42a,b). For positive damping, the free oscillation component of the response dies out with time, and the long-time response is a oscillation at the forcing frequency like in the linear system.

5.3.2.2 Case 2, superharmonic resonance: $\Omega \approx 1/3$

In this case, to express the closeness of the excitation frequency to one third of the natural frequency of the oscillator, a detuning parameter is introduced as

$$3\Omega = 1 + \varepsilon\sigma \tag{5.3.44a}$$

and it is noted that

$$3\Omega T_0 = (1+\varepsilon\sigma)T_0 = T_0 + \sigma T_1 \tag{5.3.44b}$$

Making use of Equations (5.3.44a) and (5.3.44b) in Equation (5.3.40), and collecting the source of the secular terms and setting it to zero leads to

$$2j(A' + \bar{\zeta}A) + 3\bar{\gamma}A^2 1-\Omega^2 + 6\bar{\gamma}A\Lambda^2 + \bar{\gamma}\Lambda^3 e^{j\sigma T_1} = 0 \tag{5.3.45}$$

Again, making use of the polar form of the complex amplitude given by Equation (5.3.16) in Equation (5.3.45), introducing the phase $\phi(T_1) = -(\sigma T_1 - \beta)$ leads to the following modulation equations:

$$\begin{aligned} a' &= -\bar{\zeta}a + \bar{\gamma}\Lambda^3 \sin\phi, \\ a\phi' &= -\left(\sigma a - 3\bar{\gamma}\Lambda^2 a - \frac{3}{8}\bar{\gamma}a^3 - \bar{\gamma}\Lambda^3\cos\phi\right) \end{aligned} \tag{5.3.46a,b}$$

Then, to a first approximation, the response of the oscillator (5.3.36) driven by the superharmonic resonance excitation can be expressed as

$$y(t) = a(t)\cos(3\Omega t + \phi(t)) + \frac{F}{1-\Omega^2}\cos\Omega t + O(\varepsilon) \tag{5.3.47}$$

where the time evolutions of $a(t)$ and $\phi(t)$ are given by Equation (5.3.46a,b). Thus, the system response contains a response component at three times the excitation frequency due to the nonlinearity in addition to the response component at the excitation frequency. Due to the presence of the $3\,\Omega$ component, which is at an overtone of the input frequency, the system is said to exhibit a superharmonic resonance. In order to further examine the response component at $3\,\Omega$, the fixed points of Equations (5.3.46a,b) are considered, which pertain to constant amplitude and phase. These fixed points are solutions of

$$\bar{\zeta}a = \bar{\gamma}\Lambda^3 \sin\phi, \qquad (\sigma - 3\bar{\gamma}\Lambda^2)a - \frac{3}{8}\bar{\gamma}a^3 = \bar{\gamma}\Lambda^3 \cos\phi \tag{5.3.48a,b}$$

After squaring and adding Equations (5.3.48a,b), the result is the frequency-response equation

$$\left(\bar{\zeta}^2 + \left((\sigma - 3\bar{\gamma}\Lambda^2) - \frac{3}{8}\bar{\gamma}a^2\right)^2\right)a^2 = \bar{\gamma}^2\Lambda^6 \tag{5.3.49}$$

Equation (5.3.49) can be solved to determine the detuning parameter as

$$\sigma = 3\bar{\gamma}\Lambda^2 + \frac{3}{8}\bar{\gamma}a^2 \pm \left(\frac{\bar{\gamma}^2\Lambda^6}{a^2} - \bar{\zeta}^2\right) \tag{5.3.50}$$

From Equation (5.3.50), the peak amplitude of the $3\,\Omega$ component and the corresponding frequency location are determined to be

$$a_{\mathrm{p}} = \frac{\bar{\gamma}\Lambda^3}{\bar{\zeta}}, \qquad \sigma_{\mathrm{p}} = 3\bar{\gamma}\Lambda^2 + \frac{3}{8}\bar{\gamma}\left(\frac{\bar{\gamma}\Lambda^3}{\bar{\zeta}}\right)^2 = 3\bar{\gamma}\Lambda^2\left(1 + \frac{\bar{\gamma}^2\Lambda^4}{8\bar{\zeta}^2}\right) \tag{5.3.51a,b}$$

Unlike in the case of the primary resonance, the peak amplitude is seen to depend on the strength of the nonlinearity. The stability of the fixed points of Equations (5.3.46a,b) can be studied in the same way as for the primary resonance described in Section 5.3.1.

5.3.2.3 Case 2, subharmonic resonance: $\Omega \approx 3$

In this case, to express the nearness of the excitation frequency to three times the natural frequency of the oscillator, a detuning parameter is introduced as

$$\Omega = 3 + \varepsilon\sigma \tag{5.3.52a}$$

and it is noted that

$$\Omega T_0 = (3 + \varepsilon\sigma)T_0 = 3T_0 + \sigma T_1 \tag{5.3.53b}$$

Making use of Equations (5.3.52a) and (5.3.53b) in Equation (5.3.40), and collecting the source of the secular terms and setting it to zero leads to

$$2j(A' + \bar{\zeta}A) + 3\bar{\gamma}A^2A^* + 6\bar{\gamma}A\Lambda^2 + 3\bar{\gamma}\Lambda A^{*2}e^{j\sigma T_1} = 0 \tag{5.3.54}$$

Again, making use of the polar form of the complex amplitude given by Equation (5.3.16) in Equation (5.3.54), introducing the phase $\phi(T_1) = -(\sigma T_1 - 3\beta)$ leads to the following the modulation equations:

$$\begin{aligned} a' &= -\bar{\zeta} a + \frac{3\bar{\gamma}\Lambda}{4} a^2 \sin\phi, \\ a\phi' &= -\left(\sigma a - 9\bar{\gamma}\Lambda^2 a - \frac{9}{8}\bar{\gamma} a^3 - \frac{9\bar{\gamma}\Lambda}{4} a^2 \cos\phi\right) \end{aligned} \tag{5.3.55a,b}$$

Then, to the first approximation, the response of the oscillator (5.3.36) driven by the subharmonic resonance excitation can be expressed as

$$y(t) = a(t)\cos\left(\frac{1}{3}(\Omega t + \phi(t))\right) + \frac{F}{1-\Omega^2}\cos\Omega t + O(\varepsilon) \tag{5.3.56}$$

where the time evolutions of $a(t)$ and $\phi(t)$ are given by Equation (5.3.55a,b). Thus, the system response contains a response component at 1/3 third of the excitation frequency due to the nonlinearity in addition to the response component at the excitation frequency. Due to the presence of the $\Omega/3$ component, the system is said to exhibit a subharmonic resonance. In order to further examine the response component at $\Omega/3$, the fixed points of Equations (5.3.55a,b) are considered, which pertain to constant amplitude and phase. These fixed points are solutions of

$$\begin{aligned} \bar{\zeta} a &= \frac{3\bar{\gamma}\Lambda}{4} a^2 \sin\phi, \\ \sigma a - 9\bar{\gamma}\Lambda^2 a - \frac{9}{8}\bar{\gamma} a^3 &= \frac{9\bar{\gamma}\Lambda}{4} a^2 \cos\phi \end{aligned} \tag{5.3.57a,b}$$

After squaring and adding Equations (5.3.57a,b), the result is the frequency-response equation

$$\left(9\bar{\zeta}^2 + \left((\sigma - 9\bar{\gamma}\Lambda^2) - \frac{9}{8}\bar{\gamma} a^2\right)^2\right) a^2 = \frac{81}{16}\bar{\gamma}^2\Lambda^2 a^4 \tag{5.3.58}$$

From Equation (5.3.58), it follows that either $a = 0$ or $a \neq 0$. The amplitude in the latter case can be determined from the quartic equation in a or quadratic equation in a^2

$$9\bar{\zeta}^2 + \left((\sigma - 9\bar{\gamma}\Lambda^2) - \frac{9}{8}\bar{\gamma} a^2\right)^2 - \frac{81}{16}\bar{\gamma}^2\Lambda^2 a^2 = 0 \tag{5.3.59}$$

The different solutions of this system can be studied as illustrated in [2]. Again, the stability of the fixed points of can be studied as illustrated for the primary resonance in Section 3.3.1.

For the fully nonlinear system, different types of *coexisting attractors* may exist, with the solutions undergoing various bifurcation scenarios. In the following sections, different qualitative changes and associated nonlinear phenomena are discussed to provide an illustration of these possibilities.

5.4 Periodic solutions, Poincaré sections, and bifurcations

5.4.1 Periodic solutions

The steady-state forced vibrations of the system discussed in the preceding section are periodic solutions. In general, the n-dimensional nonautonomous system

$$\dot{\mathbf{x}} = \mathbf{F}(\mathbf{x}, t) \tag{5.4.1}$$

where $\mathbf{x} \in \mathbb{R}^n$, $t \in \mathbb{R}$, is said to have a periodic solution (*orbit*) $\mathbf{X}$ of least period P if this solution satisfies $\mathbf{X}(\mathbf{x}_0; t_0) = \mathbf{X}(\mathbf{x}_0; t_0 + P)$ for all initial conditions $\mathbf{x} = \mathbf{x}_0$ on this orbit at $t = t_0$. If the vector field $\mathbf{F}$ is periodic in time with period T, then periodic solutions of Equation (5.4.1) with periods that are either an integer multiple or integer submultiple of the period T are called superharmonic and subharmonic solutions, respectively.

While approximate periodic solutions of the Duffing equation have been constructed in the previous section, there are rigorous theorems for the existence of periodic solutions for n-dimensional weakly nonlinear systems of the form (see, for example [13], Chapter 6).

$$\dot{\mathbf{x}} = \mathbf{A}\mathbf{x} + \varepsilon\mathbf{G}(\mathbf{x}, t; \varepsilon) \tag{5.4.2}$$

where $\mathbf{A}$ is a constant matrix with some zero eigenvalues; $\mathbf{G}(\mathbf{x}, t; \varepsilon)$ is periodic in t and satisfies certain smoothness conditions. To transform the Duffing equation (5.3.7) into this form, it is first recast as a system of first-order equations

$$\begin{aligned} \dot{y}_1 &= y_2, \\ \dot{y}_2 &= -y_1 - \varepsilon(2\bar{\zeta}\, y_2 + \bar{\gamma}\, y_1^3) - \bar{F}\cos\Omega t \end{aligned} \tag{5.4.3a,b}$$

The following transformations, motivated by the method of variations of parameters,

$$\begin{aligned} y_1 &= x_1\cos\Omega t + x_2\sin\Omega t, \\ y_2 &= \Omega(-x_1\sin\Omega t + x_2\cos\Omega t), \end{aligned} \tag{5.4.4a,b}$$

lead to

$$\begin{aligned} \dot{x}_1 &= -\frac{\varepsilon}{\Omega}\big(\sigma y_1 - \bar{\zeta} y_2 - \bar{\gamma} y_1^3 + \bar{F}\cos\Omega t\big)\sin\Omega t, \\ \dot{x}_2 &= \frac{\varepsilon}{\Omega}\big(\sigma y_1 - \bar{\zeta} y_2 - \bar{\gamma} y_1^3 + \bar{F}\cos\Omega t\big)\cos\Omega t \end{aligned} \tag{5.4.5a,b}$$

This system is of the form (5.4.2) with $\mathbf{A}$ being the 2×2 zero matrix. Since $\mathbf{G}$ is sufficiently smooth, it can be concluded that Equation (5.3.7) has periodic solutions.

5.4.2 Poincaré section and Poincaré map

As established in Section 5.1, the second-order nonautonomous Duffing equation (5.1.1) can be converted to the autonomous system

$$\begin{aligned}\dot{y}_1 &= y_2,\\ \dot{y}_2 &= -2\zeta y_2 - y_1 - \gamma y_1^3 + F\cos\Omega t,\\ \dot{t} &= 1\end{aligned} \tag{5.4.6a,b}$$

Note that the Duffing equation (5.4.6a–c) is invariant under the transformation $y_1 \to -y_1$, $y_2 \to -y_2$, $t \to t - \pi/\Omega$. The state space of this system (the so-called extended state space) is the three-dimensional Euclidean space $\mathbb{R}\times\mathbb{R}\times\mathbb{R} = \mathbb{R}^3$. Since the forcing is periodic with period $T = 2\pi/\Omega$, the solutions are invariant to a translation in time by T. This observation can be utilised to introduce an essential tool of nonlinear dynamics, the *Poincaré section*. Starting at an initial time $t = t_0$, the points on a suitable surface (Σ, the Poincaré section) can be collected by stroboscopically monitoring the state variables at intervals of the period T, as illustrated in Figure 5.11.

Alternatively, the Duffing equation (5.1.1) can be recast in the following form:

$$\begin{aligned}\dot{y}_1 &= y_2,\\ \dot{y}_2 &= -2\zeta y_2 - y_1 - \gamma y_1^3 + F\cos\theta,\\ \dot{\theta} &= \Omega\end{aligned} \tag{5.4.7a–c}$$

where $\theta = 2\pi t/T(\bmod 2\pi)$. Since the response at $t = 0$ and $t = T$ can be considered to be identical, the state space of Equation (5.4.7a–c) is the cylinder $\mathbb{R}^2 \times S^1$ as illustrated in Figure 5.11b. This topology results from the state space (y_1, y_2, t) with the points $t = 0$ and $t = T$ 'glued together'.

A Poincaré section in this cylindrical space can be defined as

$$\Sigma = \left((y_1, y_2, \theta) \in \mathbb{R}\times\mathbb{R}\times S | \theta = \frac{2\pi(t-t_0)}{T}(\bmod 2\pi)\right) \tag{5.4.8}$$

A section that is transverse to the flow everywhere in the state space is called a global section, as opposed to a local section that is transverse to the flow only locally.

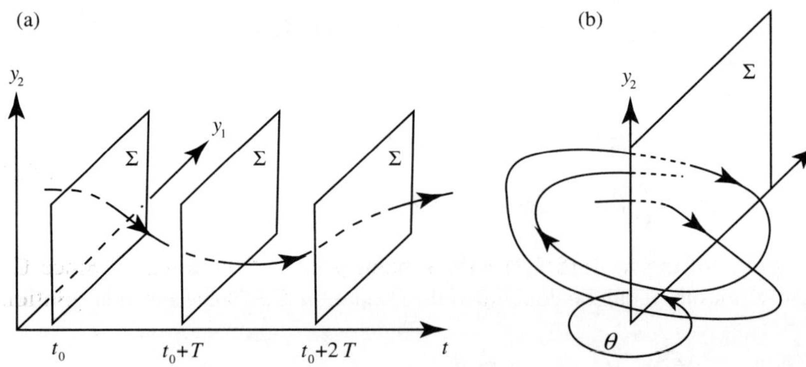

Figure 5.11 (a) Poincaré section Σ of an orbit of a two-dimensional nonautonomous system in extended state space; (b) Cylindrical phase space topology.

The normal vector $\mathbf{n}$ to this surface Σ is given by

$$\mathbf{n} = (0 \quad 0 \quad 1)^{\mathrm{T}} \tag{5.4.9}$$

and the positivity of the dot product

$$(0 \quad 0 \quad 1).\begin{pmatrix} y_2 \\ -2\zeta y_2 - y_1 - \gamma y_1^3 + F\cos\theta \\ 2\pi/T \end{pmatrix} = \frac{2\pi}{T} \tag{5.4.10}$$

implies that the section Σ is transverse to the flow everywhere in the cylindrical state space.

With the introduction of the Poincaré section, the study of the continuous system can be reduced to that of a Poincaré (or return) map $\Sigma \to \Sigma$.

In the next two subsections, the two different versions of the forced Duffing equation are studied by using the tools described previously. One of these versions is the Ueda oscillator (see Chapter 2) which has been studied extensively. The numerical simulations in Sections 5.4.3 and 5.4.4 are intended to provide an idea of the rich landscape of nonlinear dynamics possible in these systems.

5.4.3 The Ueda oscillator

A Duffing-type oscillator that has been studied extensively in the literature is the Ueda oscillator [14,15]. This oscillator is a special case of Equation (5.1.1) (see Chapter 2), and the corresponding description has a hardening nonlinearity, no linear stiffness term, and an external excitation with the frequency $\Omega = 1$; that is,

$$\ddot{y} + 2\zeta\dot{y} + y^3 = F\cos t \tag{5.4.11}$$

In Figure 5.12, following [14,15], an incomplete map of different possible types of motions over a selected parameter window is shown. In regions I, II, II$'$, II$''$, III, and IV, period-one attractors exist. In the regions marked by m/n, subharmonic or ultrasubharmonic motions (a *periodic motion* with principal frequency m/n times that of the external forcing) of order m/n are found ($n < 3$) (see Section 5.3.2 for the description of secondary resonances). Hatching indicates the existence of unique *chaotic attractors*, while in the shaded regions chaotic attractors coexist with periodic attractors.

Some representative y time series, phase plots, Poincaré sections, and power spectra of responses of the Ueda oscillator are shown for $\zeta = 0.01$ and different values of F in Figure 5.13. For $F = 0.1$ and $F = 0.9$ the responses are periodic (Figures 5.13(a) and (b)), while for $F = 0.7$ the motion is aperiodic (Figure 5.13(c)). The fractal nature of the Poincaré section of the aperiodic motion reveals the complexity of the response. This plot is known as Ueda's attractor.

5.4.4 Bifurcations and chaos in the Duffing oscillator with a softening spring

In this subsection, the Duffing oscillator with a softening nonlinear spring $\gamma = -1$ is considered with $\zeta = 0.2$, i.e.,

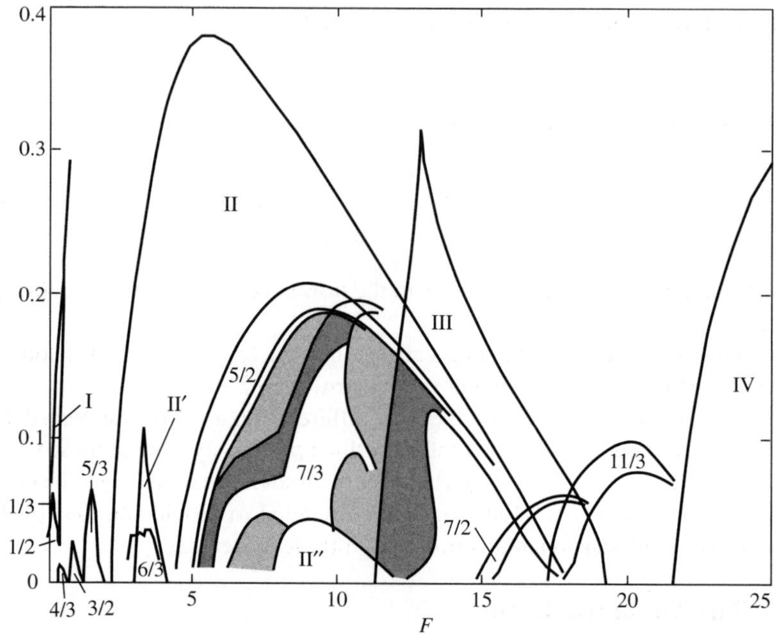

Figure 5.12 Rich dynamics exhibited by the Ueda oscillator (5.4.11) in different regions of the (F,ζ) parameter space. Period-one attractors are found in regions I, II, II′, II″, III, and IV. Subharmonic and ultrasubharmonic responses exist in the regions marked by m/n. Based on [15].

$$\ddot{y} + 0.4\,\dot{y} + y - y^3 = F\cos\Omega\, t \tag{5.4.12}$$

In Figure 5.14(a), the phase space projection of the response and the corresponding *y power spectrum* P_s are shown for $F = 0.35$ and $\Omega = 0.8$. A dominant peak at Ω can be observed in the spectrum, and the corresponding periodic attractor has a period equal to the forcing period. As F is gradually increased to 0.388, this symmetric periodic orbit undergoes a supercritical *pitchfork* (*symmetry breaking*) *bifurcation*, resulting in three asymmetric solutions (2 stable, 1 unstable). In Figure 5.14(b), one of the stable asymmetric solutions is shown along with the corresponding response spectrum. The second harmonic is discernible in the response spectrum indicating that a certain symmetry property has been broken due to the presence of the even harmonic.

This type of symmetry breaking has been observed to precede *period-doubling bifurcations*, as generally only asymmetric orbits can undergo period doubling (see, for example [7,16]). A further increase in the driving force results in period-doubled solutions, which results in *subharmonics* in the power spectrum. A gradual increase of F leads to a full *period-doubling cascade* and *chaotic solutions*. This is indicated by the complex phase plot, the fractal structure of the Poincaré section, and the

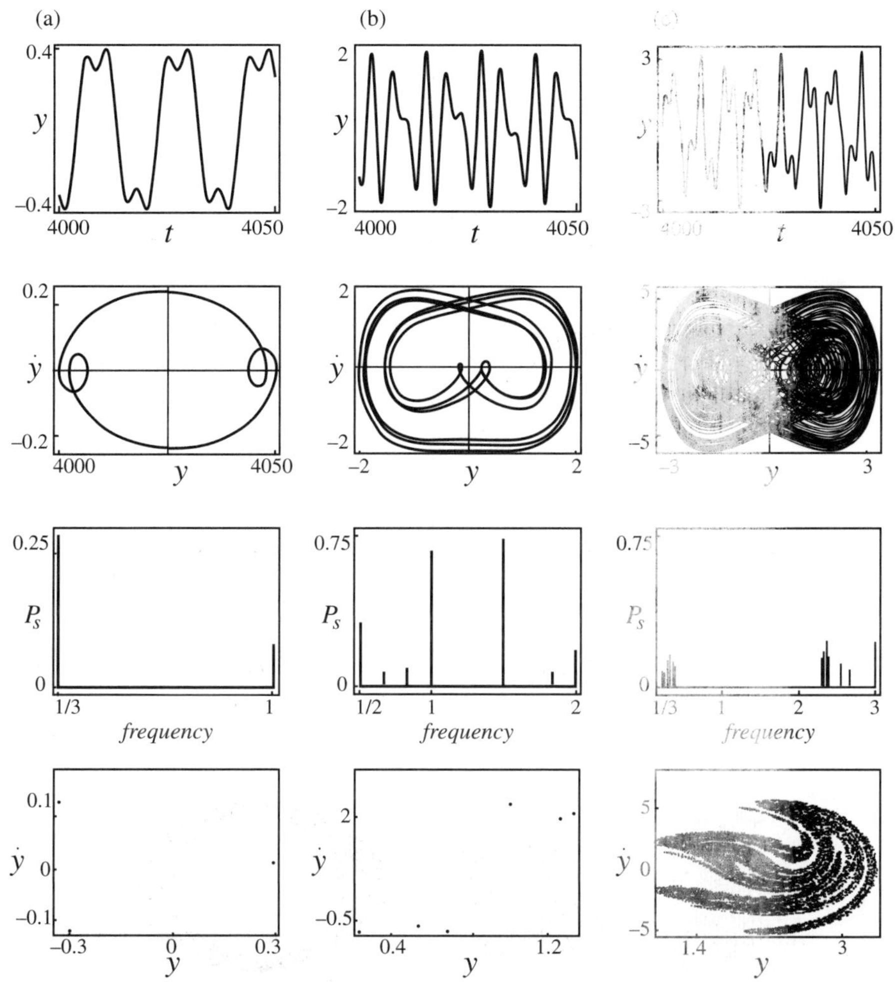

Figure 5.13 Representative y time series, phase plots, Poincaré sections, and power spectra P_s of the responses of the Ueda oscillator with $\zeta = 0.01$ *and (a)* $F = 0.1$, *(b)* $F = 0.9$, *(c)* $F = 0.7$.

broadband characteristic of the response spectrum (Figure 5.15). These results are illustrative of a sequence of period-doubling bifurcations leading to a *chaotic motion*, and this is an example of chaotic motions resulting from a local bifurcation sequence (see, for example [16]). A *bifurcation diagram* for the Duffing oscillator with parameter values $\zeta = 0.2, \gamma = -1$ and $\Omega = 0.8$, is shown in Figure 5.16(a). This diagram was constructed by splicing together intersections on the Poincaré section corresponding to a quasistatic variation with increasing values of F in the range [0.35, 0.392]. This bifurcation diagram is very similar to that exhibited by solutions of the

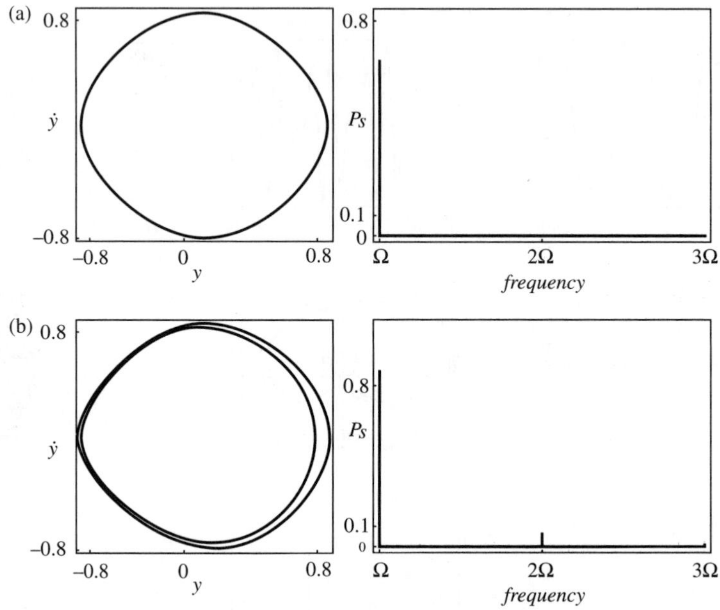

Figure 5.14 $(y, \dot{y})$ *plots of Equation (5.4.7) and corresponding y power spectra* P_s*: (a) before symmetry breaking and (b) after symmetry breaking.*

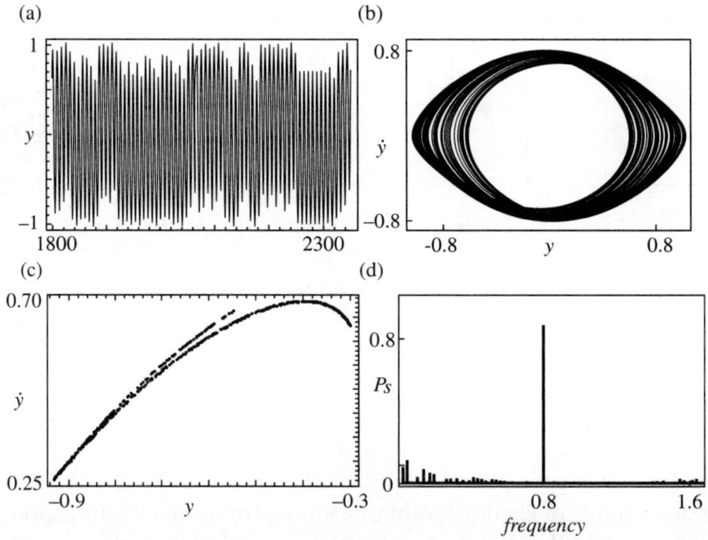

Figure 5.15 Aperiodic response of the Duffing oscillator for parameter values $\zeta = 0.2, \gamma = -1, \ \Omega = 0.8,$ *and* $F = 0.393$*: (a) y time series, (b) phase portrait projection, (c) Poincaré section, and (d) y power spectrum* P_s*.*

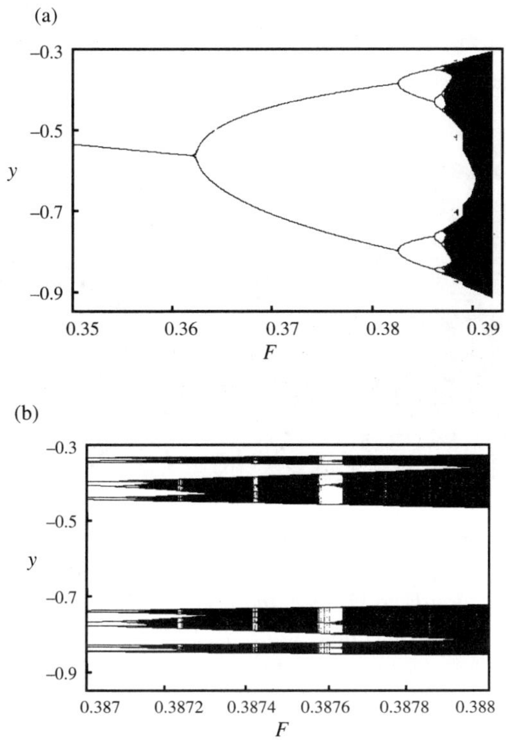

Figure 5.16 Bifurcation diagrams for the Duffing oscillator for parameter values $\zeta = 0.2$, $\gamma = -1$ *and* $\Omega = 0.8$.

logistic map (see, for example [7]), which is a one-dimensional noninvertible map. Figure 5.16(b) depicts the bifurcation diagram for the range [0.387, 0.388]. Here, windows of periodic solutions can be observed. In general, solutions of all periods that are integer multiples of the period of the forcing exist for specific parameter values.

In references [17–19], the complex dynamics of the hardening Duffing oscillator

$$\ddot{y} + 0.2\dot{y} + y + y^3 = F\cos\Omega\, t \tag{5.4.13}$$

has been explored through bifurcation diagrams, resonance curves, and phase diagrams. A phase diagram is a two-dimensional chart of the parameter space depicting the occurrence of various bifurcations. For Equation (5.4.13) such a diagram, reprinted from [18], is shown in Figure 5.17. Asymmetric period-1 solutions exist in the orange-coloured regions, while the yellow regions contain period-doubling cascades and chaotic motions. The observable 'superstructure' is due to the nonlinear resonances of the Duffing equation. The periodically recurring fine structure of the *bifurcation set* is emphasised by the use of logarithmic axes.

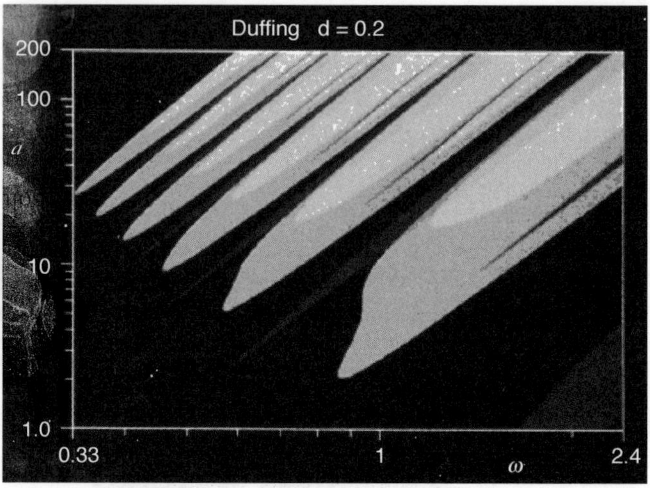

Figure 5.17 Different possible responses in the parameter space of excitation amplitude F and excitation frequency Ω of a hardening Duffing oscillator (note that ω in the figure corresponds to Ω here and d is 2ζ) The orange and yellow regions correspond to asymmetric period-1 solutions and higher-period/chaotic motions, respectively. Reprinted from [18], Copyright 1993, with permission from World Scientific Publishing Co. Pte. Ltd. See Plate 1 for the coloured version of this figure.

5.5 Global dynamics

To briefly discuss the global dynamics of the Duffing oscillator, the undamped, unforced Duffing oscillator $\ddot{y}+y-y^3=0$ is first examined, whose phase portrait is shown in Figure 5.18.

As described in Chapter 3, the potential energy of the system has a minimum at $C:(y,\dot{y})=(0,0)$, and two maxima at $S_1:(y,\dot{y})=(-1,0)$ and $S_2:(y,\dot{y})=(1,0)$. Point C is a *centre*, while S_1 and S_2 are *saddle points*. The forward orbits ($t\to\infty$) connecting S_1 with S_2, and S_2 with S_1 are called *heteroclinic orbits*, and these orbits are denoted by Γ_1 and Γ_2, respectively. The nonisolated periodic orbits inside the heteroclinic loop ($\Gamma_1\cup\Gamma_2$) of this system now become cylinders in the extended phase space (y_1,y_2,t) and tori in the (y_1,y_2,θ) space that densely fill out (foliate) the inside of the heteroclinic loop. These cylinders or tori are invariant manifolds because any solution starting on them will remain on them. In Figure 5.19, Poincaré sections of periodic responses of the undamped and forced Duffing oscillator are shown in a neighbourhood of the centre C.

Here, the structure on the Poincaré section is similar to the phase portraits of Figure 5.18, except that each 'continuous curve' is a sequence of successive intersections of the trajectory with the Poincaré plane. The Poincaré map of the unforced, undamped Duffing oscillator is area preserving (as the divergence of the

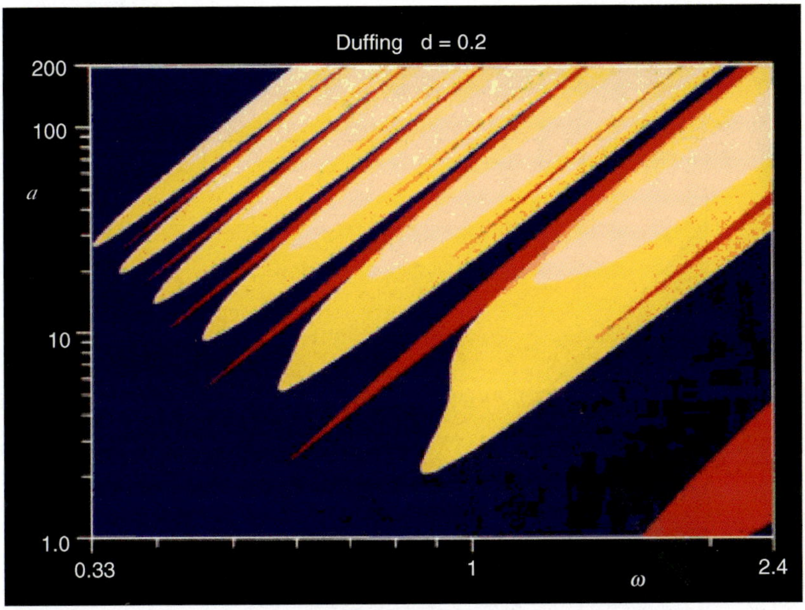

Plate 1 Different possible responses in the parameter space of excitation amplitude F and excitation frequency of a hardening Duffing oscillator (note that ω in the figure corresponds to Ω here and d is 2ζ) The orange and yellow regions correspond to asymmetric period-1 solutions and higher-period/chaotic motions, respectively. Reprinted from U. Parlitz, Common dynamical features of periodically driven strictly dissipative oscillators. International Journal of Bifurcation Chaos 3, 703–715, 1993. Copyright 1993, reprinted with permission from World Scientific Publishing Co. Pte. Ltd.

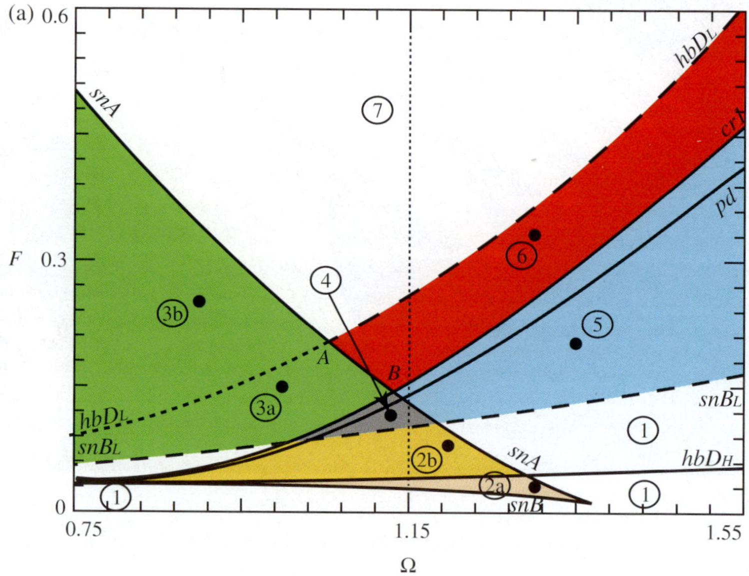

Plate 2 (a) The behaviour chart for $\zeta = 0.025$ and $\gamma = 1$. The dotted vertical line corresponds to the bifurcation diagram of Figure 7.7. The points correspond to the basins of attraction reported in Section 7.5.1.

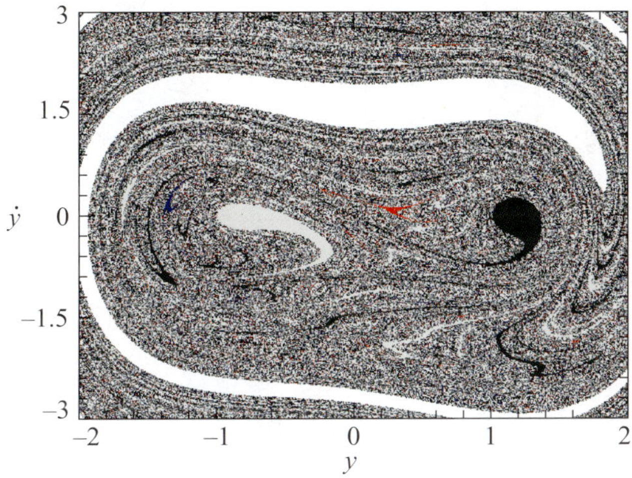

Plate 3 Overall attractor-basin phase portraits for $\Omega = 1.13$ and $F = 0.115$, point 4 in Fig. 7.11 for $\zeta = 0.025$, $\gamma = 1$.

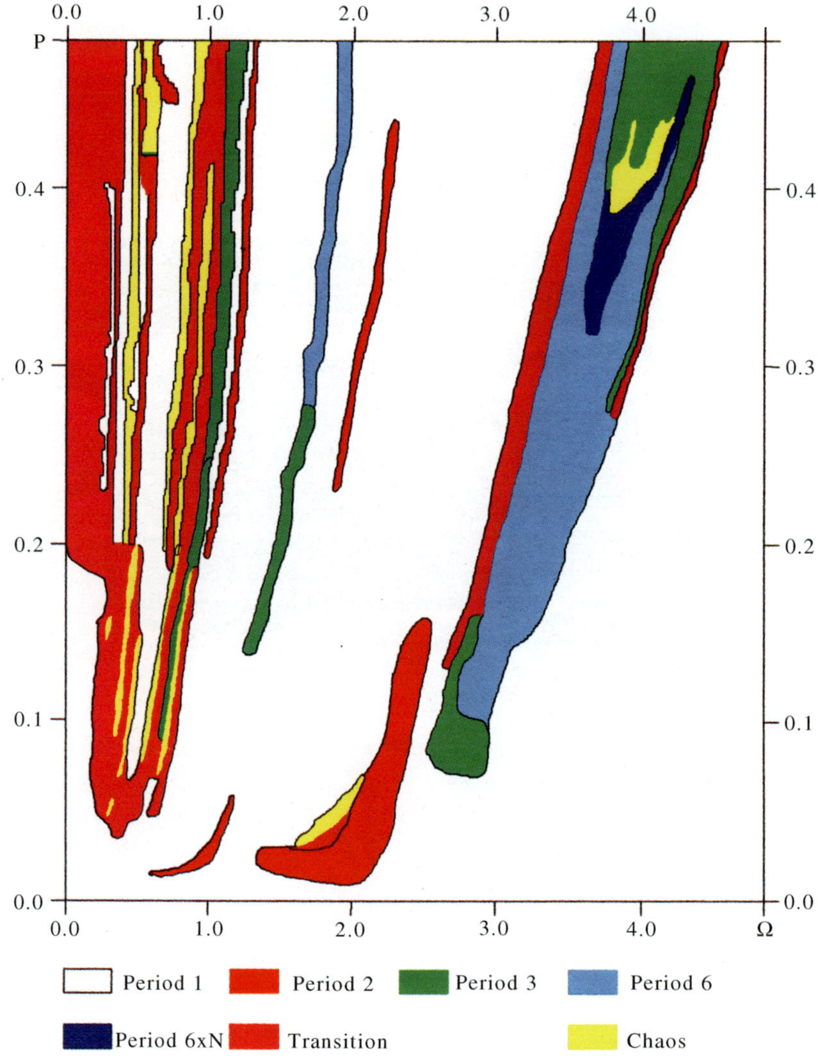

Plate 4 Periodic and chaotic response in the excitation frequency-magnitude of the excitation force plane. Note that $P \equiv F$ in Equation (8.2.11) and $\zeta = 0.05$. Reprinted from G. Rega, F. Benedettini, A. Salvatori, Periodic and chaotic motions of an unsymmetrical oscillator in nonlinear structural systems. Chaos, Solitons and Fractals, 1, 39–54, Copyright 1991, with reprinted permission from Elsevier.

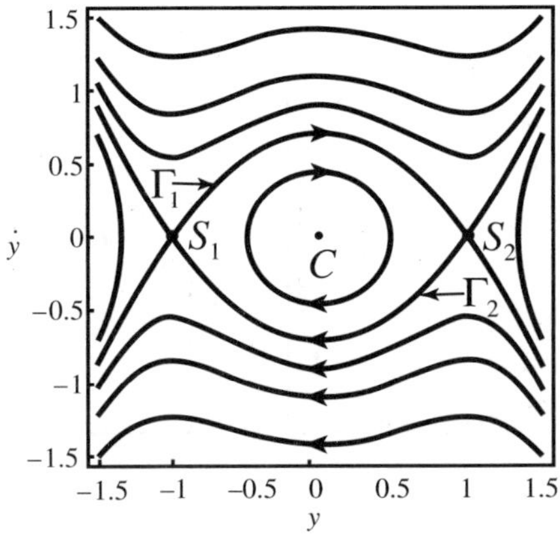

Figure 5.18 Phase portrait for the undamped, unforced Duffing oscillator $\ddot{y}+y-y^3=0$.

vector field computed in Equation (5.1.3a) is zero). The centre C corresponds to an elliptic fixed point surrounded by closed invariant curves. These are cross-sections of nearby tori.

In Figure 5.20, the influence of the increasing excitation amplitude is shown. The introduction of the slightest damping changes the centre into a *stable focus* and destroys the heteroclinic saddle-saddle connections (Figure 5.21). The system is not structurally stable to a damping perturbation. The stable manifolds of the two saddles divide the phase space into three regions. Initial conditions to the left of the stable manifold of S_1 and to the right of the stable manifold of S_2 are attracted to infinity. The

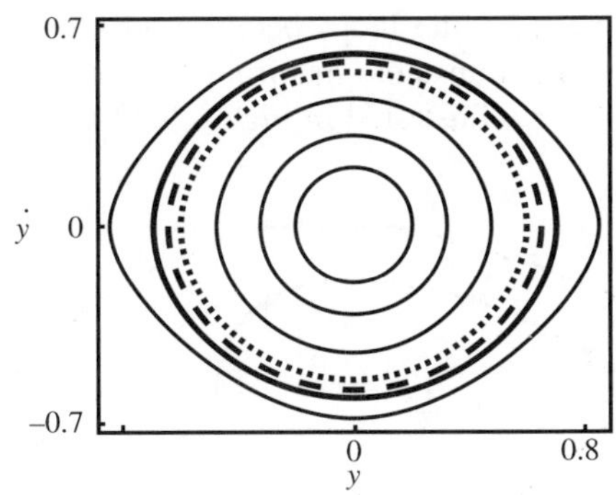

Figure 5.19 Poincaré sections of the periodic responses of the undamped, unforced Duffing oscillator. The closed thick curves of increasing enclosed area correspond to the initial conditions $y(0)=-0.6$ *(dotted line),* $y(0)=-0.65$ *(dashes line),* $y(0)=-0.7$ *(solid line) and* $\dot{y}(0)=0$.

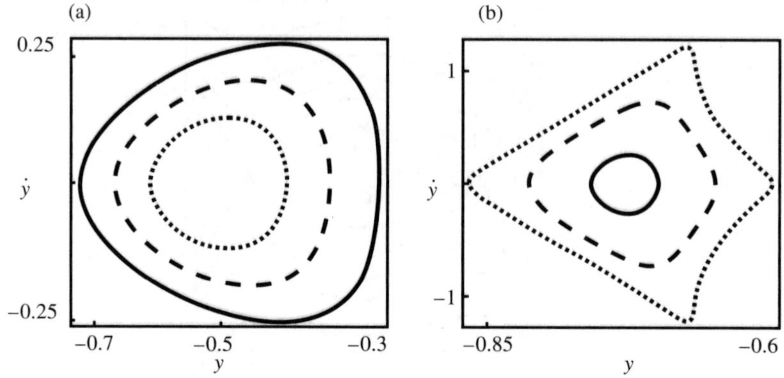

Figure 5.20 Poincaré sections for increasing forcing amplitude: (a) $F = 0.1$; (b) $F = 0.3$. The initial conditions corresponding to the closed thick curves with increasing enclosed area are $y(0) = -0.6$ (dotted line), $y(0) = -0.65$ (dashed line), $y(0) = -0.7$ (solid line) and $\dot{y}(0) = 0$. The excitation frequency is $\Omega = 1$.

third region is the *basin of attraction* of the focus. When the forcing is small, there is still a cylinder/torus – albeit slightly deformed – close to that of the corresponding one of the unforced system. As the forcing amplitude is increased, the stable and unstable manifolds of the two saddle points move closer to one another. As F reaches a critical value F_C, the two manifolds graze each other (Figure 5.22(a)) and with further increase of F they intersect transversally (Figure 5.22(b)) (see, for example [7]).

Due to the invariance of the sets W^S and W^U, if they have an intersection point I_0, they must intersect infinitely many times because $I_0 \in W^S \cap W^U \Rightarrow P^m(I_0) \in W^S \cap W^U$, $\forall m \in Z$. Each point of transversal intersection (I_0, I_1, I_2, ...) is called a transversal heteroclinic point, and the orbit of such a point under the Poincaré map produces a transversal heteroclinic orbit that oscillates wildly. As the unstable manifold of S_1 approaches S_2, the loops between adjacent heteroclinic points are stretched parallel to the local unstable manifold W^U and contracted parallel to the local stable manifold W^S, as illustrated in Figure 5.22(c).

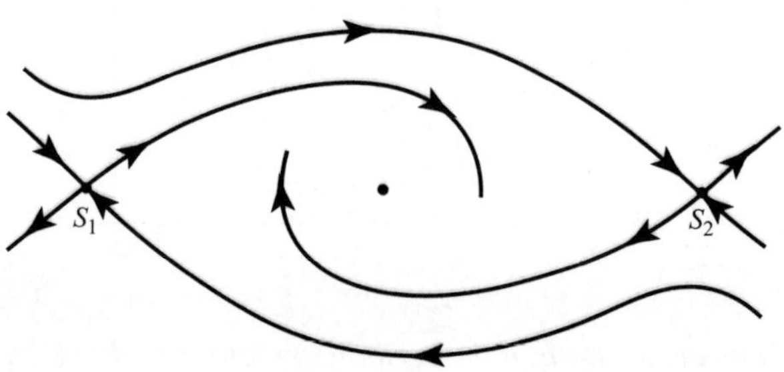

Figure 5.21 Destruction of the heteroclinic connections of the Duffing oscillator by damping.

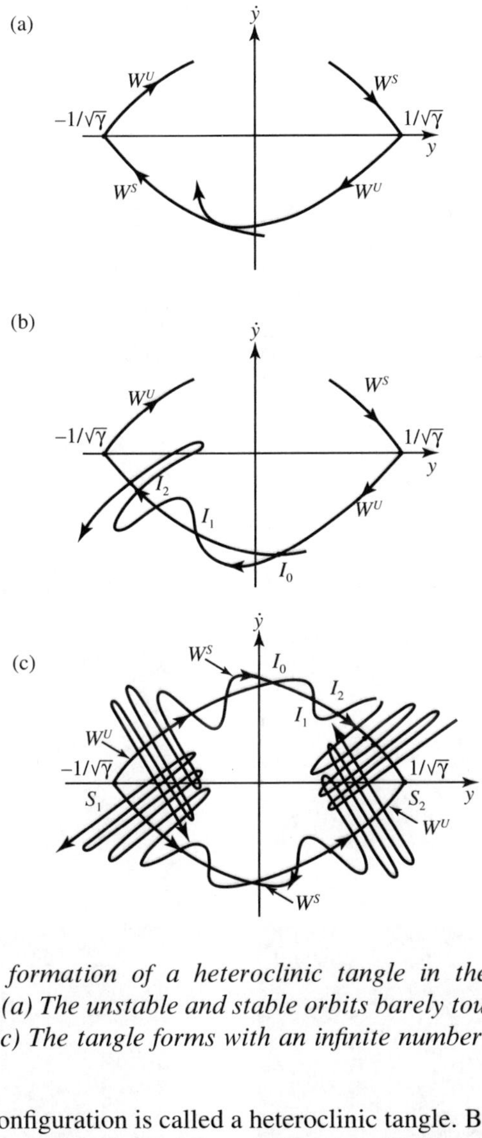

Figure 5.22 The formation of a heteroclinic tangle in the Poincaré map of Equation (5.4.12): (a) The unstable and stable orbits barely touch; this is the onset of chaos; (b) and (c) The tangle forms with an infinite number of intersections.

The resulting configuration is called a heteroclinic tangle. Because of the strong folding and stretching near the saddle points, two initially nearby points may be mapped far apart, resulting in sensitive dependence on initial conditions. The heteroclinic tangle structure also has a profound influence on the basins of attraction for bounded and unbounded motions. In the neighbourhood of the transverse heteroclinic points these basins are delicately intertwined and exhibit a fractal structure. Basin boundaries can sometimes undergo metamorphoses, transforming a smooth basis boundary into a fractal structure [20–22]. Following [20], in Figure 5.23, a series of basin-boundary metamorphoses is shown as the excitation amplitude F is increased for $\Omega = 0.8$. As F increases from 0.3, the once smooth basin boundary develops fingers

(Figures 5.23(a) and (b)). The white and dark regions represent the set of initial conditions that results in bounded or unbounded solutions, respectively. The interface of the regions becomes increasingly complicated, as shown in Figures 5.23(c)–(e), and gradually the basin of attraction for bounded solutions fades away (Figure 5.23(f)). The

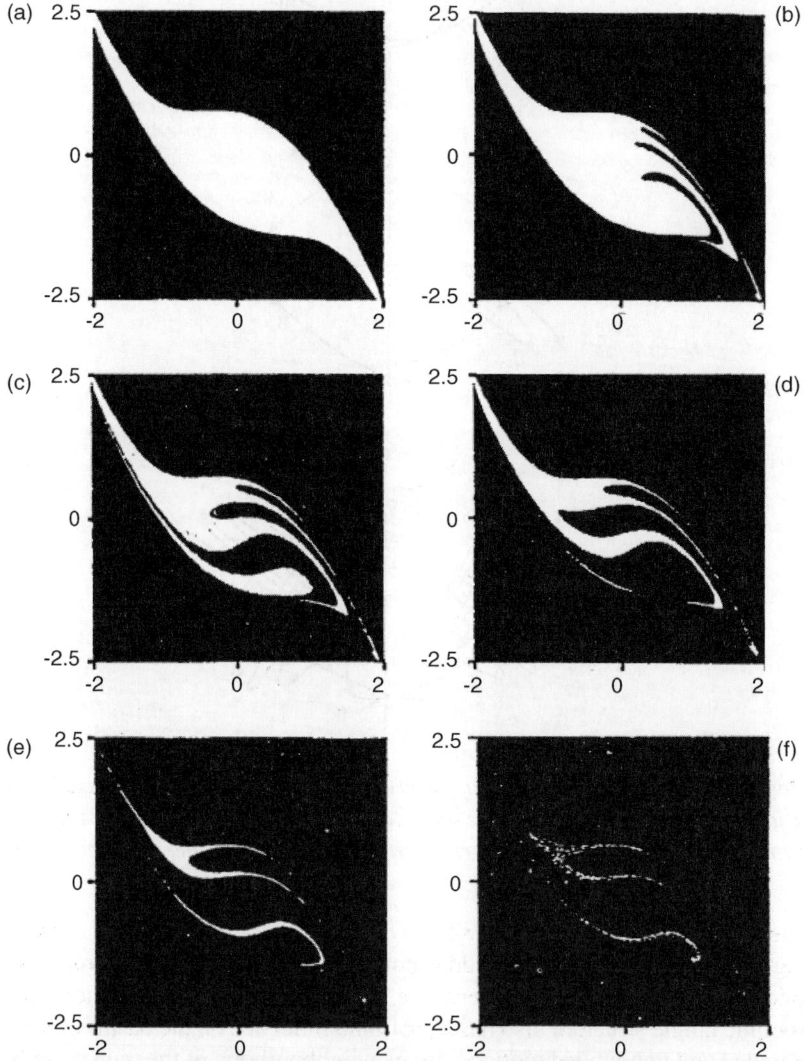

Figure 5.23 Basin-boundary metamorphoses for $\Omega = 0.8$. Increasing values of F correspond to decreasing white basin area: (a) $F = 0.30$; (b) $F = 0.32$; (c) $F = 0.33$; (d) $F = 0.345$; (e) $F = 0.38$; (f) $F = 0.394$. Reprinted from [20], Copyright 1989, with permission from Elsevier.

disappearance of the basin is related to the point of escape from the potential well of the Duffing oscillator. This erosion of the basin has been quantified in reference [22] and other studies.

5.6 Summary

In this chapter, some characteristic nonlinear behaviours exhibited by the externally excited Duffing oscillator have been explored. For weak nonlinearity and weak damping, the use of perturbation analysis to obtain an analytical approximation for the forced response has been illustrated. Stability analysis of the steady-state solutions has also been carried out to explain the dramatic jump behaviour between solutions in the primary resonance. A glimpse into the rich variety of nonlinear phenomena for this system has been provided through numerical exploration of qualitative changes. Some aspects of the global dynamics have also been discussed in this chapter. Based on the discussion presented, the following observations can be made: i) the presence of the cubic nonlinearity in a damped oscillator allows only for construction of approximate solutions; ii) in addition to the primary resonance, there are also others, the so-called, secondary resonances; iii) the oscillator displays hardening or softening behaviour, depending on the sign of the nonlinearity; iv) the presence of the nonlinearity results in multiple solutions for the same parameter values; v) the system exhibits a plethora of dynamically interesting solutions, including *limit cycles* (isolated periodic motions) and aperiodic motions (*strange attractors*) with simple periodic motions coexisting with exotic attractors in some parameter windows, and vi) the basins of attraction of responses can exhibit a fractal structure.

References

[1] J.J. Stoker, *Nonlinear Vibrations*, Interscience, New York, 1950.

[2] A.H. Nayfeh, D. Mook, *Nonlinear Oscillations*, Wiley, New York, 1979.

[3] S.W. Shaw, B. Balachandran, A review of nonlinear dynamics of mechanical systems in year 2008. *Journal of System Design and Dynamics*, 2, 611–640, 2008.

[4] R. Lifshitz, M. C. Cross, Nonlinear dynamics of nanomechanical and micromechanical resonators. *Reviews of Nonlinear Dynamics and Complexity*, Volume 1, Wiley, 2008.

[5] J. Guckenheimer and P. Holmes, *Nonlinear Oscillations, Dynamical Systems, and Bifurcations of Vector Fields*, Springer, New York, 1983.

[6] F.C. Moon, *Chaotic Vibrations*, Wiley, New York, 1987.

[7] A.H. Nayfeh, B. Balachandran, *Applied Nonlinear Dynamics: Analytical, Computational, and Experimental methods*, Wiley, 1995.

[8] A.H. Nayfeh, *Perturbation Methods*, Wiley, New York, 1973.

[9] A.H. Nayfeh, *Introduction to Perturbation Techniques*, Wiley, New York, 1981.

[10] M.J. Brennan, I. Kovacic, A. Carrella, T.P. Waters, On the jump-up and the jump-down frequencies of the Duffing oscillator. *Journal of Sound and Vibration*, 318, 1250–1261, 2008.

[11] R. Thom, *Structural Stability and Morphogenesis*, W.A. Benjamin, New York, 1973.

[12] P.J. Holmes, D.A. Rand, The bifurcations of Duffing's equation: an application of catastrophe theory. *Journal of Sound and Vibration*, 44, 237–253, 1976.

[13] J.K. Hale, *Oscillations in Nonlinear Systems*, McGraw-Hill, New York, 1963.

[14] Y. Ueda, Randomly transitional phenomena in the system governed by Duffing's equation. *Journal of Statistical Physics*, 20, 181–196, 1979.

[15] Y. Ueda, *Steady Motions Exhibited by Duffing's Equation: A Picture Book of Regular and Chaotic Motions*, in Hao Bai-Lin, D. H. Feng, and J.-M. Yuan, (eds.), *New Approaches to Nonlinear Problems*. SIAM, Philadelphia, 1980.

[16] D. D'Humieres, M.R. Beasley, B.A. Huberman, A. Libchaber, Chaotic states and routes to chaos in the forced pendulum. *Physical Review A*, 26, 3483–3496, 1982.

[17] U. Parlitz, W. Lauterborn, Superstructure in the bifurcation set of the Duffing equation $\ddot{x}+d\,\dot{x}+x+x^3=f\cos(\omega t)$. *Physics Letters A*, 107, 351–355, 1985.

[18] U. Parlitz, Common dynamical features of periodically driven strictly dissipative oscillators. *International Journal of Bifurcation and Chaos* 3, 703–715, 1993.

[19] U. Parlitz, *Complex Dynamics of Nonlinear Systems, Oscillations, Waves and Interactions*, pp. 405–434 T. Kurz, U. Parlitz, and U. Kaatze, (eds.) Universitatsverlag Gottingen, 2007.

[20] A.H. Nayfeh, N.E. Sanchez, Bifurcations in a forced softening Duffing oscillator. *International Journal of Non–Linear Mechanics*, 24, 483–497, 1989.

[21] C. Grebogi, E. Ott, J.A. Yorke, Metamorphoses of basin boundaries in nonlinear dynamical systems. *Physical Review Letters*, 56, 1011–1014, 1986.

[22] M.S. Soliman, J.M.T. Thompson, Integrity measures quantifying the erosion of smooth and fractal basins of attraction. *Journal of Sound and Vibration*, 135, 453–475, 1989.

6

Forced harmonic vibration of a Duffing oscillator with different damping mechanisms

Asok Kumar Mallik
Department of Mechanical Engineering, Indian Institute of Technology Kanpur, India,

6.1 Introduction

As discussed in Chapter 2, the stiffness force in one form of a Duffing oscillator is expressed by a combination of a linear and a cubic nonlinear term. For small amplitude vibrations, this nonlinear term may be negligible compared to the linear term. The damping force, on the other hand, is assumed to be linear (viscous damping) only for mathematical simplification. Pippard [1] writes 'There is something of a tendency among physicists to try to reduce everything to linearity . . ., reality may not conform to what we may wish, rather more so with the damping forces than with the restoring force in small amplitude vibration.' In real life, the damping force, representing various dissipative mechanisms present in a system, may be a distinctly nonlinear function of velocity [2–5]. In some situations, a combination of linear and nonlinear terms may be necessary to model the damping satisfactorily. In this chapter,

The Duffing Equation: Nonlinear Oscillators and their Behaviour, First Edition. Edited by Ivana Kovacic and Michael J. Brennan.

the response of a harmonically excited Duffing oscillator with different damping mechanisms is discussed.

6.2 Classification of nonlinear characteristics

Nonlinear characteristics, present in both stiffness and damping forces, of the Duffing oscillator, are classified in this section. Some facts related to the stiffness force and the corresponding geometric nonlinearity shown in Chapter 2 and Chapter 3 are summarised and given here for the convenience of the reader.

6.2.1 Stiffness force

The nonlinear stiffness force in the Duffing equation is expressed in nondimensional form as

$$F_{\mathrm{r}}(y) = \alpha\, y + \gamma\, y^3 \tag{6.2.1}$$

where y denotes the nondimensional displacement and α and γ represent, respectively, the linear stiffness parameter (linear stiffness) and nonlinear cubic stiffness parameter (cubic/nonlinear stiffness).

If α and γ both are positive, then the system has only one *equilibrium point* at $y = 0$, which is stable. This is a hardening system, since the stiffness, implied by the slope of the restoring force curve, increases with increasing $|y|$.

When $\alpha > 0$ and $\gamma < 0$, the system has one stable equilibrium point at $y = 0$ and two unstable equilibrium points at $y = \pm\sqrt{-\alpha/\gamma}$. In this case, the slope of the force–deflection curve decreases with increasing $|y|$ and the system is softening. For a pendulum, the force–deflection characteristic, in which the force is proportional to $\sin y$, can be approximately modelled so that the force is proportional to $y - (y^3/6)$. Thus, the nonlinear characteristic of a pendulum is that of a softening Duffing oscillator.

If $\alpha < 0$ and $\gamma > 0$, then such an oscillator has one unstable equilibrium point at $y = 0$ and two stable equilibrium points at $y = \pm\sqrt{-\alpha/\gamma}$. This system is referred to as a negative linear-positive cubic stiffness oscillator or also as a *two/double/twin-well potential* Duffing oscillator. The oscillations of a buckled beam can be modelled by such an oscillator (see Chapter 2).

6.2.2 Damping force

As in the previous chapters, the linear viscous damping force in the nondimensional form of the Duffing equation is expressed as

$$F_{\mathrm{d}}(\dot{y}) = 2\zeta\, \dot{y} \tag{6.2.2}$$

where ζ is the viscous damping ratio.

A nonlinear damping term is written as a velocity nth power model as

$$F_{\mathrm{d}}(\dot{y}) = 2\zeta_n\, \dot{y}|\dot{y}|^{n-1}, \quad n \geq 0, \quad \zeta_1 = \zeta \tag{6.2.3a–c}$$

The modulus, rather than a simple exponent, is used to maintain the dissipative characteristic of the damping force for all values of n. The value of the index n is unity for linear viscous damping. Such viscous damping is encountered in low velocity oscillation of a solid in a fluid medium. Dry friction between two solid interfaces or *Coulomb damping* can be modelled with $n=0$. The value of the exponent $n=2$ for quadratic damping is encountered at high Reynold's number [2]. If the damping is provided by an oscillating fluid flow through an orifice, then the value of n can be taken as 1.5 [3]. Sometimes a combination of viscous and a nonlinear damping term is needed to satisfactorily model the dissipative mechanisms. For example, the damping moment encountered during the rolling motion of a ship is expressed as $2\zeta_1\dot{y}+2\zeta_2\dot{y}|\dot{y}|$ [4]. Similarly, the damping capacity of a metal-bonded rubber isolator is also modelled by a combination of viscous and quadratic damping terms [5].

6.2.3 Equivalent viscous damping

So far as the mathematical treatment of the nonlinear damping force is concerned, the most common method is to use an *equivalent viscous damping* coefficient (ratio, factor). A harmonic response of the system is assumed and the equivalence is based on equal energy dissipation per cycle.

Let the harmonic motion across the damper be expressed as

$$y = Y\cos(\Omega t-\varphi) \tag{6.2.4}$$

Then, with the linear damping term $2\zeta\dot{y}$, the energy dissipated per cycle is given by

$$D = 2\zeta\pi\,\Omega Y^2 \tag{6.2.5}$$

With a velocity nth power damping, the energy dissipated per cycle is found to be [2]

$$D = 4\zeta_n Y^{n+1}\Omega^n\sqrt{\pi}\,\frac{\Gamma\left(\dfrac{n+2}{2}\right)}{\Gamma\left(\dfrac{n+3}{2}\right)} \tag{6.2.6}$$

where Γ represents the Gamma function. Combining Equations (6.2.5) and (6.2.6), the equivalent viscous damping ratio is given by [2]

$$\zeta_{\text{eq}} = \zeta_n Y^{n-1}\,\Omega^{n-1}\beta_n \tag{6.2.7}$$

where

$$\beta_n = \frac{2}{\sqrt{\pi}}\frac{\Gamma\left(\dfrac{n+2}{2}\right)}{\Gamma\left(\dfrac{n+3}{2}\right)} \tag{6.2.8}$$

It may be noted that for Coulomb damping ($n=0$), there is a discontinuity in the damping force at $\dot{y}=0$. Consequently, there exists the possibility of *stick-slip motion*, which needs special consideration. In subsequent sections, cases with $n=0$ and

$n \neq 0$ are discussed separately. It should be mentioned that the equivalent viscous damping ratio, given by Equation (6.2.7), depends on the amplitude and frequency of oscillation.

6.3 Harmonically excited Duffing oscillator with generalised damping

Consider a harmonically excited Duffing oscillator governed by the following nondimensional equation of motion

$$\ddot{y} + 2\zeta_n \dot{y}|\dot{y}|^{n-1} + \alpha y + \gamma y^3 = F \cos \Omega t \tag{6.3.1}$$

Depending on the nature of the nonlinearity in the stiffness force, values of the damping parameter and exponent and the level and frequency of excitation, the response may be taken as harmonic with the same frequency as the excitation. The response may also be periodic containing *superharmonics* and *subharmonics*. The response may even be chaotic in some situations. In the sections to follow, various types of response characteristics for different damping mechanisms are discussed.

6.4 Viscous damping

In this section, $n = 1$ is considered and Equation (6.3.1) is rewritten as

$$\ddot{y} + 2\zeta \dot{y} + \alpha y + \gamma y^3 = F \cos \Omega t \tag{6.4.1}$$

This equation is also considered in Chapter 5 for the case when $\alpha = 1$. The amplitude response is obtained there by using the method of multiple scales and discussed in terms of the magnification factor. The *stability* of the solution is investigated by considering the eigenvalues of the *Jacobian*. Some characteristics of the response are demonstrated, such as its multivaluedness and the *jump phenomenon*. Parts of this analysis are revisited here, but are given in a different way, which will be used later for the case when Equation (6.3.1) is studied for the case when $n \neq 1$. Thus, the amplitude response is derived by using the harmonic balance method and plotted as a *frequency-response curve*. The stability of the solutions is investigated by adding a small disturbance (perturbation) to the harmonic solution and analysing the corresponding differential equation, which has the form of the *Mathieu equation*. The analogy between the results derived in this chapter and those from Chapter 5 are emphasised. The intention is to illustrate some of the possible, but different approaches that can be used for the same problem.

6.4.1 Harmonic solution for a hardening system

Assume the steady-state harmonic solution of Equation (6.4.1) as

$$y = Y \cos(\Omega t - \theta) \tag{6.4.2}$$

Substituting Equation (6.4.2) in Equation (6.4.1) and equating the coefficients of $\cos \Omega t$ and $\sin \Omega t$ from both sides, the solution for Y can be determined as

$$Y^2 = \frac{F^2}{\left(4\zeta^2\Omega^2 + \left(\Omega^2 - \alpha - \frac{3}{4}\gamma Y^2\right)^2\right)}. \tag{6.4.3}$$

Introducing the following parameters

$$\omega = \frac{\Omega}{\sqrt{\alpha}}, \quad \xi = \frac{\zeta}{\sqrt{\alpha}}, \quad A^2 = \frac{\gamma}{\alpha} Y^2, \quad f^2 = \frac{\gamma}{\alpha^3} F^2 \tag{6.4.4a–d}$$

Equation (6.4.3) can be rewritten as

$$A^2 = \frac{f^2}{\left(4\xi^2\omega^2 + \left(\omega^2 - 1 - \frac{3}{4}A^2\right)^2\right)} \tag{6.4.5}$$

It may be noted that Equation (6.4.5) is cubic in A^2. Therefore, there may be either one or three real roots (with $A > 0$), i.e., either one or three values of the amplitude are possible at any given frequency, as discussed in Chapter 5, which is also related to Figures 5.3–5.5. Figure 6.1 shows some typical plots of A versus ω for a given value of the magnitude of excitation $f = 1$ with different amounts of damping, which are analogous to the *response curve* shown in Figure 5.4. When three amplitude values are obtained, the intermediate one is unstable. This will be shown in the next section in a different way than in Chapter 5. Of the two stable values, which one will be obtained

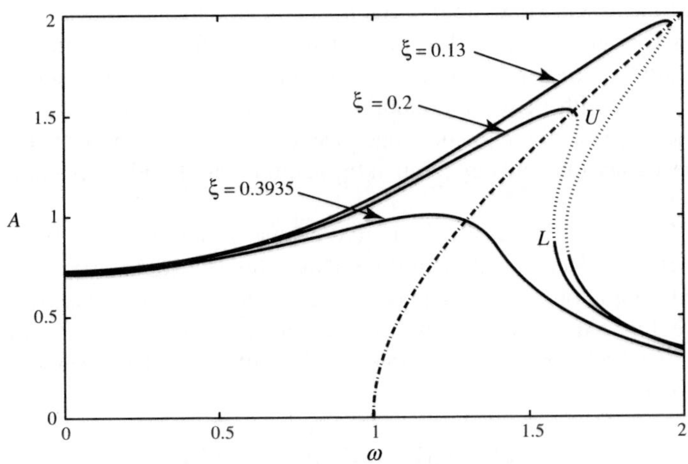

Figure 6.1 Frequency-response curves for a hardening system with $f = 1$ and different amounts of viscous damping; stable (solid line); unstable (dotted line); backbone curve (dashed-dotted line); U, L – upper and lower jump points.

in numerical simulation or observed during an experiment depends on the initial disturbance. Thus, the steady-state amplitude depends on the initial conditions, even in the presence of damping. It may be pointed out that Figure 6.1 may be drawn also by considering Equation (6.4.5) as a quadratic equation in ω^2 for a given value of the amplitude A [6]. There may be none, one or two values of ω depending on the value of A.

In Figure 6.1, the unstable solutions are indicated by dotted lines and the stable solutions by solid lines. The dashed-dotted line is the *backbone curve*, which shows the dependence of the nonlinear 'natural' frequency on the amplitude of motion, and is defined by

$$\omega^2 = 1 + \frac{3}{4}A^2 \tag{6.4.6}$$

It is interesting to note that if Equations (6.4.4) are used, then for $\alpha = 1$ then $\Omega^2 = 1 + 3\gamma Y^2/4$. For a small nonlinearity, this expression can be approximated to the one given in Equation (5.3.26a) $\Omega = 1 + 3\gamma Y^2/8$, where $Y \equiv a_p$.

In Figure 6.1, the frequency-response curve has vertical tangents at the points U and L: these are the *jump points*. The part of the frequency-response curve between the jump points is unstable. If the frequency of excitation is gradually increased from a low value, then at point U (the jump-down point)) the response jumps from the *resonance* to nonresonance branch undergoing a jump or *saddle-node bifurcation* as discussed in Chapter 5, related to Figure 5.6 (note that the points U and L correspond, respectively, to the points S_1 and S_3 labelled therein). Thus, starting at a high frequency on the nonresonance branch of the response curve, if the frequency of excitation is gradually reduced, then again the amplitude jumps to the resonance branch at the lower jump point L (jump-up point). After a jump occurs, the system takes more time to settle into the *steady state* compared to that along the resonance or nonresonance branch. The settling time depends on the sweep rate of excitation frequency and the amount of damping. It is clearly seen that the width of the jump region decreases with increasing damping and eventually for a critical amount of damping the jump phenomenon no longer occurs. As shown later, the value of critical damping necessary to avoid any jump in the response amplitude depends on the level of excitation.

Figure 6.2 shows the results for a given value of damping $\xi = 0.1$ at different levels of excitation. Here, the dashed-dotted lines show the loci of the two sets of jump points for different values of f. The equations of these two loci can be obtained as discussed below. At the points of vertical tangency in Figure 6.1, $dA/d\omega \to \infty$. Differentiating both sides of Equation (6.4.5) with respect to ω with ξ and f as constants, gives

$$\frac{dA}{d\omega} = \frac{-A\left(4\xi^2\omega^2 + 2\omega\left(\omega^2 - 1 - \frac{3}{4}A^2\right)\right)}{\left(4\xi^2\omega^2 + \left(\omega^2 - 1 - \frac{3}{4}A^2\right)^2 - \frac{3}{2}A^2\left(\omega^2 - 1 - \frac{3}{4}A^2\right)\right)} \tag{6.4.7}$$

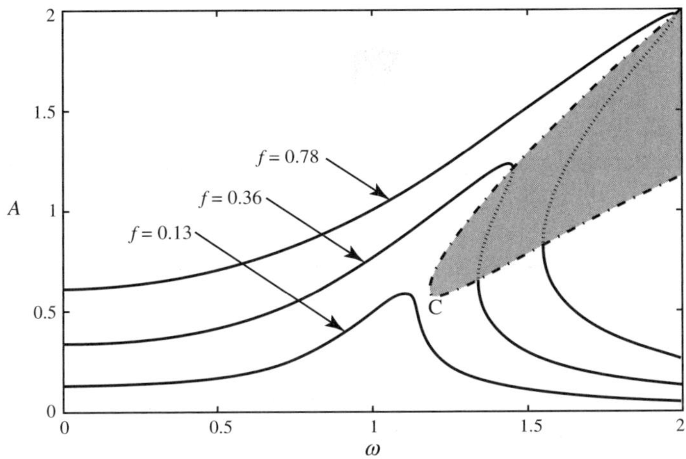

Figure 6.2 Frequency-response curves for a hardening system with $\xi = 0.1$ and various levels of excitation; stable (solid line); unstable (dotted line); boundary (dashed-dotted line) of the unstable region (hatched area) in the frequency-response plane; C – the point of intersection of the loci of upper and lower jump points.

Setting the denominator of the right-hand side of Equation (6.4.7) to zero, gives

$$4\xi^2\omega^2 + \left(\omega^2 - 1 - \frac{3}{4}A^2\right)^2 - \frac{3}{2}A^2\left(\omega^2 - 1 - \frac{3}{4}A^2\right) = 0 \qquad (6.4.8)$$

Solving the above quadratic equation in A^2, two roots are obtained as

$$A^2 = \frac{8}{9}(\omega^2 - 1) \pm \frac{4}{9}\sqrt{(\omega^2 - 1)^2 - 12\xi^2\omega^2} \qquad (6.4.9a,b)$$

Thus, Equations (6.4.9) describe the two loci of the jump points. The shaded area bounded by these two loci indicates the unstable region in the frequency-response plane. An interesting fact is revealed by the frequency-response plots for different levels of excitation (Figure 6.2). With increasing level of excitation at the same frequency, the amplitude of the response increases if it is stable but decreases if it is unstable. The two loci are seen to meet at a point C shown in Figure 6.2, where the response curve for the corresponding value of f would show a cusp. For the given damping, with any excitation lower than this critical value, the response remains single-valued at all frequencies. Likewise for a given force amplitude, there is a critical damping value beyond which again the response remains single-valued at all frequencies. These critical values can be easily determined as shown below.

Equating the two roots given by Equation (6.4.9), the coordinates where the two loci meet can be obtained from the following equations:

$$\omega_C^2 = 1 + \frac{9}{8}A_C^2 \qquad (6.4.10)$$

and

$$\left(\omega_C^2 - 1\right)^2 = 12\xi^2\omega_C^2 \tag{6.4.11}$$

Equation (6.4.10) implies $\omega_C^2 > 1$. Therefore, taking the proper root of Equation (6.4.11) (the other being less than unity), after neglecting terms of order ξ^2, gives

$$\omega_C^2 \approx 1 + 2\sqrt{3}\xi \tag{6.4.12}$$

From Equations (6.4.10) and (6.4.12), the critical value of A is given by

$$A_C^2 \approx \frac{16}{3\sqrt{3}}\xi \tag{6.4.13}$$

It should be noted that this critical value slightly differs from the one stemming from Equation (5.3.30). When combined with Equations (6.4.4b,c), this critical value is proportional to $\sqrt{\zeta/\gamma}$ with the coefficient 1.75, approximately, while this coefficient, according to Equation (5.3.30) is approximately 1.63.

Further, for small damping, using Equations (6.4.12) and (6.4.13) in Equation (6.4.5), the critical (minimum) value of f required to give a multivalued response is

$$f_{\min} \approx \frac{16}{3}\left(\frac{\xi^3}{\sqrt{3}}\right)^{1/2} \tag{6.4.14}$$

The maximum amount of damping beyond which no jump phenomenon occurs is, thus, given by

$$\xi_{\max} \approx \left(\frac{9\sqrt{3}f^2}{256}\right)^{1/3} \tag{6.4.15}$$

6.4.1.1 Stability analysis

In the previous section, it was mentioned that if there are three values for the amplitude at a particular excitation frequency, then the intermediate value is unstable. Consequently an unstable zone in the frequency-response plane occurs. This is shown as a shaded area in Figure 6.2. Stability analysis of the harmonic solution given by Equation (6.4.2) is now carried out. For this, the harmonic solution is perturbed by η and the time development of this disturbance is tracked. Thus, $y = Y\cos(\Omega t - \theta) + \eta$ is substituted into Equation (6.4.1), Equation (6.4.2) is then the assumed solution of Equation (6.4.1), and higher (than linear) terms in η are neglected, to give

$$\ddot{\eta} + 2\zeta\dot{\eta} + \left[\alpha + 3\gamma Y^2\cos^2(\Omega t - \theta)\right]\eta = 0$$

or

$$\ddot{\eta} + 2\zeta\dot{\eta} + \left(\left(\alpha + \frac{3}{2}\gamma Y^2\right) + \frac{3}{2}\gamma Y^2\cos 2(\Omega t - \theta)\right)\eta = 0 \tag{6.4.16}$$

Using $z = \Omega t - \theta$ and the parameters defined in Equations (6.4.4a–d), Equation (6.4.16) can be rewritten as

$$\frac{d^2\eta}{dz^2} + \frac{2\xi}{\omega}\frac{d\eta}{dz} + \left(\frac{1 + \frac{3}{2}A^2}{\omega^2} + \frac{3}{2}\frac{A^2}{\omega^2}\cos 2z \right)\eta = 0 \tag{6.4.17}$$

Comparing Equation (6.4.17) with the standard form of the Mathieu equation with a viscous term given by [7]

$$\frac{d^2\eta}{dz^2} + 2\mu\frac{d\eta}{dz} + (\delta + 2\varepsilon\cos 2z)\,\eta = 0 \tag{6.4.18}$$

the following variables can be defined

$$\mu = \frac{\xi}{\omega}, \quad \delta = \frac{1 + \frac{3}{2}A^2}{\omega^2}, \quad \varepsilon = \frac{3}{4}\frac{A^2}{\omega^2} \tag{6.4.19a–c}$$

It is known that the boundary curves of the primary unstable region of Equation (6.4.18) in the $\delta - \varepsilon$ plane are approximately given by [7]

$$\delta = 1 \pm \sqrt{\varepsilon^2 - 4\mu^2} + O(\varepsilon^2) \tag{6.4.20a}$$

or

$$(\delta - 1)^2 = \varepsilon^2 - 4\mu^2 \tag{6.4.20b}$$

Using the parameters given in Equations (6.4.19a–c), the boundary curves described by Equation (6.4.20b) can be transferred to the amplitude frequency $A - \omega$ a plane as

$$\left(\frac{1}{\omega^2} + \frac{3}{2}\frac{A^2}{\omega^2} - 1 \right)^2 = \frac{9}{16}\frac{A^4}{\omega^4} - 4\frac{\xi^2}{\omega^2}$$

which are shown in Figure 6.2. Solving the above quadratic in A^2, gives

$$A^2 = \frac{8}{9}(\omega^2 - 1) \pm \frac{4}{9}\sqrt{(\omega^2 - 1)^2 - 12\xi^2\omega^2} \tag{6.4.21}$$

Equations (6.4.21) and (6.4.9a,!b) are identical, hence demonstrating that the shaded area in Figure 6.2 is, in fact, the unstable region. For different values of ξ, using Equation (6.4.21) the primary unstable regions can be drawn as shown in Figure 6.3.

As expected, the extent of the unstable region shrinks with increasing damping. Also, the minimum amplitude required to cause instability increases with increasing damping.

Returning to the standard form of the Mathieu equation, with a viscous damping term, given by Equation (6.4.18), it should be noted that there also exists higher unstable regions in the $\delta - \varepsilon$ plane. The transition curves bounding the secondary unstable region are given by [7]

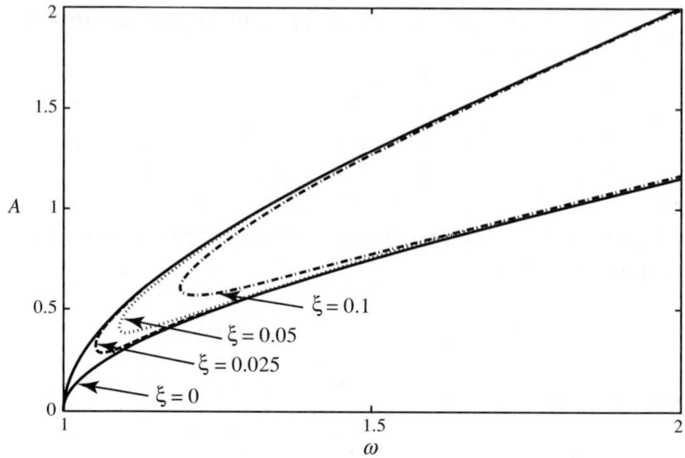

Figure 6.3 The primary unstable regions in the frequency-response plane for different values of viscous damping.

$$\delta = 4 + \left(\varepsilon^2/6\right) \pm \sqrt{(\varepsilon^4/16) - 16\mu^2} + O\left(\varepsilon^3\right) \tag{6.4.22a}$$

or

$$\left(\delta - 4 - \frac{\varepsilon^2}{6}\right)^2 = \frac{\varepsilon^4}{16} - 16\mu^2 \tag{6.4.22b}$$

Using Equations (6.4.19a–c) and (6.4.22b) a quartic equation in A^2 is obtained, which can be solved for given values of ω and ξ [8]. Taking the feasible roots, the boundary curves of the secondary unstable region in the amplitude–frequency plane can be obtained. Such secondary unstable regions, shown in Figure 6.4, are seen to be quite narrow. Regions of still higher-order instability exist, which are of progressively narrower extent.

It should be noted that the higher-order unstable regions imply that in narrow frequency ranges, even the unique amplitudes predicted using Equation (6.4.5) may be unstable. Furthermore, by neglecting higher-order terms in ε, the boundary curves obtained above, are accurate only in narrow frequency ranges. With the viscous damping term in the Mathieu equation, it is well known that the unstable regions in the δ–ε plane start only beyond a critical value of ε. This is evident from Equation (6.4.20a) and Equation (6.4.22a). This fact is reflected in the shifting of the unstable regions from the frequency axis in the frequency-response plane, as seen in Figures 6.3 and 6.4.

As stated earlier, the primary unstable region signifies that the intermediate value of the amplitude of the harmonic solution is unstable. This can also be shown by carrying out a linear stability analysis of the assumed harmonic form of the

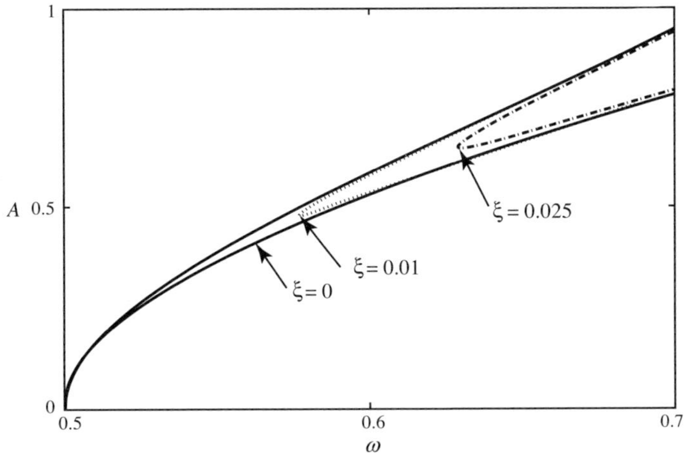

Figure 6.4 The secondary unstable regions in the frequency-response plane for different amounts of damping.

solution [7,9]. First, the solution of Equation (6.4.1) is written as

$$y = Y(t)\cos(\Omega\, t - \theta(t)) \tag{6.4.23}$$

Comparing this with Equation (6.4.2), it is observed that the same form of the harmonic solution is retained but now the amplitude and phase are assumed to vary with time. Assuming slow variation of these two quantities, the second time derivatives of these two quantities are neglected. Substituting Equation (6.4.23) into Equation (6.4.1) and equating the coefficients of $\cos\Omega\, t$ and $\sin\Omega\, t$ from both sides, the so-called slow flow equations or a two-dimensional flow are obtained

$$\begin{aligned} \dot{Y} &= g(Y,\theta), \\ \dot{\theta} &= h(Y,\theta) \end{aligned} \tag{6.4.24a,b}$$

The equilibrium points of this flow are the values of Y and θ for which $\dot{Y} = 0 = \dot{\theta}$ They correspond to the steady-state solution given by Equation (6.4.2). The phase plane $Y-\theta$ of this two-dimensional flow is known as the van der Pol plane. When the *phase portrait* of the flow given by Equation (6.4.24a,b) exhibits one equilibrium point, it is always a stable *focus* implying a stable harmonic amplitude. However, at some frequencies three equilibrium points are obtained; the one with the intermediate value of Y is a *saddle* while the other two are stable foci, as also discussed in Chapter 5, and shown in Figure 5.9. This confirms that the intermediate value of the amplitude is unstable and the other two are stable. This linear stability analysis cannot reveal the higher-order unstable regions obtained by using the Mathieu equation.

6.4.2 Harmonic solution for a softening system

Consider Equation (6.4.1) with $\alpha > 0$ and $\gamma < 0$. In this case, Equation (6.4.5) is replaced by

$$A^2 = \frac{f^2}{\left(4\xi^2\omega^2 + \left(\omega^2 - 1 + \frac{3}{4}A^2\right)^2\right)} \tag{6.4.25}$$

with

$$A^2 = -\frac{\gamma}{\alpha}Y^2, \quad f^2 = -\frac{\gamma}{\alpha^3}F^2 \tag{6.4.26a,b}$$

The *backbone curve* is given by

$$\omega^2 = 1 - \frac{3}{4}A^2 \tag{6.4.27}$$

Figure 6.5 shows the frequency-response curves for a given level of excitation with different values of damping. The backbone curve, leaning to the left, is shown by the dashed-dotted line. The stable and unstable values are shown by solid and dotted lines, respectively. It can be seen that that up to the value of damping $\xi^{(2)}$, the response curve has no jump-down frequency only a jump-up frequency, for the lower to the upper branch. Beyond this particular value of damping, the response curve splits into two parts and a jump-down in the response can be observed from the higher to the lower branch with increasing frequency. This has been referred to as an 'anomalous' jump in

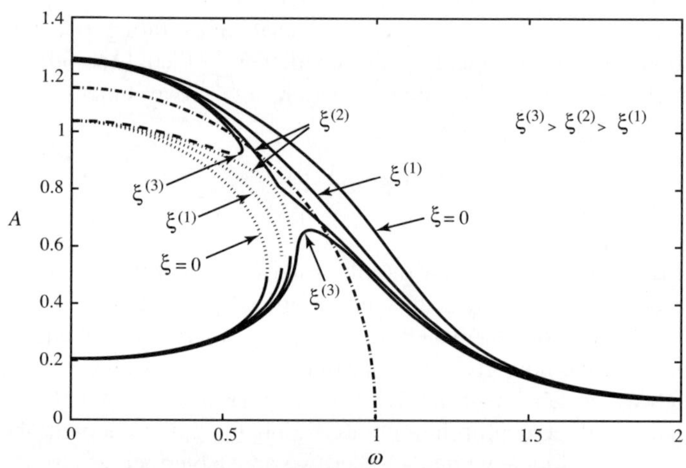

Figure 6.5 Frequency-response curves for a softening system with $f = 0.2$ and different amounts of viscous damping; stable (solid line); unstable (dotted line); backbone curve (dashed-dotted line).

the literature. Besides this anomalous jump, the normal jump phenomena with decreasing and increasing frequency persist up to some value of damping. As the damping is further increased, then, as observed in the hardening system, these normal jumps disappear but the anomalous jump persists. Of course, inclusion of superharmonics and subharmonics in the solution makes the response curve much more complicated. A series of *symmetry-breaking* and *period-doubling bifurcations* occur, as illustrated in Chapter 5. Such softening systems with viscous damping have been investigated extensively both numerically and experimentally [10–13]. In this chapter, the focus is first on the hardening system, which has been relatively less well explored. Thereafter, softening systems with nonlinear damping are discussed.

6.4.3 Superharmonic and subharmonic response

So far it has been assumed that if the excitation is harmonic then the response is also harmonic, having the same frequency as the excitation, given by Equation (6.4.2). However, this is true only approximately in some ranges of frequency. The periodic solution of Equation (6.4.1) can be sought involving only odd superharmonics of the form

$$y = Y_1\cos(\Omega\, t - \theta_1) + Y_3\cos(3\Omega\, t - \theta_3) + \cdots \tag{6.4.28}$$

It should be noted that Equation (6.4.1) has the symmetry $y \to -y$ as $t \to t + (\pi/\Omega)$. This implies that if $y(t)$ is a solution of Equation (6.4.1), then so is $-y[t + (\pi/\Omega)]$. These two solutions may or may not be distinct. The solution is symmetric if

$$y(t) = -y[t + (\pi/\Omega)] \tag{6.4.29}$$

It is clear that the solution given by Equation (6.4.28) maintains this symmetry. The closed *orbit* in *phase space*, depicting this periodic solution, is symmetric about both the y- and $\dot{y}$- axes. At a given frequency, as the level of excitation is increased, the symmetric solution given by Equation (6.4.28) loses its stability and generates even superharmonics [14,15]. With both odd and even superharmonics, the periodic solution of Equation (6.4.1) may be written as

$$y = \sum_{N=1,2,3,\ldots} Y_N \cos(N\Omega\, t - \theta_N) \tag{6.4.30}$$

In the presence of even superharmonics, $y(t) \neq -y[t + (\pi/\Omega)]$ and thus, the symmetry of the solution is lost. In such situations, dual solutions occur for different sets of initial conditions. The closed orbits corresponding to each of the dual solutions are not symmetric in phase space. However, the closed orbits of the dual solutions are mirror images of each other with mirrors placed along both $\dot{y}$ and y axes. In this way, the dual solutions together maintain symmetry. The appearance of these dual solutions is the result of a *symmetry-breaking bifurcation*. The time period of the solution is still the same as that of the excitation. Numerical results presented in a subsequent section illustrate such dual solutions.

Besides superharmonics, the response may also contain subharmonics of some specific order. For example, if the forcing frequency is about three times the linearised natural frequency, then the response contains a one-third subharmonic, i.e., a frequency one-third of that of the excitation. Detailed analytical treatment of the one-third subharmonic response and its stability analysis are available in textbooks on nonlinear vibrations [9,16]. For the one-third subharmonic to be excited, the damping must be less than some critical value depending on the system parameters and level of excitation. Three unstable and three stable solutions exist for this subharmonic. Linear stability analysis of the subharmonic solutions is carried out using slow flow equations and *phase portraits* in the van der Pol plane, mentioned earlier in the context of a harmonic solution. Each of the stable solutions has its own *basin of attraction* in the Van der Pol plane. In the presence of a one-third subharmonic, the time period of the solution is three times that of the excitation. Such solutions are referred to as period-3 solutions. In the low-frequency regime, with increasing excitation level, numerical results reveal period two solutions also, thereby confirming the presence of a half subharmonic.

6.4.4 Chaotic and other types of responses

For a hardening Duffing oscillator, with varying levels of excitation and damping, numerical simulations have revealed chaotic and other types of complicated responses. These include simultaneous coexistence of different types of *attractors*, i.e., the steady-state behaviour depends on the initial conditions. Each distinct attractor is associated with its own basin of attraction defined by the set of initial conditions. As well as periodic attractors having different periods, even periodic and *chaotic attractors* can coexist [17,18]. Besides the normal jump in the response, indeterminate or unsafe jumps have been reported [19]. Unpredictable response involving *Wada basin*s has also been found [20]. In the next subsection, some of the response features of a hardening Duffing oscillator, discussed in this and previous subsections, are illustrated through experimental and numerical results.

6.4.5 Experimental and numerical results

Following Pippard [21], an experimental setup consisting of a torsional oscillator, governed by the Duffing equation with hardening characteristics, was designed [22,23]. Figure 6.6 shows a permanent bar magnet of rectangular cross-section, which was held at the centre by two thin Mylar strips. The ends of these strips were clamped inside a fixed circular ring made of perspex. This constituted a torsional oscillator capable of oscillations about a vertical axis. The torsional stiffness of the system was obtained by measuring the applied static torque M_s and the resulting rotation θ. The plot of M_s (N mm) versus θ (rad), shown in Figure 6.7 behaves in accordance with

$$M_s(\theta) = k_1\,\theta + k_2\,\theta^3 \tag{6.4.31}$$

where $k_1 = 60.14$ N mm/rad, $k_2 = 62.46$ N mm/rad^3.

Figure 6.6 Experimental setup of a hardening Duffing torsional oscillator.

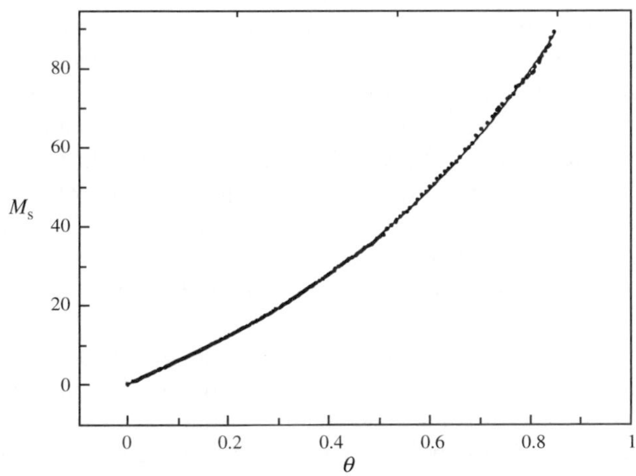

Figure 6.7 Experimental results of static torque versus angular rotation; experimental results (dots); fit of Equation (6.4.31) (solid line).

An electric coil was wound along the periphery of the circular ring. When a suitably amplified harmonic current from a signal generator was passed through this coil, the system was subjected to harmonic excitation. The resulting response was sensed by a small piezoelectric accelerometer. The output of this transducer was amplified and processed so that a voltage proportional to the rotational response was obtained. This voltage was treated as the rotation of the magnet to an arbitrary scale. The excitation torque was also not calibrated. Thus, without any quantitative estimate, only the qualitative nature of the response could be investigated. The damping present in the system was measured by the decay rate of free oscillation and was found to be independent of the amplitude. Hence, the damping could be satisfactorily modelled as viscous damping. The damping ratio was found to be 0.015. The centroidal moment of inertia J of the bar magnet was measured as 4.69e–05 kg m^2. Therefore, the linearised natural frequency of this system was $(1/2\pi)\sqrt{(k_1/J)} = 5.7$ Hz. The equation of motion of this oscillator under harmonic excitation could be written as

$$J\ddot{\theta} + C\dot{\theta} + k_1\,\theta + k_3\,\theta^3 = M\cos 2\pi\,\bar{\omega}\,t \tag{6.4.32}$$

where the excitation torque amplitude M was taken to be proportional to the amplitude of current flowing through the coil. The excitation frequency $\bar{\omega}$ (Hz) was controlled by the signal generator. Some experimental results obtained from this setup with varying excitation amplitude and frequency and numerically simulated results from Equation (6.4.32) are presented in this section. The numerical simulation was carried out with different sets of initial conditions after rendering Equation (6.4.32) into a suitable nondimensional form [23].

Figure 6.8 shows the peak-to-peak response θ_{pp}, obtained experimentally to an arbitrary scale, with three different levels of excitation. For each level of excitation, the response curves are plotted with both increasing and decreasing sequences of excitation frequency. The jump phenomenon discussed in Section 6.4.1 as well as in Section 5.3.1. is clearly discernible. The size of the *hysteresis* loop, as expected, is seen to increase with increasing level of excitation.

In the low-frequency regime with increasing level of excitation, higher-order unstable regions interact in a complicated manner. This is manifested in a nonsmooth variation of the peak-to-peak response in this frequency range with high excitation amplitude. Figure 6.9(a) clearly shows this phenomenon. The response also contained other harmonics that was confirmed by the frequency analysis of the signal [23]. Numerical results, presented in Figure 6.9b, also demonstrate the same phenomenon.

To investigate the one-third subharmonic response, numerical results were obtained for a rather high level of excitation at frequencies greater than 17 Hz, i.e., just above three times the linearised natural frequency. Figure 6.10 shows these results for three sets of initial conditions mentioned in the figure. At frequencies above 18 Hz, the higher response values in this figure correspond to period-three ($N = 3$) solutions and the lower ones to the nonresonance branch of harmonic solution ($N = 1$). With initial conditions $(0, 0)$, only the harmonic solution is obtained. However, with another two sets of initial conditions, period-three and harmonic

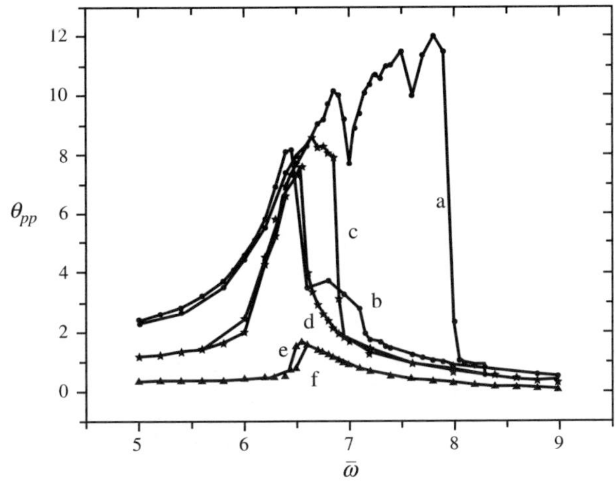

Figure 6.8 *Experimental frequency-response curve s for different levels of excitation ($M = M_3 > M_2 > M_1$) exhibiting jump phenomena with increasing and decreasing frequencies; curve a: $M = M_3, \bar{\omega}$ increasing; curve b: $M = M_3, \bar{\omega}$ decreasing; curve c: $M = M_2, \bar{\omega}$ increasing; curve d: $M = M_2, \bar{\omega}$ decreasing; curve e: $M = M_1, \bar{\omega}$ increasing; curve f: $M = M_1, \bar{\omega}$ decreasing.*

solutions are obtained at different frequencies. Sometimes, a small change in the frequency causes the response to jump back and forth between these two solutions. This is indicated by joining two such solutions given by the solid lines in Figure 6.10. Moreover, in a narrow frequency range around 18 Hz, even period-two solutions ($N = 2$) are also observed with initial conditions (1,1).

The time-series $\theta - t$ and power spectra $S_{\theta\theta} - \bar{\omega}$ of the response at three narrowly separated frequencies in this range, respectively, at 18.01 Hz, 18.02 Hz and 18.03 Hz obtained are shown in Figure 6.11. Figures 6.10 and 6.11 correspond to the same level of excitation. The power spectra clearly indicate $N = 1$, 2 and 3 solutions. Experimental results for the time series obtained at 18 Hz for three different time windows are shown in Figure 6.12. These are taken when the system is supposed to reach a steady state. But it can clearly be seen that the response keeps on changing without showing the expected steady-state behaviour. Of course, there was some drift in the frequency, however small, during the experiment. The system response was highly sensitive to the excitation frequency and thus prevented attainment of the steady state.

Experimental results obtained at 1.9 Hz with three different levels of excitation are shown in Figure 6.13. Both the time series and the corresponding *power spectrum* are shown in this figure. With the moderate (Figure 6.13(a)) and high (Figure 6.13(b)) excitation levels, the appearance of superharmonics and subharmonics is clearly seen. At the highest level of excitation (Figure 6.13(c)), all the lower frequencies extending down to zero appear and the response becomes nonperiodic or chaotic. Numerical

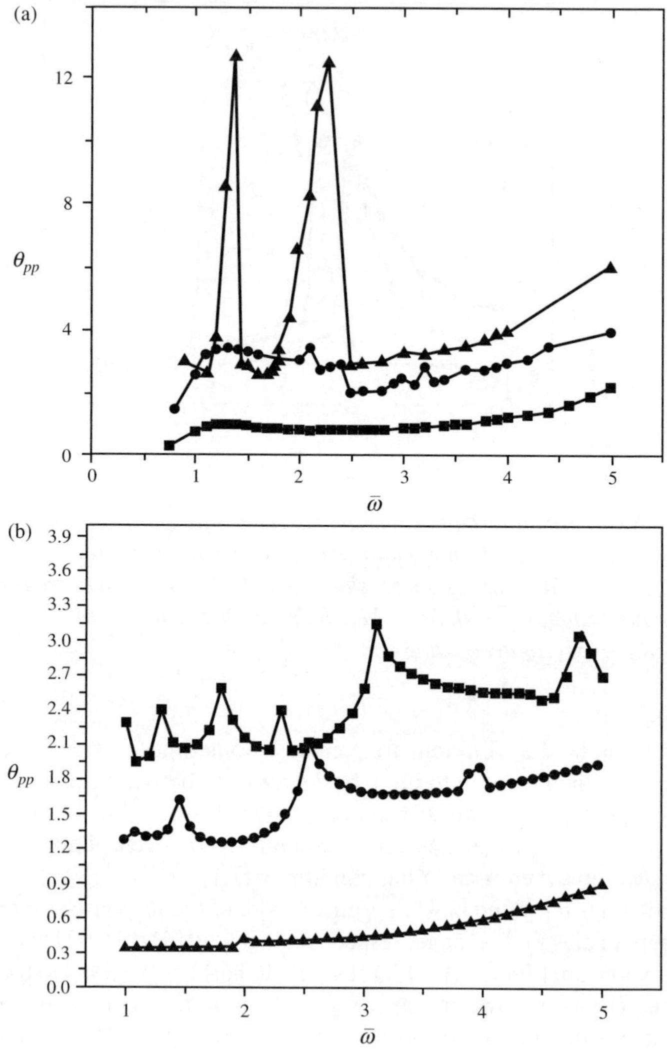

Figure 6.9 Peak-to-peak response in the low-frequency regime with different levels of excitation exhibiting nonsmooth variation: (a) experimental results; (b) numerical results.

results obtained at 1.8 Hz with increasing level of excitation are shown in Figure 6.14. The plots show the time series, phase paths or the *stroboscopic maps*. In this set, different types of solutions can be found that include a symmetric periodic solution (Figure 6.14(a)), dual solutions after symmetry-breaking (Figure 6.14(b)), period-three and period-two solutions which eventually turn chaotic (Figures 6.14(c) and (d)) with a slight change in the excitation level.

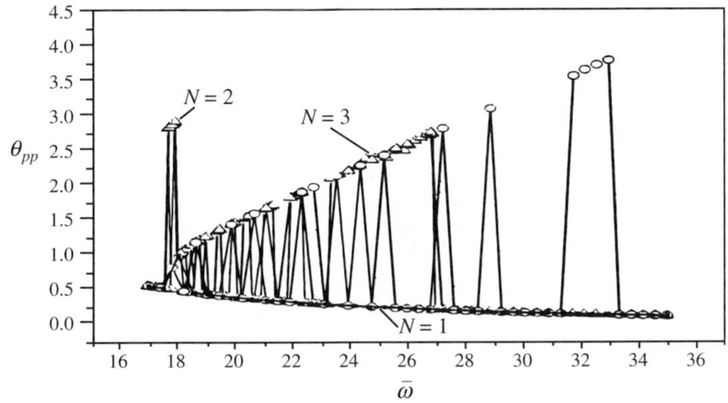

Figure 6.10 Numerically obtained frequency-response curves showing period-three, two and one solutions with sets of different initial conditions; △ (1,1); ○ (1,3); ● (0,0).

6.5 Nonlinear damping in a hardening system

Now consider Equation (6.3.1) for the case of $n \neq 1$. For the reasons already mentioned in Section 6.2.3, two cases, namely, $n > 0$ and $n = 0$ are discussed separately. Harmonic response and its stability analysis are discussed. Numerical results revealing other types of response are also included.

6.5.1 Harmonic solution

In this section the case $n > 0$ is considered, when the harmonic solution of Equation (6.3.1) can be obtained by considering the following equation similar to that with viscous damping

$$\ddot{y} + 2\zeta_{\text{eq}}\dot{y} + \alpha y + \gamma y^3 = F\cos\Omega t \tag{6.5.1}$$

where ζ_{eq} is given by Equation (6.2.7). Consequently, the amplitude parameter A can be obtained using Equation (6.4.5) as [24]

$$A^2 = \frac{f^2}{\left(4\xi_{\text{eq}}^2\omega^2 + \left(\omega^2 - 1 - \frac{3}{4}A^2\right)^2\right)} \tag{6.5.2}$$

where $\xi_{\text{eq}} = \beta_n \xi_n A^{n-1}\omega^{n-1}$, with

$$\xi_n = \frac{\zeta_n \alpha^{(2n-3)/2}}{\gamma^{(n-1)/2}} \tag{6.5.3}$$

Various quantities used in Equation (6.5.3) have been defined in Equations (6.2.7), (6.2.8) and (6.4.4a–d). However, it must be noted that unlike ζ in case of viscous

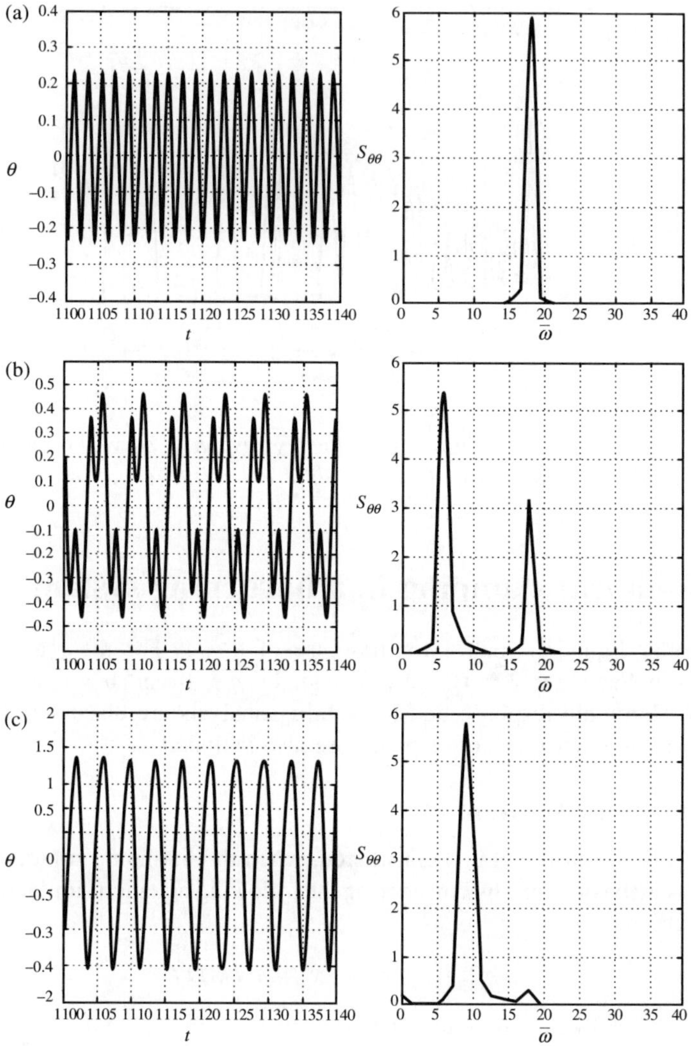

Figure 6.11 Numerically obtained time series and power spectrum of the response for three narrowly separated frequencies around 18 Hz with initial conditions (1,1)*; (a)* $\bar{\omega} = 18.03$ *Hz,* $N=1$*; (b)* $\bar{\omega} = 18.02$ *Hz,* $N=3$*; (c)* $\bar{\omega} = 18.01$ *Hz,* $N=2$.

damping, the equivalent viscous damping ratio ζ_{eq}, and also ξ_{eq} are not constant system parameters. Rather they depend on the amplitude of response, the frequency of excitation and the damping exponent n, as mentioned in Section 6.2.3. It has been shown [24,25] that for all values of n, the frequency-response curve leans to the right with jump frequencies for low values of damping as in the case $n=1$. The intermediate solution between the jump frequencies is unstable. The points on

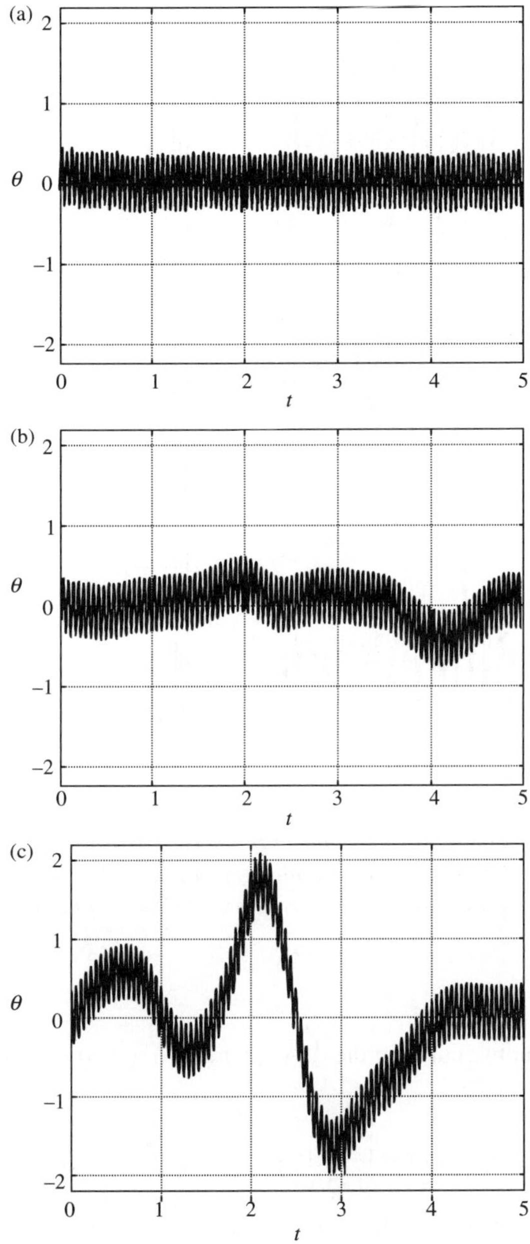

Figure 6.12 Experimentally obtained time series at 18 Hz showing three different time windows indicating unsteady behaviour after long interval: (a) negligible low-frequency content; (b) appearance of low-frequency content; (c) predominant low-frequency content.

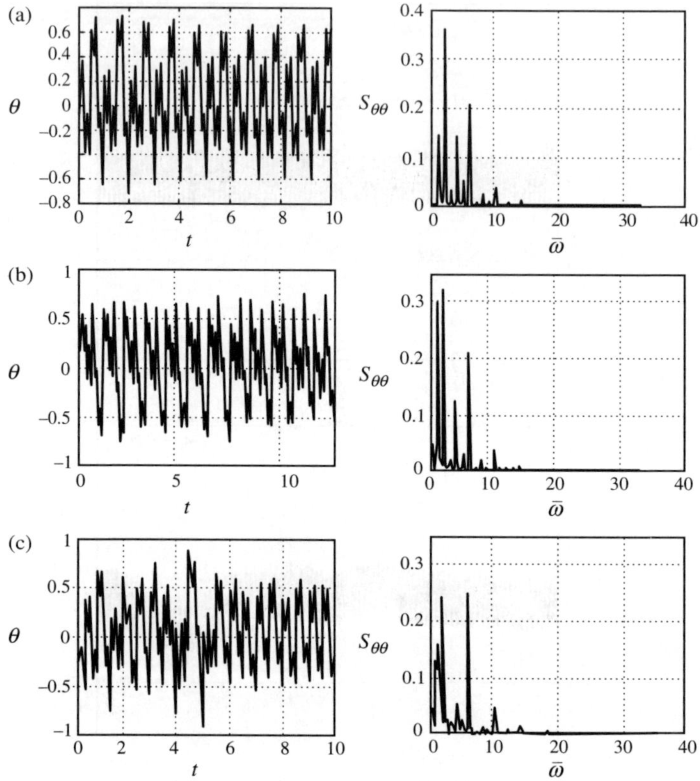

Figure 6.13 Experimentally obtained time series and power spectrum at 1.9 *Hz exhibiting superharmonics, subharmonics and eventually chaotic motion with increasing levels of excitation: (a) moderate excitation; (b) high excitation; (c) very high excitation.*

the frequency-response curve in the $A{-}\omega$ plane having vertical tangents are seen to satisfy

$$4n\xi_{\mathrm{eq}}^2\omega^2 + \left(\omega^2 - 1 - \frac{3}{4}A^2\right)^2 - \frac{3}{2}A^2\left(\omega^2 - 1 - \frac{3}{4}A^2\right) = 0 \qquad (6.5.4)$$

However, due to dependence of ξ_{eq} on A and ω, Equation (6.5.4) cannot be easily transferred to the $A{-}\omega$ plane for $n \neq 1$.

To study the variation of the excitation frequencies corresponding to the jump points with nonlinear damping, the Duffing equation without the linear stiffness force is considered. Moreover, besides the harmonic excitation of constant amplitude, the so-called frequency-squared excitation is also considered. In the latter case, the amplitude of excitation is proportional to the square of the excitation frequency as

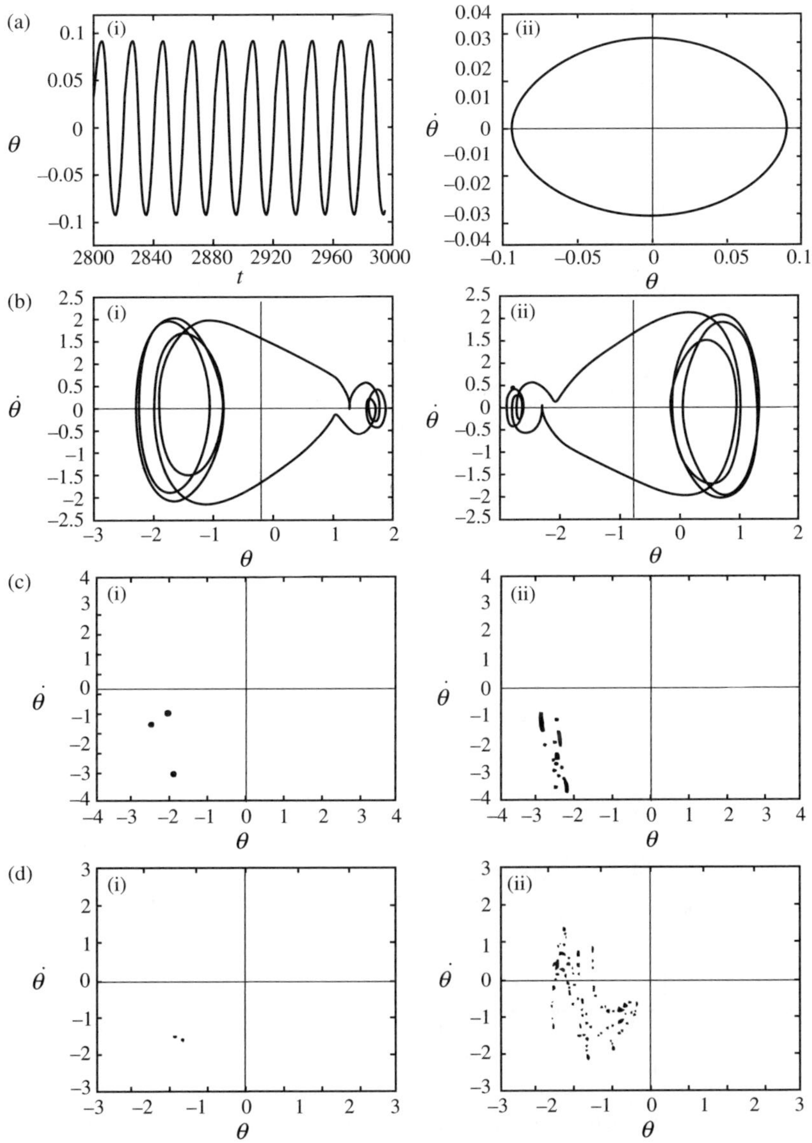

Figure 6.14 Numerically obtained response at 1.8 *Hz with increasing level of excitation; (a) period-one symmetric solution (i) time series, (ii) phase trajectory; (b) symmetry breaking and phase trajectories of dual solutions with initial conditions (i)* [1,1], *(ii)* [0,0]; *(c) stroboscopic maps of (i) period-three solution leading to (ii) chaotic motion; (d) stroboscopic maps of (i) period-two solution leading to (ii) chaotic motion.*

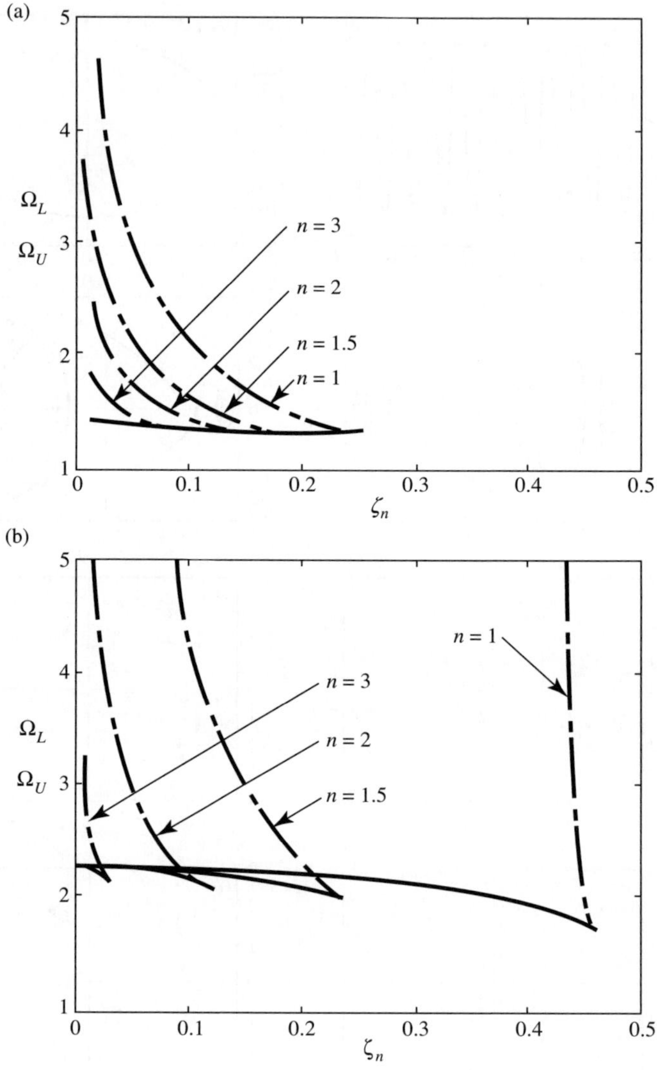

Figure 6.15 The variation of upper Ω_U and lower Ω_L jump point frequencies with damping parameters, Ω_L (solid line); Ω_U (dashed-dotted line): (a) for Equation (6.5.5); (b) for Equation (6.5.6).

provided by a rotary unbalance, a frequent source of excitation. Thus, the following two equations are considered:

$$\ddot{y} + 2\zeta_n \dot{y}|\dot{y}|^{n-1} + y^3 = \cos \Omega t \tag{6.5.5}$$

and

$$\ddot{y} + 2\zeta_n \dot{y}|\dot{y}|^{n-1} + y^3 = \Omega^2 \cos \Omega t \tag{6.5.6}$$

The responses in both cases, obtained through the equivalent viscous damping, show the usual jump phenomena. The variations of the jump frequencies, Ω_U and Ω_L, with the damping parameters are shown in Figures 6.15(a) and (b) for Equations (6.5.5) and (6.5.6), respectively. It is interesting to note that in both cases, the lower jump frequency is rather insensitive to the damping exponent n. The upper jump frequency, on the other hand, decreases sharply with increasing ζ_n for higher values of n. When the upper and lower frequencies become the same, the jump phenomenon vanishes. Thus, with increasing value of the damping exponent n, the jump phenomenon and consequently the primary instability of the harmonic solution disappear at a smaller value of the damping ratio ζ_n.

6.5.2 Stability analysis

The method for stability analysis of the harmonic solution outlined in Section 6.4.1.1 cannot be extended for any general value of n. However, it will be shown later in this section that for cubic damping, i.e., for $n=3$, a similar approach is possible. Before that, recast Equation (6.5.4) is rewritten as

$$\left(\frac{1+(3/2)A^2}{\omega^2}-1\right)^2=\left(\frac{3}{4}\frac{A^2}{\omega^2}\right)^2-\left(\frac{2\sqrt{n}\xi_{\text{eq}}}{\omega}\right)^2 \tag{6.5.7}$$

Similar to Equations (6.4.19a–c), the following variables can be defined

$$\mu_n=\frac{\sqrt{n}\,\xi_{\text{eq}}}{\omega},\qquad \delta=\frac{1+(3/2)A^2}{\omega^2},\qquad \varepsilon=\frac{3}{4}\frac{A^2}{\omega^2} \tag{6.5.8a–c}$$

when Equation (6.5.7) can be written as

$$(\delta-1)^2=\varepsilon^2-4\mu_n^2 \tag{6.5.9}$$

or

$$\delta=1\pm\sqrt{\varepsilon^2-4\mu_n^2} \tag{6.5.10}$$

For $n=1$, Equation (6.5.10) reduces to Equation (6.4.22a) defining the boundary curves of the primary unstable region. However, attention must be drawn to the appearance of the term $\sqrt{n}$ in the parameter μ_n and dependence of ξ_{eq} on A, ω and the damping exponent n. Considering a small disturbance η, for the case $n=3$, to the steady-state harmonic solution (as in Section 6.4.1.1), the following differential equation governs the growth of η [25]:

$$\frac{d^2\eta}{dz^2}+\left(\frac{4\mu_3}{\sqrt{3}}-\frac{4\mu_3}{\sqrt{3}}\cos 2z\right)\frac{d\eta}{dz}+(\delta+2\varepsilon\cos 2z)\eta=0 \tag{6.5.11}$$

where $z=\Omega\,t-\theta$.

It should be noted that $\eta=0, d\eta/dz=0$ is an equilibrium point of the two-dimensional flow given by Equation (6.5.11). The instability of this equilibrium solution implies growth of the small disturbance η. According to *Floquet theory* [7],

η is periodic with time period π or 2π along the transition curves, in the $\delta-\varepsilon$ plane, separating the stable and unstable solutions. First, setting $\mu_3 = \varepsilon\hat{\mu}_3$ and treating ε as the perturbation parameter the Lindstedt–Poincaré method is used to obtain the periodic solutions. To this end, the periodic solution and the value of δ along the transition curve are expanded as

$$\eta(z,\varepsilon) = \eta_0(z) + \varepsilon\eta_1(z) + \varepsilon^2\eta_2(z) + \cdots \tag{6.5.12}$$

and

$$\delta(\varepsilon) = \delta_0 + \varepsilon\delta_1 + \varepsilon^2\delta_2 + \cdots \tag{6.5.13}$$

Now, for η to be periodic, $\eta_0,\ \eta_1,\ \eta_2, \ldots$ all have to be periodic. Assuming the period to be 2π, the boundary curves of the primary unstable region are found to be [25]

$$\delta_0 = 1, \quad \delta_1 = \pm\sqrt{1 - 4\hat{\mu}_3^2} \tag{6.5.14a,b}$$

Using Equations (6.5.13) and (6.5.14a,b), gives

$$\delta = 1 \pm \sqrt{\varepsilon^2 - 4\mu_3^2} + O(\varepsilon^2) \tag{6.5.15}$$

Comparing Equation (6.5.15) with Equation (6.5.10), which is same as Equation (6.5.4), it can be concluded that Equation (6.5.10) gives the boundaries of the primary unstable region for both $n = 1$ and $n = 3$. As mentioned earlier, for a general value of n such analytical treatment is not possible. However, it has been shown that Equation (6.5.10) can be used for all values of n in the range $0 < n \leq 3$ [25]. This has been done by numerical verification of the results obtained in this way with those obtained by other approximate method using the Van der Pol plane as outlined in reference [26]. For example, with $n = 2, f = 1.0$, $\omega = 2$ and $\xi_2 = 0.025$, three roots of the amplitude parameter A are obtained as $A_1 = 0.34339$, $A_2 = 1.847$ and $A_3 = 2.1022$. Figure 6.16 shows the primary unstable regions obtained from Equation (6.5.10). The values of μ_2, δ and ε are calculated for the three values of A and the corresponding points are shown by a cross in Figure 6.16. It can be seen that the intermediate value of A_2 is unstable, whereas the lowest value of A_1 and the highest value of A_3 are stable. This approximate method, of course, cannot be applied for $n = 0$.

6.5.3 Chaotic motion

The chaotic motion in the presence of nonlinear damping has been investigated in [27,28]. Here again the linear term has been omitted as in Equation (6.5.5). To study both symmetric and asymmetric systems, the following equation is considered:

$$\ddot{y} + 2\zeta_n\,\dot{y}|\dot{y}|^{n-1} + y^3 = F_0 + F\cos\Omega\,t \tag{6.5.16}$$

The presence of the term F_0 makes the system asymmetric, i.e., with $F_0 = 0$, the system becomes symmetric. It has been found that the bifurcation structure associated with the resonance of the system remains unaffected by the value of the damping

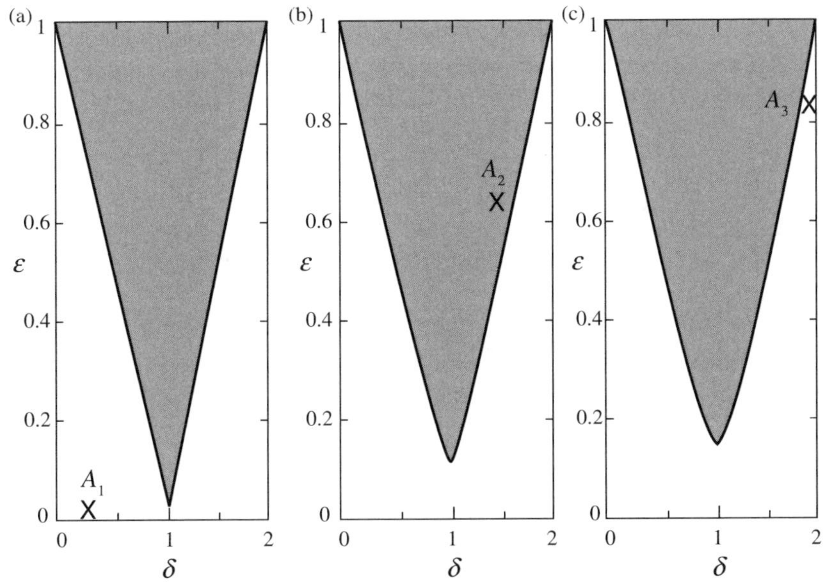

Figure 6.16 Primary unstable region in the δ–ε plane from Equation (6.5.10) for $n=2$ with: (a) $A=A_1$; (b) $A=A_2$; (c) $A=A_3$.

exponent n. However, the critical values of the parameters for the onset of different types of bifurcations depend on the values of both ζ_n and n. Numerical simulation is carried out at a particular frequency with increasing level of excitation.

As shown in Chapter 5, for the symmetric system, the symmetry-breaking bifurcation occurs before the period-doubling bifurcation, but it is important to emphasise here that it takes place irrespective of the value of n.

Figures 6.17(a)–(c) show the variation of the critical values of F, labelled by F_{cr}, for the onset of different bifurcations, with changing damping ratio for $n=1$, 2 and 3. These results are obtained for $\Omega=1$. It can be seen that the basic structure of bifurcation is independent of the value of n. For all values of n, a minimum value of ζ_n is necessary for symmetry-breaking. This minimum value decreases with increasing value of n. Furthermore, with increasing n, the cascade of period-doubling bifurcation ceases at lower values of ζ_n. For the asymmetric system, the period doubling occurs at a much lower value of F as compared to the symmetric system.

At $\Omega=1$, with $n=2$ and $\zeta_2=0.025$, chaotic solutions are obtained at $F=6$ for the symmetric system as shown in Figure 6.18. Such chaotic solutions at the same value of forcing persist with a little asymmetry $F_0=0.03$. However, with this small asymmetry, another chaotic attractor following a period-doubling route is observed for a very low value of excitation $F=0.08$ [28]. This chaotic attractor is obtained with a different set of initial conditions and has rather small values of maximum

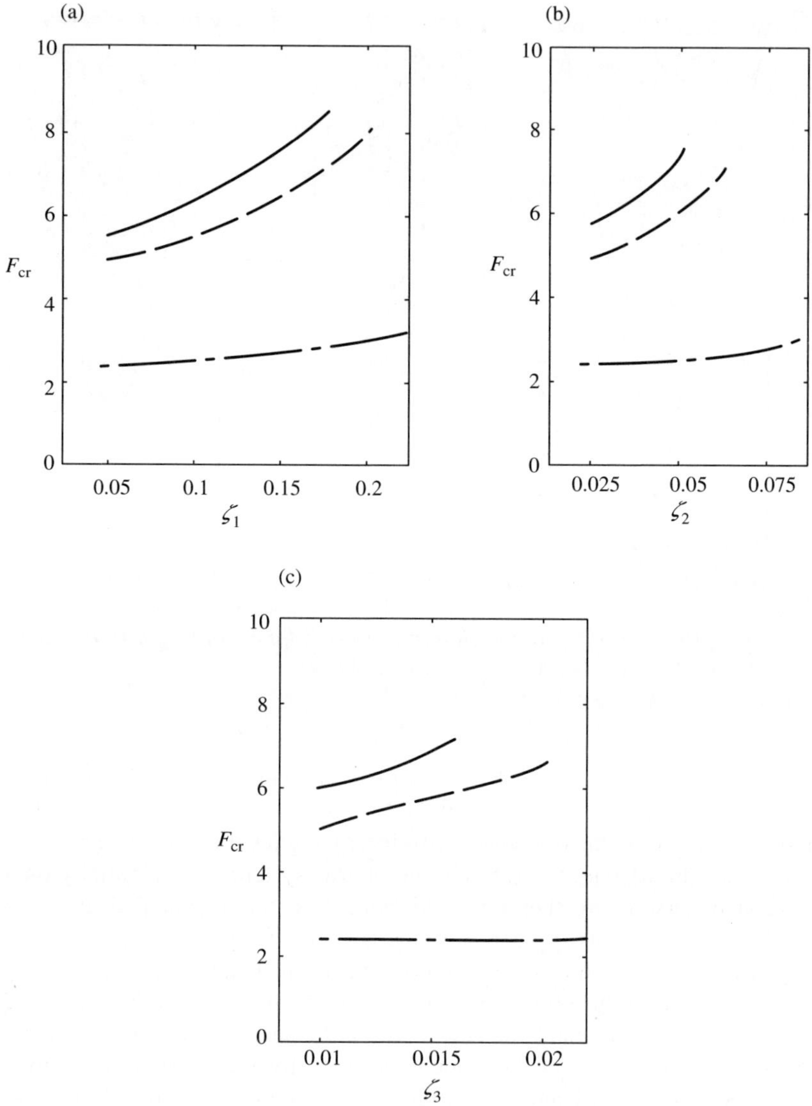

Figure 6.17 Variation of critical excitation level for onset of typical bifurcations with damping parameters; symmetry breaking (dashed-dotted line); period-two (dashed line); period-four (solid line); (a) $n = 1$; (b) $n = 2$; $n = 3$.

displacement as compared to the one obtained with $F = 6$. Figure 6.19 shows the stroboscopic maps indicating a sharp transition to *chaos* following a period-three solution for the symmetric system. This route and the critical values are found to be insensitive to the presence of small asymmetry.

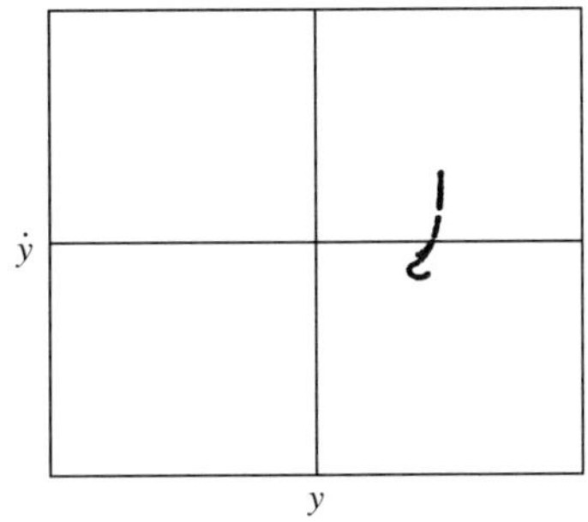

Figure 6.18 Stroboscopic map at the onset of chaotic solution with $F = 6$.

6.5.4 Coulomb damping

Some interesting features in the response of the hardening Duffing oscillator with Coulomb damping $n = 0$ have been reported in [29]. Again, the linear term in the restoring force is omitted and a frequency-squared harmonic excitation is considered. Accordingly the following equation of motion is considered:

$$\ddot{y} + 2\zeta_0 \frac{\dot{y}}{|\dot{y}|} + y^3 = \Omega^2 \cos \Omega\, t \tag{6.5.17}$$

which can also be written as

$$\ddot{y} + 2\zeta_0 \operatorname{sgn}(\dot{y}) + y^3 = \Omega^2 \cos \Omega\, t \tag{6.5.18}$$

Substituting the assumed harmonic solution $y = A\cos(\Omega\, t - \theta)$, in the equation of motion and then following either the method of harmonic balance or the method of equivalent linearisation [9], one gets the following amplitude–frequency equation:

$$k_1^2 A^6 - 2k_1 \Omega^2 A^4 + \Omega^4 A^2 + \left(4k_2^2 \zeta_0^2 - \Omega^4\right) = 0 \tag{6.5.19}$$

with $k_1 = 3/4$ and $k_2 = 4/\pi$.

For the oscillator with Coulomb damping, no motion across the damper is possible unless the external force overcomes the friction force. The frequency at which motion across the damper just starts is called the *break-loose frequency* [30]. For the excitation under consideration, the break-loose frequency Ω_b, is obtained, by equating the exciting force amplitude with the magnitude of the maximum friction force, as

$$\Omega_b = \sqrt{2\zeta_0} \tag{6.5.20}$$

When the term within the parenthesis in Equation (6.5.19) is equated to zero, one value of the amplitude A turns out to be zero. In this way an approximate value of the break-loose frequency is determined to be

$$\Omega_b \approx \sqrt{2k_2\zeta_0} \tag{6.5.21}$$

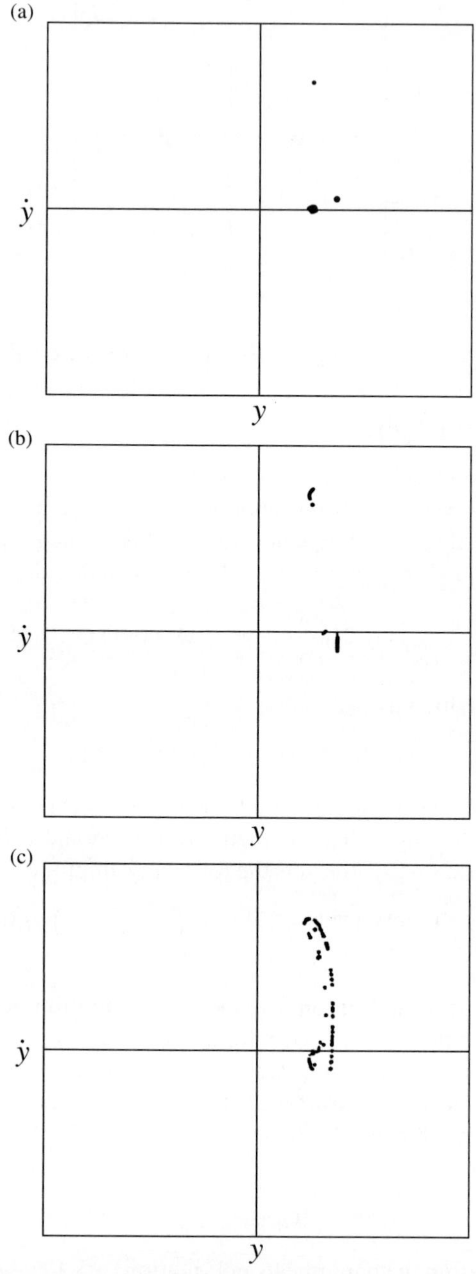

Figure 6.19 Stroboscopic maps showing transition to chaos from a period-three solution: (a) period-three solution, $F = 9.5$; (b) prechaotic solution, $F = 10$; (c) chaotic solution, $F = 10.5$.

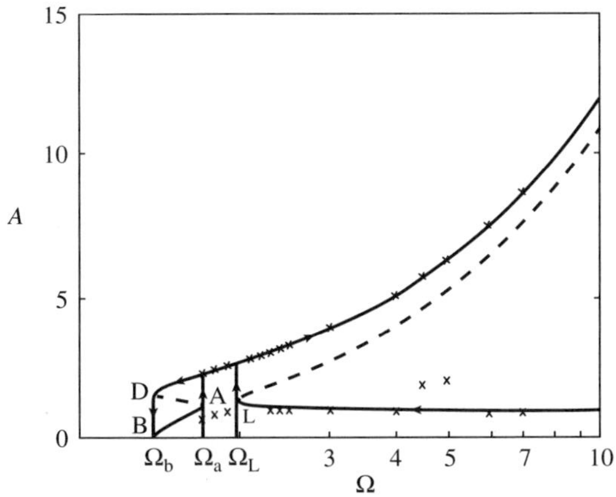

Figure 6.20 Frequency-response curve with Coulomb damping showing various jumps; $\zeta_0 = 0.75$; stable (solid line); unstable (dashed line); numerical results (crosses).

Comparing Equations (6.5.20) and (6.5.21), it can be noted that an error of about 11% in the break-loose frequency is caused by the assumed continuous harmonic movement of the oscillator. It may be concluded that Equation (6.5.20) should not be used for $\Omega < \Omega_b$, when the amplitude is zero. In fact, for Ω just greater than Ω_b, there exists rather long intervals of stick-slip motion. In this regime, special numerical treatment is necessary to obtain the response accurately [31]. For Ω greater than Ω_b, it is assumed that the amplitude given by Equation (6.5.19) is sufficiently accurate.

Solving for A from Equation (6.5.19), one or three real roots are obtained. Figure 6.20 shows the frequency-response curve for $\zeta_0 = 0.75$. In this figure, the point B corresponds to the break-loose frequency.

The stable solutions are indicated by the solid lines and the unstable solution by the dashed line. The stability analysis is carried out following the method outlined in reference [26]. A linear system with Coulomb damping, if treated by equivalent viscous damping, behaves like an undamped system at resonance, i.e., the amplitude at resonance tends to infinity. The nonlinear system also does not exhibit any upper jump frequency as can be seen in Figure 6.20. The normal jump occurs at the point L, with frequency Ω_L, from the nonresonance to the resonance branch. In the low-frequency regime, with increasing frequency a jump in the response occurs at $\Omega = \Omega_a$ from the branch BA to the resonance branch. This jump is referred to as an anomalous jump. From the resonance branch, with decreasing frequency another jump can be observed from the point D to A at the break-loose frequency Ω_b.

The critical frequencies, Ω_L, Ω_a and Ω_b can be obtained from the conditions of repeated roots of Equation (6.5.19), which is cubic in A^2. First, setting $x = A^2$ Equation (6.5.19) can be rewritten as

$$x^3 + a_1 x^2 + a_2 x + a_3 = 0 \tag{6.5.22}$$

with

$$a_1 = -\frac{2\Omega^2}{k_1}, \quad a_2 = \frac{\Omega^4}{k_1^2}, \quad a_3 = \frac{4k_2^2\zeta_0^2 - \Omega^4}{k_1^2} \tag{6.5.23a–c}$$

It is easy to see that at the break-loose frequency, when $a_3 = 0$, one value of the amplitude is zero and the other two roots of A^2 are the same as $a_1^2 = 4a_2$. For a general treatment, first Equation (6.5.22) is reduced to the standard form by putting $x = z-(a_1/3)$. Thus, Equation (6.5.22) is rewritten as

$$z^3 + b_1\, z + b_2 = 0 \tag{6.5.24}$$

where

$$b_1 = -\frac{a_1^2}{3} + a_2, \quad b_2 = \frac{2a_1^3}{27} - \frac{a_1 a_2}{3} + a_3 \tag{6.5.25a,b}$$

Applying the condition of repeated roots for Equation (6.5.24) [32], gives

$$\frac{b_2^2}{4} + \frac{b_1^3}{27} = 0 \tag{6.5.26}$$

Using Equations (6.5.25a,b) and (6.5.23a–c) in Equation (6.5.26), gives (besides $a_3 = 0$, which has already been considered)

$$4k_2^2\zeta_0^2 - \Omega^4 + \frac{4}{27k_1}\,\Omega^6 = 0 \tag{6.5.27}$$

For a given value of ζ_0 with $\Omega > \Omega_b$, Equation (6.5.27) can be solved to give the other two critical frequencies, respectively, Ω_a and Ω_L. Figure 6.21 shows the variation of the three critical frequencies with ζ_0. The value of ζ_0 at which Ω_a and Ω_L coincide, which is depicted by point M in Figure 6.21, is indicated by $(\zeta_0)_m$.

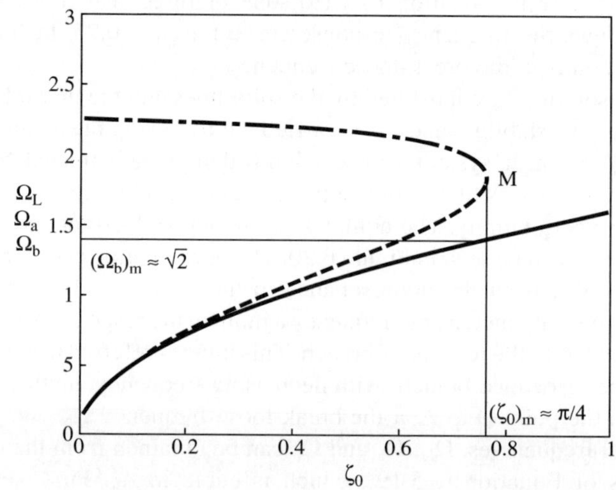

Figure 6.21 Variation of critical frequencies with Coulomb damping ratio ζ_0; Ω_b (solid line); Ω_a (dashed line); Ω_L(dashed-dotted line).

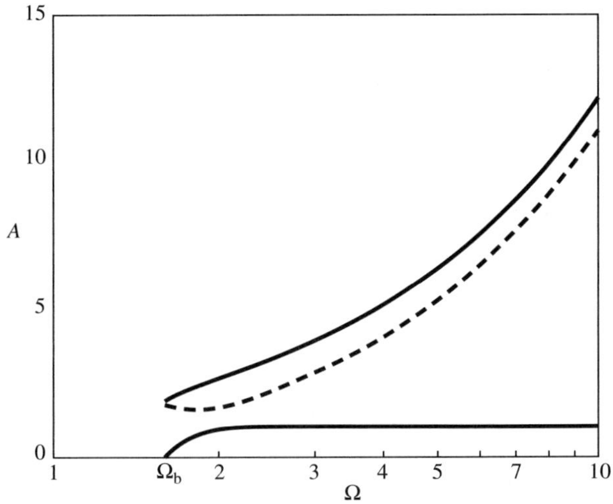

Figure 6.22 Frequency-response curve after disappearance of jump phenomena with $\zeta_0 = 1 > (\zeta_0)_m$; stable (solid line); unstable (dashed line).

This value can again be obtained from the condition of repeated roots of Equation (6.5.27), which is a cubic equation in Ω^2. Thus, substituting for k_1 and k_2 gives

$$(\zeta_0)_m = \frac{3^{5/2}}{2^6}\pi \approx \frac{\pi}{4} \tag{6.5.28}$$

with the corresponding break-loose frequency

$$(\Omega_b)_m = \left(\frac{8}{\pi}(\zeta_0)_m\right)^{1/2} \approx \sqrt{2} \tag{6.5.29}$$

For $\zeta_0 > (\zeta_0)_m$, both the anomalous jump and the usual jump phenomena disappear. The frequency-response curve appears as shown in Figure 6.22, which has been obtained with $\zeta_0 = 1$.

Numerical simulation results for $\zeta_0 = 0.75$ are also shown in Figure 6.20 and are indicated by crosses. As usual, multiple steady-state solutions are obtained with different sets of initial conditions. The discrepancy with the analytical results around $\Omega = 5$ is due to the presence of a one-third subharmonic in the response not accounted for in the analytical treatment. Such period-three solutions were confirmed by the stroboscopic maps of the numerically obtained response. As already mentioned, the response at frequencies in the neighbourhood of the break-loose frequency can be obtained accurately only with special treatment.

In some systems, the damping needs to be modelled by a combination of viscous and Coulomb damping. The equation of motion in such circumstances can

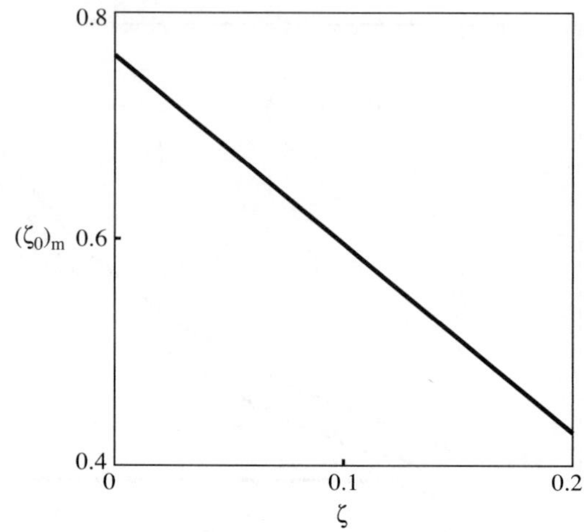

Figure 6.23 Variation of critical Coulomb damping ratio $(\zeta_0)_m$ in the presence of viscous damping.

be written as

$$\ddot{y} + 2\zeta\dot{y} + 2\zeta_0 \mathrm{sgn}(\dot{y}) + y^3 = \Omega^2 \cos\Omega\, t \tag{6.5.30}$$

For this case also, the value of $(\zeta_0)_m$ at which the frequencies Ω_a and Ω_L coincide can be obtained as before. The variation of $(\zeta_0)_m$ in the presence of ζ is shown in Figure 6.23. Surprisingly, the value of $(\zeta_0)_m$ decreases almost linearly with increasing ζ.

Numerical results [28] indicate that chaotic motion occurs via symmetry breaking at $\Omega = 0.2$ with $\zeta = 0.01$ in the absence of Coulomb damping. At a still lower value of Ω, a different intermittency *route to chaos* is also observed. With the same value of viscous damping, if a small amount of Coulomb damping is added with $\zeta_0 = 0.005$, then symmetry breaking occurs at $\Omega = 0.4$. However, no further bifurcation leading to chaos is seen. Thus, it can be concluded that addition of a small amount of Coulomb damping can quench some chaotic oscillations in a viscously damped, hardening Duffing oscillator.

6.6 Nonlinear damping in a softening system

In this section, a softening system with velocity-dependent nonlinear damping is discussed. Accordingly the following equation of motion is considered [33]:

$$\ddot{y} + 2\zeta_n\, \dot{y}|\dot{y}|^{n-1} + y - \varepsilon\, y^3 = F\cos\Omega\, t, \quad \varepsilon > 0, n > 0 \tag{6.6.1}$$

First, the effects of nonlinear damping on the harmonic response and bifurcation set are illustrated.

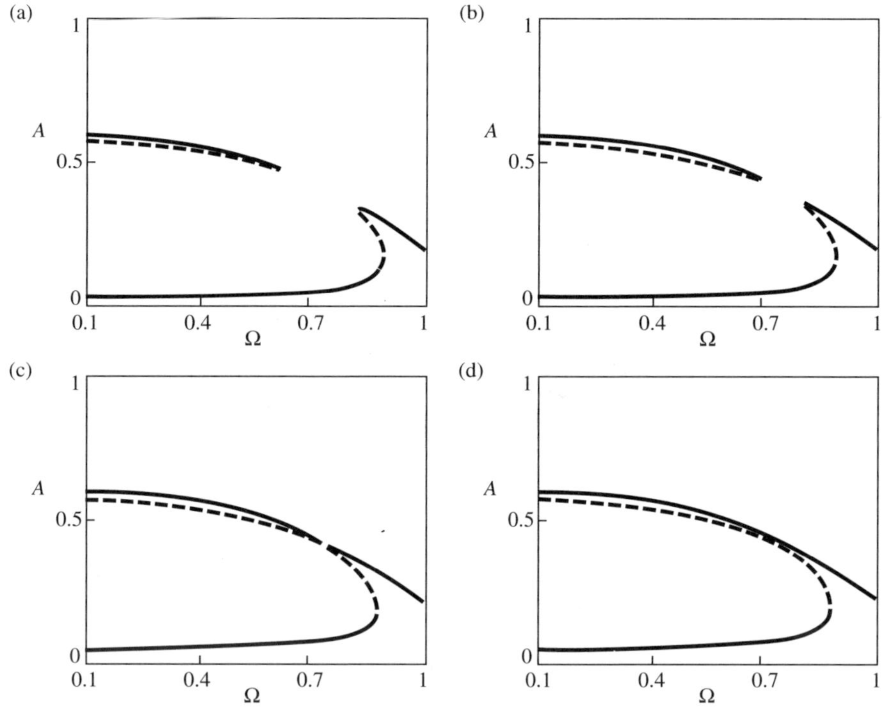

Figure 6.24 Frequency-response curves with increasing levels of excitation with $\zeta_2 = 0.2, \varepsilon = 4$; stable (solid line); unstable (dashed line); (a) $F = 0.025$; (b) $F = 0.027$; (c) $F = 0.028$; (d) $F = 0.029$.

Substituting the assumed harmonic solution$y = A\cos(\Omega t - \theta)$ into Equation (6.6.1) and using the method of harmonic balance gives the amplitude–frequency equation

$$A^2\left(1-\Omega^2\right)^2 + \frac{9}{16}\varepsilon^2 A^6 + \frac{3}{2}\left(\Omega^2 - 1\right)\varepsilon A^4 + \left(2\zeta_n A^n \Omega^n \beta_n\right)^2 = F^2 \tag{6.6.2}$$

It should be mentioned that Equation (6.6.2) is obtained by using the concept of equivalent viscous damping outlined in Section 6.2.3 and β_n is defined by Equation (6.2.8).

Typical frequency-response curves for $n = 2$ with different levels of excitation and a given value of ζ_2 are shown in Figures 6.24(a)–(d). It should be noted that apart from the usual jumps, an anomalous jump also occurs with increasing frequency in the low-frequency regime. With increasing level of excitation, finally the two separate branches of the frequency-response curve merge (Figure 6.24(d)) and only the lower jump point exists. This behaviour is similar to what is observed with viscous damping and is related to the symmetry-breaking bifurcation.

To obtain the bifurcation set depicting the unstable region in the $F-\Omega$ plane, the following are defined

$$\rho = 1 - \Omega^2, \quad \sigma_n = 2\zeta_n\beta_n\Omega^n \tag{6.6.3a,b}$$

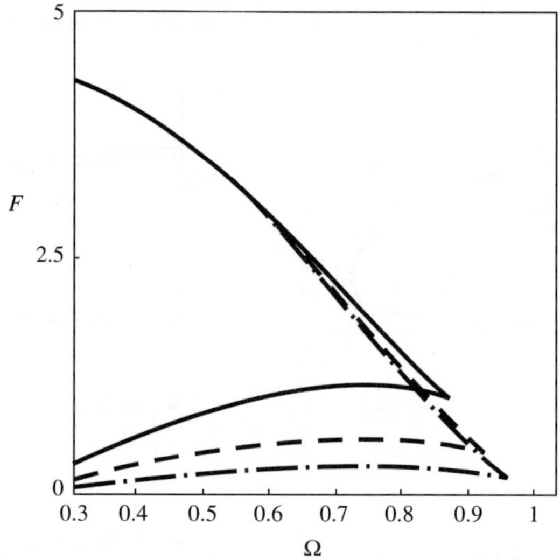

Figure 6.25 Bifurcation sets with different values of linear viscous damping; $\varepsilon = 0.01$; $\zeta_1 = 0.01$ *(solid line);* $\zeta_1 = 0.05$ *(dashed line);* $\zeta_1 = 0.025$ *(dashed-dotted line).*

and Equation (6.6.2) is rewritten as

$$9\varepsilon^2 A^6 - 24\rho\varepsilon A^4 + 16\rho^2 A^2 + 16(\sigma_n A^n)^2 - 16F^2 = 0 \tag{6.6.4}$$

Equation (6.6.4) remains cubic in A^2 for $n = 1$, 2 and 3. For these values of n, the conditions for repeated roots can be used to get the loci of the jump points in the F–Ω plane. The bifurcation sets obtained in this manner for $n = 1$ and 2 are shown in Figures 6.25 and 6.26, respectively.

The value of the damping exponent n does not seem to alter the structure of the bifurcation set. It may be mentioned that the softening Duffing oscillator acquires a negative stiffness beyond some value of the amplitude and the results obtained so far do not remain valid beyond that amplitude.

For a driven pendulum (with periodic potential), the *Melnikov criterion* can be used to determine the critical value of Frequired for the onset of chaos [34]. It has been pointed out that the Melnikov criterion depicts the occurrence of fractal basin boundaries rather than the onset of chaos [35]. Following reference [34], for the driven pendulum, the critical value of forcing for $n = 1$ is given by,

$$F_{cr} = \frac{8\zeta}{\pi}\cosh(\pi\,\Omega/2) \tag{6.6.5}$$

and for $n = 3$ by,

$$F_{cr} = \frac{64}{3\pi}\zeta_3\cosh(\pi\,\Omega/2) \tag{6.6.6}$$

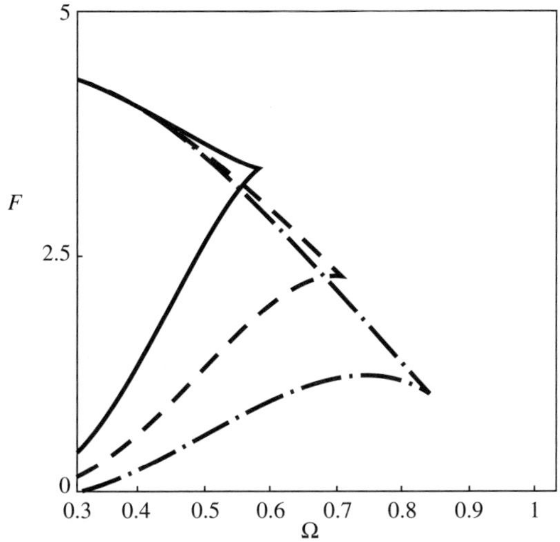

Figure 6.26 Bifurcation sets with different values of quadratic damping; $\varepsilon = 0.01$*;* $\zeta_1 = 0.01$ *(solid line);* $\zeta_1 = 0.05$ *(dashed line);* $\zeta_1 = 0.025$ *(dashed-dotted line).*

As mentioned in Section 6.2.1, a softening system behaves like a pendulum. From Equations (6.6.5) and (6.6.6) it can be concluded that for a softening system, the critical value of forcing necessary is higher for $n = 3$ than that for $n = 1$ with the same value of ζ_n. In both cases this critical value depends linearly on the value of the damping ratio.

6.7 Nonlinear damping in a double-well potential oscillator

In this section the equation of motion with the negative linear-positive nonlinear stiffness, i.e. with the double-well potential is considered:

$$\ddot{y} + 2\zeta_n \dot{y}|\dot{y}|^{n-1} + \varepsilon\left(-y + y^3\right) = F\cos\Omega\, t, \quad \varepsilon > 0, \quad n > 0 \tag{6.7.1}$$

Following the procedure outlined in reference [36], an analytical criterion for the period-doubling bifurcation for $n = 1$ and 3 is determined. First, Equation (6.7.1) is rewritten for $n = 1$ and 3 separately, as

$$\ddot{y} + 2\zeta\dot{y} + \varepsilon\left(-y + y^3\right) = F\cos\Omega\, t \tag{6.7.2}$$

and

$$\ddot{y} + 2\zeta_3\dot{y}^3 + \varepsilon\left(-y + y^3\right) = F\cos\Omega\, t \tag{6.7.3}$$

For ease of comparison with previously published results [36], the parameter ε is set to 0.5. For this asymmetric system, substituting assumed harmonic solution of the form

$$y = A_0 + A\cos(\Omega t - \theta) \tag{6.7.4}$$

in Equation (6.7.2) and carrying out the harmonic balance, gives

$$A_0^2 = \frac{2-3A^2}{2} \tag{6.7.5}$$

and

$$\left(\frac{15}{8}\right)^2 A^6 - \frac{15}{4}(1-\Omega^2)A^4 + \left(1-\Omega^2\right)^2 A^2 + 4\zeta^2 A^2\Omega^2 - F^2 = 0 \tag{6.7.6}$$

Similarly, for Equation (6.7.3), Equation (6.7.5) is determined, but Equation (6.7.6) is replaced by

$$\left(\frac{15}{8}\right)^2 A^6 - \frac{15}{4}(1-\Omega^2)A^4 + \left(1-\Omega^2\right)^2 A^2 + \frac{9}{4}\zeta_3^2 A^6\Omega^6 - F^2 = 0 \tag{6.7.7}$$

To analyse the stability of the harmonic solution obtained, a small disturbance η is considered and the procedure outlined in Section 6.4.1.1 is followed. In this way, for $n = 1$,

$$\ddot{\eta} + 2\zeta\dot{\eta} + (\lambda_0 + \lambda_1\cos\Omega t + \lambda_2\cos 2\Omega t)\eta = 0 \tag{6.7.8}$$

and for $n = 3$:

$$\ddot{\eta} + \zeta_3 A^2\Omega^2(2-3\cos 2\Omega t)\dot{\eta} + (\lambda_0 + \lambda_1\cos\Omega t + \lambda_2\cos 2\Omega t)\eta = 0 \tag{6.7.9}$$

where

$$\lambda_0 = \frac{3}{2}A_0^2 + \frac{3}{4}A^2 - \frac{1}{2}, \quad \lambda_1 = 3A_0A, \quad \lambda_2 = \frac{3A^2}{4} \tag{6.7.10a–c}$$

For the onset of period-doubling bifurcation, it is assumed that

$$\eta = \eta_{1/2}\cos\left(\frac{\Omega t}{2} - \varphi\right) \tag{6.7.11}$$

Substituting Equation (6.7.11) in Equations (6.7.8) and (6.7.9) and carrying out harmonic balance, for nontrivial solution, some conditions must be satisfied depending on the value of n. For $n = 1$, it is found that

$$\left(\lambda_0 - \frac{\Omega^2}{4}\right)^2 + \zeta^2\Omega^2 - \frac{\lambda_1^2}{4} = 0 \tag{6.7.12}$$

and for $n = 3$:

$$\left(\lambda_0 - \frac{\Omega^2}{4}\right)^2 + \frac{9}{4}\zeta_3^2 A^4\Omega^6 - \frac{\lambda_1^2}{4} = 0 \tag{6.7.13}$$

Equation (6.7.12) in conjunction with Equation (6.7.6) determines the boundary of the period-doubling bifurcation in the F–Ω plane for $n = 1$. Likewise

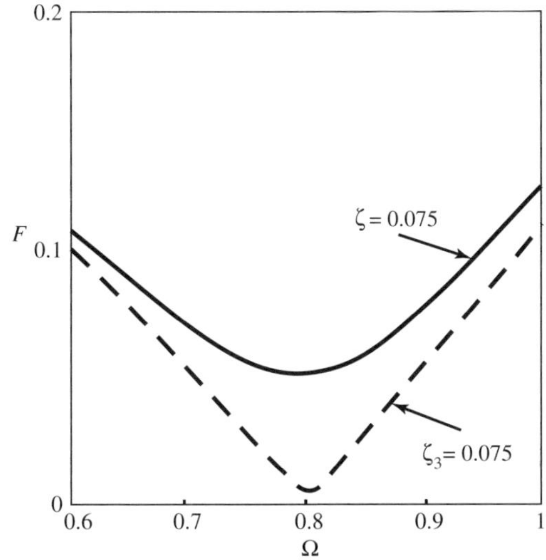

Figure 6.27 Critical values of excitation for period-doubling with the same value of linear and cubic damping.

Equations (6.7.13) and (6.7.7) together determine the similar boundary for $n=3$. These boundaries are shown in Figure 6.27 for the same value of ζ and ζ_3. The results for $n=1$ are in complete agreement with that reported in reference [36]. It may be noted that with cubic damping, the critical forcing level required to generate the period-doubling bifurcation is less than that with linear viscous damping. This result is just opposite of what had been observed in Figure 6.17 for a hardening system.

Numerical simulations for Equation (6.7.1) with $\varepsilon = 1$, $\Omega = 1.5$ are carried out for $n=1$, 2 and 3 to obtain the critical forcing level required for period-doubling bifurcation. These results are shown in Table 6.1. The trend of decreasing the critical excitation level with increasing value of the damping exponent, observed in Figure 6.27, is also confirmed in Table 6.1. Numerical results [28] suggest that both period-doubling and intermittency routes to chaotic motion exist independently of the value of the damping exponent. The stroboscopic maps of a chaotic attractor for $n=1$, 2 and 3 are shown in Figure 6.28.

Table 6.1 Critical values of F required to generate period two solution with $\varepsilon = 1$, $\Omega = 1.5$.

ζ_n	$n=1$	$n=2$	$n=3$
0.10	0.41	0.39	0.38
0.15	0.44	0.40	0.39
0.20	0.49	0.42	0.40
0.25	0.54	0.45	0.41

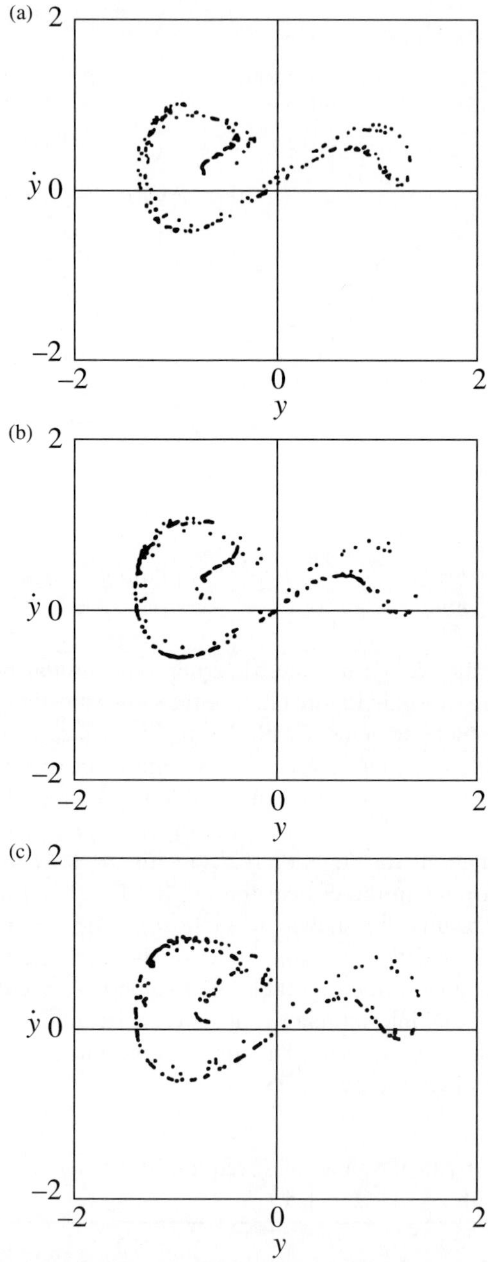

Figure 6.28 Stroboscopic maps of chaotic attractors with different damping exponents for $\varepsilon = 1$, $\Omega = 1$, $F = 0.4$, $\zeta_n = 0.125$*: (a)* $n = 1$*; (b)* $n = 2$*; (c)* $n = 3$.

It can been seen that the structure of the chaotic attractor, including the *fractal dimension*, is rather insensitive to the value of n.

Following reference [37], the Melnikov criterion for the critical value of forcing, with $\varepsilon = 0.5$, is obtained for $n = 1$:

$$F_{\text{cr}} = \frac{8\zeta}{3\sqrt{2}\pi\,\Omega}\cosh(\pi\,\Omega/2) \tag{6.7.14}$$

and for $n = 3$:

$$F_{\text{cr}} = \frac{32\zeta_3}{35\sqrt{2}\pi\,\Omega}\cosh(\pi\,\Omega/2) \tag{6.7.15}$$

From Equations (6.7.14) and (6.7.15) it is concluded that just like the period-doubling bifurcation, the chaotic motion also occurs at a lower critical value of the forcing F_{cr} for cubic damping as compared to linear viscous damping with $\zeta = \zeta_3$. Thus, the nature of variation of F_{cr} with the damping exponent in this case is just opposite to what was mentioned for a pendulum at the end of Section 6.6.

6.8 Summary

The response of three types of Duffing oscillators; softening, hardening and with negative linear-positive nonlinear stiffness, i.e., with a double-well potential, have been considered under harmonic excitation with different damping mechanisms. Harmonic response and its stability analysis have been presented. The analytical treatment was similar for linear viscous and nonlinear cubic damping. Experimental results, exhibiting chaotic motion for the hardening Duffing oscillator with linear viscous damping, were included. Velocity-dependent nonlinear damping has been treated in terms of equivalent viscous damping. The break-loose frequency in the presence of Coulomb damping was introduced. Besides the normal jump phenomena, an anomalous jump in the response was observed. The effects of nonlinear damping on the onset of period-doubling bifurcation and chaotic motion have also been discussed. The value of the damping exponent has been found to change various critical parameters differently, depending on the type of the oscillator. Some numerical results have been presented to confirm the analytical and experimental results or to give some results, at least, when no such results were available.

Acknowledgments

Thanks are due to Dr. B. Ravindra and Dr. A. Banerjee for their help in the preparation of this chapter.

References

[1] A.B. Pippard, *The Physics of Vibration*, Cambridge University Press, Cambridge, 1989.

[2] J.E. Ruzicka, T. F. Derby, *Influence of Damping in Vibration Isolation*, The Shock and Vibration Center, Naval Research Laboratory, Washington, D.C., 1971.

[3] W. Szemplinska–Stupnicka, J. Bajkowski, The 1/2 subharmonic resonance and its transition to chaotic motion in a nonlinear oscillator. *International Journal of Non-Linear Mechanics*, 21, 401–419, 1986.

[4] J.B. Roberts, P.D. Spanos, *Random Vibration and Statistical Linearization*, John Wiley, Chichester, 1990.

[5] A.K. Mallik, V. Kher, M. Puri, H. Hatwal, On the modelling of non-linear elastomeric vibration isolators. *Journal of Sound and Vibration*, 219, 239–253, 1999.

[6] K. Magnus, *Vibrations*, Blackie, London, 1965.

[7] A.H. Nayfeh, D.T. Mook, *Nonlinear Oscillations*, John Wiley & Sons, New York, 1979.

[8] A.K. Mallik, B. Ravindra, Nonlinear Mechanical Vibrations, Chapter 2 in *Modelling of Complex Systems* – eds. J.K. Bhattacharjeeand A.K. Mallik, Narosa Publishing House, New Delhi, 1997.

[9] D.W. Jordan, P. Smith, *Nonlinear Ordinary Differential Equations*, Clarendon Press, Oxford, 2nd edn, 1987.

[10] A.Y.T. Leung, Rigorous studies of a Duffing oscillator. *Journal of Sound and Vibration*, 149, 147–149, 1991.

[11] W. Szemplinska-Stupnicka, Bifurcations of harmonic solution leading to chaotic motion in the softening type Duffing's oscillator. *International Journal of Non-Linear Mechanics*, 23, 257–277, 1988.

[12] J. Miles, Resonance and symmetry-breaking for the pendulum. *Physica D*, 31, 252–268, 1988.

[13] G.L. Baker, J.P. Gollub, *Chaotic Dynamics: An Introduction*, Cambridge University Press, Cambridge, 2nd edn, 1996.

[14] S. Novac, R.G. Frehlich, Transition to chaos in the Duffing oscillator. *Physical Review A*, 26, 3660–3663, 1982.

[15] K. Wiesenfeld, Calculation of period doubling in a Josephson junction. *Physical Review A*, 29, 2102–2109, 1984.

[16] C. Hayashi, *Nonlinear Oscillations in Physical Systems*, Mc-Graw Hill Book Co., New York, 1964.

[17] Y. Ueda, Randomly transitional phenomena in the system governed by Duffing's equation. *Journal of Statistical Physics*, 20, 181–196, 1979.

[18] J.M.T. Thompson, H. B. Stewart, *Nonlinear Dynamics and Chaos, Geometric Methods for Engineers and Scientists*, John Wiley & Sons, New York, 1986.

[19] M.S. Soliman, Non-linear vibrations of hardening systems:chaotic dynamics and unpredictable jumps to and from resonance. *Journal of Sound and Vibration*, 207, 383–392, 1997.

[20] J. Aguirre, M.A.F. Sanjuan, Unpredictable behaviour in the Duffing oscillator: Wada basins. *Physica D*, 171, 41–51, 2002.

[21] A.B. Pippard, *Response and Stability: An Introduction to the Physical Theory*, Cambridge University Press, Cambridge, 1985.

[22] N.S. Patil, A.K. Mallik, Experimental investigation of the response of harmonically excited hard Duffing oscillator. *Pramana (Journal of Physics)*, 68, 99–104, 2007.

[23] N.S. Patil,*Forced vibrations of a hard Duffing oscillator – numerical and experimental investigations*, M. Tech Thesis, Indian Institute of Technology Kanpur, 2004.

[24] B. Ravindra, A.K. Mallik, Performance of nonlinear vibration isolators under harmonic excitation. *Journal of Sound and Vibration*, 170, 325–337, 1994.

[25] B. Ravindra, A.K. Mallik, Stability analysis of a non-linearly damped Duffing oscillator. *Journal of Sound and Vibration*, 171, 708–716, 1994.

[26] D. Capecchi, F. Vestroni, Periodic response of a class of hysteretic oscillators. *International Journal of Non-Linear Mechanics*, 25, 309–317, 1990.

[27] B. Ravindra, A.K. Mallik, Chaotic response of a harmonically excited mass on an isolator with non-linear stiffness and damping characteristics. *Journal of Sound and Vibration*, 182, 345–353, 1995.

[28] B. Ravindra, *Harmonic and Chaotic Responses with Nonlinear Vibration Isolators*, Ph.D Thesis, Indian institute of Technology Kanpur, 1994.

[29] B. Ravindra, A.K. Mallik, Hard Duffing-type vibration isolator with combined Coulomb and viscous damping. *International Journal of Non-Linear Mechanics*, 28, 427–440, 1993.

[30] A.K. Mallik, *Principles of Vibration Control*, Affiliated East-West Press (P) Ltd., New Delhi, 1990.

[31] T. K. Pratt, R. Williams, Non-linear analysis of stick-slip motion. *Journal of Sound and Vibration*, 74, 531–542, 1981.

[32] A.G. Kurosh, *Algebraic Equations of Arbitrary Degrees*, Mir Publishers, Moscow, 1977.

[33] B. Ravindra, A.K. Mallik, Role of nonlinear damping in soft Duffing oscillators. *Physical Review E*, 49, 4950–4954, 1994.

[34] P. Holmes, Poincaré, celestial mechanics, dynamical-systems theory and chaos. *Physics Reports (Review section of Physics Letters)*, 193, 137–163, 1990.

[35] F. C. Moon, *Chaotic and Fractal Dynamics*, Wiley, New York, 1992.

[36] K.R. Asfar, K.K. Masoud, On the period-doubling bifurcations in the Duffing's oscillator with negative linear stiffness. *ASME Journal of Vibration and Acoustics*, 114, 489–494, 1992.

[37] J. Guckenheimer, P.J. Holmes, *Nonlinear Oscillations, Dynamical Systems and Bifurcation of Vector Fields*, Springer, New York, 1983.

7

Forced harmonic vibration in a Duffing oscillator with negative linear stiffness and linear viscous damping

Stefano Lenci[1] **and Giuseppe Rega**[2]
[1]*Department of Architecture, Buildings and Structures, Polytechnic University of Marche, Italy*
[2]*Department of Structural and Geotechnical Engineering, Sapienza University of Rome, Italy*

7.1 Introduction

In this chapter systems described by the following nondimensional equation are studied

$$\ddot{y} + 2\zeta\,\dot{y} - \alpha\,y \pm \gamma\,y^3 = F\cos\Omega t \qquad (7.1.1)$$

where y, ζ, α, γ, F and Ω are, respectively, the displacement, damping ratio, linear stiffness parameter, nonlinear stiffness parameter, excitation amplitude and excitation (circular) frequency; overdots denote derivatives with respect to nondimensional time t. Two points, which restrict the analysis, are first highlighted. The first is that if there is a minus sign in front of γ, the steady-state dynamics are trivial, since all solutions exponentially tend to infinity. So, the case with a plus sign is considered. The second is

The Duffing Equation: Nonlinear Oscillators and their Behaviour, First Edition. Edited by Ivana Kovacic and Michael J. Brennan.

that by an appropriate rescaling ($\bar{y} = \alpha y$, $\bar{t} = t\sqrt{\alpha}$, $\bar{\zeta} = \zeta/\sqrt{\alpha}$, $\bar{\gamma} = \gamma/\alpha^3$ and $\bar{\Omega} = \Omega/\sqrt{\alpha}$), Equation (7.1.1) can be rewritten in an equivalent form

$$\ddot{y} + 2\zeta\,\dot{y} - y + \gamma\,y^3 = F\cos\Omega t \tag{7.1.2}$$

where the overbars are omitted for simplicity.

Other variations of Equation (7.1.1) have been considered in the literature, for example with $\alpha = 1$ and $\gamma = 2$, or with $\alpha = 1$ and $\gamma = 1$. Also, equation

$$\ddot{z} + 2\zeta\,\dot{z} + z \pm \frac{3}{2}z^2 + \frac{1}{2}z^3 = F\cos\Omega t \tag{7.1.3}$$

has been considered, which can be obtained from Equation (7.1.1) by setting $\alpha = 1/2$ and $\gamma = 1/2$ and by shifting the origin of the coordinates by introducing the new variable $z = y \mp 1$.

In this chapter Equation (7.1.2) is taken as the representative equation.

The first point to be addressed is terminology, i.e., how Equation (7.1.2) is referred to in the scientific literature and how to distinguish it from the other Duffing-type equations.

From the dynamical systems point of view, Equation (7.1.2), with no damping or excitation, has three equilibrium points, as shown in Chapter 3. It has a symmetric two-well potential and a *phase portrait* with two symmetric *homoclinic orbits* (see, Figure 3.9(b)). Thus, focusing on any of these features, alternative – and equivalent – labels of the equation are *two/double/twin-well* Duffing oscillator, as discussed in Chapter 3. However, it has also been referred to in the literature as a symmetric Duffing oscillator, unstable Duffing oscillator, and even – at the same time – hardening-unstable or softening Duffing oscillator. This apparent contradiction is because hardening-unstable refers to the overall bounded cross-well behaviour of the system whereas softening refers to the local, possibly unbounded, in-well behaviour. In this chapter the term twin-well Duffing oscillator (or equation) will be used.

This chapter is organised as follows. After an extensive literature review (Section 7.2), the dynamics of the *conservative system* is discussed together with the main effects of damping and excitation, which make the system nonconservative and are considered as (small) perturbations of the conservative system (Section 7.3). Classical nonlinear periodic oscillations, and in particular the principal nonlinear *resonance*, are investigated in Section 7.4. Then, a transition to complex response, including *chaos*, is systematically discussed in Section 7.5, based on computer simulation results and analytical predictions. Finally, two nonclassical topics, namely the *control of chaos* and system *dynamical integrity*, are investigated (Section 7.6) before the chapter is closed with a summary.

7.2 Literature survey

The twin-well Duffing oscillator can basically exhibit two distinct types of periodic steady-state oscillations, namely (i) in-well, small *orbit* dynamics, where the system state remains within the *potential well* centred at a stable equilibrium point and

(ii) cross-well, large orbit dynamics, whose trajectory surrounds the three equilibrium points (see Chapter 3 for more details). In both cases, under external excitation, periodic oscillations can evolve to steady in-well or *cross-well chaotic motions*, as a control parameter is changed.

The occurrence of irregular motion consisting of random-like *jumps* from oscillations around one stable equilibrium point to oscillations around the other equilibrium point was first observed experimentally by Tseng and Dugundji in 1971 [1]. They called the motion *snap-through oscillations*, but did not pay too much attention to the phenomenon.

Awareness of the occurrence and meaning of *strange attractor* occurred at the end of the 1970s, when the twin-well equation was derived as a mathematical model for describing large transverse deflections of a buckled beam using a Galerkin-based single-mode approximation ([2]; see also [3]). It also occurred in the Ueda oscillator as discussed in Chapter 1. Relevant theoretical and numerical results were also verified by physical experiments [4,5]. The experimental apparatus consisted of a cantilever beam interacting with two magnets (the so-called Moon beam, which is shown in Figure 2.8). The beam snapped from one static equilibrium point to the other in a chaotic manner.

Since the earlier work by Holmes and Moon and the following work [6–9] making use of sophisticated dynamical systems tools, the twin-well potential system – which is also used to describe plasma oscillations [10] – has become one of the archetypal systems in nonlinear and chaotic dynamics. It has been used as a means of understanding features of chaotic dynamics and testing. It has also been used for comparing and validating methods and tools for the analysis of regular and *chaotic oscillations*, and for their control.

Bifurcation of *periodic motions* to *subharmonics* had already been addressed using second-order averaging [11]. Much later, Yagasaki [12] used second-order averaging to analyse bifurcation behaviour of the twin-well oscillator at various resonances, when comparing classical asymptotic methods and the *Melnikov method* [13]. Thereafter, in an effort to understand the effect of system parameters on the onset of chaos and to complement previous computer simulation results [14,15], the harmonic balance technique coupled with the continuation scheme was used in [16] to track the various branches of the solution to the Duffing equation. A complete picture of solution trajectories over a range of loading conditions was achieved along with a few previously undetected strange attractors arising along these trajectories.

On the experimental side, physical realisations of the twin-well Duffing oscillator included a rollercoaster-type mechanical system whereby a small cart rolling on a double-well potential energy shaped track was contrived to mimic the behaviour of the equation [17]. Details of the design of the rolling track are given in [18]. It was used to illustrate such nonlinear dynamical features as competing steady-state *attractor*, *hysteresis*, sensitivity to initial conditions, subharmonic oscillations and chaos.

In developing a successful experimental technique (the stochastic interrogation method) for determining the evolution of *basins of attraction* and *homoclinic bifurcation*, Cusumano and Kimble [19] considered a driven twin-well magnetomechanical oscillator similar to the one used by Moon and Holmes [4,5]. However, it was stiffened slightly to force the system to behave as a single degree-of-freedom even during *transient motion*. Later, the stochastic interrogation technique was used to determine basins of attraction, global transient behaviour and indeterminate bifurcation of the experimental rollercoaster [20,21]. It was also used to determine the transient behaviour near a *saddle-node bifurcation* in an electronic circuit that behaved like the twin-well Duffing equation [22].

More recently, the twin-well oscillator is being used: (i) for the global analysis and evolution of catastrophic and explosive *crises* under harmonic excitation and noise, using the generalised cell mapping with digraph method [23]; (ii) for the mathematical classification of homoclinic tangencies in multiharmonically perturbed systems using the order of zeros in the Melnikov function [24]; (iii) as the simplest type of smooth nonlinear oscillator with nonideal driving and amplitude constraints, which is useful for modelling more complex engineering applications [25]; (iv) as the reference oscillator for studying homoclinic orbits, bifurcation and chaos in a system with fractional order deflection [26].

Overall, the list of references related to nonregular phenomena in the twin-well Duffing system is quite large and the reader is referred to both a few general books on chaos [27–32] and a series of papers by various authors aimed at (i) numerically characterising in detail the features involved in the transition to in-well or, mostly, *cross-well chaotic response* and/or (ii) constructing approximate criteria to estimate critical system parameter values for which a periodic motion leaves the potential well and turns into chaotic cross-well motion.

The specific research achievements on these two general matters are discussed chronologically, and this is followed by a summary of contributions and results on the control of the nonlinear dynamics of the twin-well Duffing equation.

7.2.1 Former numerical studies and approximate criteria for chaos

Early studies in the 1980s clarified several aspects concerned with the system global response such as the occurrence of fractal basin boundaries and their relation with the necessary *Melnikov criterion* for chaos (also renamed Holmes–Melnikov, after [2]), based on the existence of homoclinic orbits [33]. This gives the condition for the intersection of *stable* and *unstable manifolds* associated with the saddle point of the *Poincaré map*, or the evaluation of the *fractal dimension* of the Duffing–Holmes twin-well potential strange attractor [34]. It is based on both numerical solutions and experimental data from chaotic vibrations of the buckled beam.

Features of the local and global dynamic response of the twin-well oscillator as obtained by numerical simulations, sometimes complemented by the comparison with experiments, were also discussed in [14,15,35]. The determination of the minimum threshold force for the occurrence of chaos in a cantilever beam was addressed in [36] both numerically and experimentally, by including up to three modes of the beam within a Galerkin procedure. It was shown that if the value of damping was chosen carefully, better estimates could be obtained, compared to those provided by the approximate threshold criteria of Holmes–Melnikov [2] and Moon [5,29], which were based on a single-mode approximation for the buckled beam.

The criteria for chaos received a great deal of attention in the 1980s and early 1990s, and was examined by analytical, numerical and experimental methods [27,29,37,38]. The first semianalytical criterion for the imminence of *cross-well chaos* in the twin-well Duffing oscillator was proposed by Moon [5,29]. The criterion is when the maximum velocity of the in-well periodic motion is near the maximum velocity on the *separatrix*. Later, Melnikov's idea of *homoclinic tangling* of *invariant manifolds* of the hilltop saddle point and the concept of fractal basin boundaries were used as the mathematically rigorous criteria for chaos to occur. The Melnikov criterion ([13], see Section 7.5.2) can be interpreted as giving the critical parameter values for which an erosion of basin boundaries (see Section 7.6.2) can be observed. This results in the possibility of transient chaos where the final *steady states* are regular and periodic.

There is also interest in critical system parameter values for which stable periodic orbits do not exist or, rather, where they become unstable with respect to infinitesimal disturbances, and are replaced by persistent, steady-state chaotic attractors. A possible necessary and sufficient condition for the occurrence of steady-state chaos was heuristically discussed in [39–41] by way of computer simulation results, again with reference to the twin-well Duffing oscillator. However, this turned out to be a nontrivial matter [42].

In contrast, within a general framework of archetypal polynomial oscillators (for example, the *single-well* Duffing oscillator with escape, see Chapter 5), Szemplinka-Stupnicka showed that approximate analytical methods can give simple, closed-form algebraic formulae providing good estimates of the critical system parameter values where steady chaos is established. This author exploited concepts and approximate techniques for nonlinear oscillations (like perturbation and harmonic balance) to gain insight into the relationships between periodic and *chaotic solutions*. Chaotic regions were located against the background of classical phenomena of principal, subharmonic or ultraharmonic resonances. Regions in system parameter space were also identified so computer simulations could be conducted to detect and characterise chaotic responses.

Considering Equation (7.1.1) with $\alpha = \gamma = 1/2$, Rudowski and Szemplinka-Stupnicka [43] proposed an approximate criterion for the change of small periodic orbit into *chaotic motion* in the region of excitation frequency close to the *primary resonance*. This was based on the first approximate harmonic solution of the equation. By complementing theoretical analysis with computer simulations they showed how chaotic motion can occur as a transition zone from either resonant to nonresonant small periodic orbit or resonant small orbit to large orbit motion, depending on the amplitude of the excitation force. The approximate criterion was then refined [44] by considering the second approximate solution for the small orbit. Hill's type of variational equation provides the *stability* loss of the resonant solution through period-doubling bifurcation that is seen to start the *route to chaos* in the computer simulations. Still, based upon the approximate analysis of local instabilities in the approximate periodic solutions, [45] dealt with predicting the occurrence of chaos in the neighbourhood of the order-two *superharmonic resonance* of the small orbit. Computer simulations revealed a scenario of transition to chaos significantly different from that at primary resonance, in which both small and large periodic orbits lose stability and jointly turn into a unique chaotic attractor. The approximate criteria for chaos were identified with the symmetry-breaking instability of the approximate large orbit symmetric periodic solution and with the first-order instability of the nonresonant branch of the superharmonic small-orbit solution. In all of these studies, local methods were used to predict a global occurrence.

The success of the analysis of local instability of approximate periodic solutions in providing satisfactory approximate criteria for the occurrence of chaos is because the regions of the system parameter values where complex chaotic responses are observed are very close to, or even border upon, regions of highly regular periodic responses. This was summarised in [46], where a picture book of different steady states – including single or coexisting, small or large orbit, periodic and chaotic attractor – was displayed, based on computer simulations. Comparison with the instability analysis of approximate periodic solutions made it possible to estimate the system parameter domains where certain types of steady-state motions occur. It also made it possible to predict boundaries of V-shaped regions where the system exhibits cross-well chaotic motion at both principal and superharmonic resonance. Along the same lines, theoretical predictions and numerical simulations of transition to cross-well chaos in the subharmonic frequency region – where studies had been conducted in [7,11,12], also including a partial analysis of basin bifurcation in the excitation amplitude-damping plane [47,48] – were accomplished in [49]. Attention was also paid to the differences in the bifurcation features in the region of primary resonance [50].

Szemplinka-Stupnicka compared three oscillators that have a softening-type restoring force characteristic [51] – including the twin-well Duffing oscillator, and extended the method to systems for which, upon the local instabilities, cross-well chaos is replaced by the escape phenomenon. Finally, within an overall framework encompassing steady-state chaos or escape phenomena in the class of Duffing-type oscillators with softening characteristics, Szemplinska-Stupnicka [52] surveyed the analytical predictive criteria by considering both the rigorous Melnikov criterion for transient chaos and the three different heuristic criteria proposed by Moon [29], Schmidt [37] and herself. All of these were based on a low-order periodic approximate solution for the in-well periodic attractor. For the twin-well Duffing oscillator, a comparison of Melnikov's, Moon's and Szemplinska-Stupnicka's criteria showed a progressively improved estimation of the bifurcational curves and of the V-shaped region of chaos, as obtained via computer simulations (see Section 7.5.1). The author concluded that the 'far distance' view associated with the crude approximation involved in the predictive criteria for steady-state chaos had to be suitably complemented by the fine details (the 'electronic microscope' view) of complex bifurcational phenomena that could be provided by systematic computer based studies. Both views are necessary for the overall understanding of chaotic dynamics.

7.2.2 Refined computational investigations

Following the numerical studies of the 1980s, refined computational investigations of the twin-well Duffing oscillator were carried out at the beginning of the 1990s [48,53]. These studies were aimed at describing the metamorphoses of basin boundaries of a *coexisting attractor* and the generation and destruction of the chaotic attractor.

In 1990, Ueda *et al.* published a pioneering paper [48] that began to explain complex bifurcation and escape from single-well to cross-well behaviour of the twin-well Duffing oscillator. They did this in the context of the intersections of stable and unstable manifold of a specific periodic saddle, the two events being shown to coincide.

Lansbury *et al.* [53] also explored basin-boundary bifurcation phenomena in the context of manifold intersections and described two different mechanisms through which a catastrophic erosion of a basin of attraction may take place via the formation of a *chaotic saddle* (whose role in the instability of a chaotic attractor of the twin-well oscillator had earlier been introduced in [54]). This is induced by the *homoclinic intersection* between the stable and unstable manifold of a principal saddle and is accompanied by a smooth-fractal basin bifurcation, and its subsequent growth. In both scenarios the basin boundary develops a fractal structure, whose evolution and growth is associated with a specific sequence of bifurcational events whose understanding is also obtained using geometric models. Later, starting from [48] and exploiting some results from [53], a discussion on the role of the hilltop saddle stable manifolds and a detailed explanation as to why the crossing of the manifolds of a specific in-well saddle implies the destruction of the single-well chaotic attractor were provided by Katz and Dowell [55]. They referred to escape scenarios that occurred when either one of two different control parameters (the excitation amplitude or frequency) was varied. Generally, it was shown how the study of the behaviour of manifolds provides a useful stepping stone in the understanding of overall system dynamics.

The possibility of global stable periodic and chaotic oscillations surrounding the unstable equilibrium point (hilltop effect) was discussed in Zakrzhevsky [56], who showed how they can occur if the damping is large.

Szemplinska-Stupnicka and associates also joined people working computationally on the twin-well Duffing oscillator [46,57–61]. They observed the zones of existence and coexistence of different attractor in the region of primary resonance and the corresponding bifurcation scenarios [46]. This enabled them to distinguish between the domains where the evolution from periodic motion to cross-well chaos is initiated by main *local bifurcation* of periodic attractor and the domains where the destruction of cross-well chaos can only be explained in terms of global homoclinic bifurcation.

Specifically, Szemplinska-Stupnicka and Tyrkiel [59] focused on the system behaviour after annihilation of the resonant attractor in the region of nonlinear resonance hysteresis occurring within each of the two potential wells. The interest was in understanding what happens after the strict loss of stability (crisis) of the attractor and which factors play a decisive role on the occurrence of 'indeterminate' or 'determinate' final outcomes [62,63]. The answer was found by an analysis of the structure of the basins of attraction of the T-periodic coexisting attractor and of their metamorphoses at the thresholds of a series of *homoclinic* and *heteroclinic bifurcation*. This led to the identification of two system parameter subdomains where the two different outcomes occur.

In contrast, Szemplinska-Stupnicka and Janicki [57] focused on the destruction of cross-well chaos upon which, for large excitation amplitudes, the resonant large orbit (cross-well periodic attractor) remains as a unique attractor of the system. In particular, upon discussing a 'hierarchy' of (hilltop, in-well small orbits, cross-well large orbit) saddles possibly playing the role of principal saddles, a codimension two bifurcation, i.e., the bifurcation defined by the intersection of the homoclinic bifurcation of the large orbit with the saddle-node bifurcation of the nonresonant small orbit, was investigated. Numerical analysis of the manifold structure of the large orbit saddle and of the basins of attraction revealed new features of the system behaviour. It also revealed a complex bifurcational structure that includes a boundary crisis of the cross-well attractor ending up with the explosion of the large-orbit basin of attraction.

The twin-well Duffing oscillator has also been used as a representative model, along with others, in comparative computational studies aimed at investigating bifurcational properties and features of transition to or destruction of chaos. Specifically, in [58], the general phenomena of regular or chaotic boundary crises in dissipative, forced oscillators, with the related concepts of regular or chaotic destroyer saddles, were investigated after the destruction of the chaotic attractor. Features of the system behaviour (fractal basin boundary, indeterminate outcome, role of periods of the destroyer and of the principal saddle of the attractor) were also studied.

Szemplinska-Stupnicka and Tyrkiel [60] investigated the common phenomenological aspects of the onset of persistent cross-well chaos. This exists in a wide region of system control parameters, in a class of oscillators with multiple wells. Finally, Tyrkiel [61] discussed the formation of chaotic saddles as the primary mechanism triggering chaotic transient motions, independently of whether single or multiple attractors exist. She also investigated levels of control parameters much lower than those corresponding to the onset of the steady-state chaos. Formation and expansion of a chaotic saddle are the result of a sequence of global (homoclinic and heteroclinic) bifurcation. Their investigation is useful for establishing criteria for the occurrence of chaotic behaviour as the control parameter changes, and for evaluating the decrease of system 'stability in the large', i.e., of the phase plane area of safe disturbances of the regular orbits.

7.2.3 Control of nonlinear dynamics

Within a more general perspective of 'system safety', global bifurcational events may induce erosion of the so-called *safe basin* of attraction, with the ensuing reduction of the system dynamical integrity. This raises the question of how to reduce and possibly control such potentially dangerous phenomena for engineering systems [64,65].

Control of nonlinear dynamics and chaos in the twin-well Duffing oscillator has been studied by many authors using various techniques (for a summary on chaos control techniques see, for example, Fradkov [66] and Rega and Lenci [67]). Sifakis and Elliott [68] compared four different strategies for chaos control in the twin-well Duffing equation: (i) the open-loop periodic perturbation method, which consists of adding a periodic perturbation to the excitation force; (ii) the continuous delayed feedback method, i.e., Pyragas' method [69]; (iii) the occasional proportional feedback method proposed by Hunt [70]; and (iv) the OGY method [71]. As to the first technique, in Sifakis and Elliott [68] the controlling perturbation was chosen empirically by a trial-and-error procedure, whereas in other works it was optimally determined on the basis of a theoretical analysis relying on system dynamical properties [72]. The possibility to reduce and even suppress chaos in the Duffing–Holmes oscillator by means of a small parametric perturbation of suitable frequency was demonstrated in [73] both analytically, via the Melnikov method, and numerically, through the computation of *Lyapunov exponents*. Later it was verified experimentally in a bistable magnetoelastic beam [74]. Inhibition of Melnikov chaos using active control in a twin-well Duffing oscillator was considered by Nana Nbendjo *et al.* [75]. They used time delay in the control system and *parametric excitation* [76].

Since the very first formulation of the OGY method [71], some kind of magnetoelastic beam has been considered as the archetypal experimental model for control of chaos in smooth mechanics. Experimental control of chaos in a continuous system was addressed in [77] by considering the same apparatus (a beam deformed by permanent magnets) already considered by the leading author in his pioneering paper [5] on experimental chaos in mechanical systems. The control method was Hunt's occasional proportional feedback control [70], which is a semiempirical variant of the OGY method [71].

Inspired by that classical experiment, Dressler *et al.* [78] demonstrated the tracking of an unstable periodic orbit in a horizontal elastic bronze ribbon cantilever, which was driven harmonically. The orbit stabilisation was performed at each tracking step via the local control method, which is a variant of the OGY method.

Many other researchers have developed techniques for the control of chaos and applied them to the twin-well oscillator considered as an archetypal system. Examples include: Chen and Dong [79], who designed a conventional feedback controller to drive a *chaotic trajectory* to one of the system inherent multiperiodic orbits; Nijmeijer [80], who compared robust and adaptive controllers; Ravindra and Mallik [81], who controlled chaos by using nonlinear dissipation; Mahmoud *et al.* [82], who applied Pyragas' method [69] to the complex Duffing oscillator; Bowong and Kakmeni [83], who applied a robust feedback controller to suppress chaos in the uncertain Duffing system; Yamasue and Hikihara [84], who used numerical simulations to investigate the issue of domain of attraction of stabilised orbits under time delayed feedback control; Lei *et al.* [85], who controlled chaos in a twin-well Duffing oscillator subject to a harmonic parametric control by adjusting the level of a random phase; Aguilar-López and Martínez-Guerra [86], who applied a state-observer-based active controller, which provides robustness against model uncertainties and noisy output measurements, for chaos suppression.

A number of recent papers have been seeking a chaos control methodology that exploits the underlying structure of the chaotic behaviour of a system, i.e., topology and dynamics.

Alvarez-Ramirez and Espinosa-Paredes [87] applied a suitable feedback control, based on the application of damping. It is related to the Pyragas' delayed feedback method [69], and aims to eliminate the transverse homoclinic orbit responsible for chaotic behaviour in the twin-well Duffing oscillator.

Still in the context of dynamical systems but in a nonfeedback context, Lenci and Rega [88,89] applied a general control method to the twin-well oscillator. It is based on eliminating, or shifting, the homoclinic intersections by optimally modifying the shape of a periodic excitation. While discussing the theoretical features of the method in [71,90] for various archetypal oscillators, the authors were concerned with reducing basin erosion and delaying onset of cross-well chaos, applied to the twin-well system. This involved either analytical control of the hilltop saddle manifolds or numerical control of other saddles playing a role in the transition from single- to cross-well chaos. A modified version of the method [91] was applied to the twin-well oscillator in [92].

7.3 Dynamics of conservative and nonconservative systems

In this section the main aspects of the dynamics of Equation (7.1.2) are discussed. The conservative case is considered first, this is similar as was done in Chapter 3. However, this consideration is revisited for the sake of the completeness of the analysis presented below and because it allows a general understanding of the multiwell potential. Damping is then added and the system is excited harmonically. The complexity of the phenomena observed is then discussed.

7.3.1 The conservative case

In the absence of damping ($\zeta = 0$) and excitation ($F = 0$), Equation (7.1.2) becomes

$$\ddot{y} - y + \gamma y^3 = 0 \tag{7.3.1}$$

and it is called conservative, i.e., unforced and undamped. It has the property of symmetry, such that if $y(t)$ is a solution, then also $y(-t)$ and $-y(t)$ are solutions.

The *equilibrium points* (*fixed points*) are obtained by setting $y(t) = \hat{y}$, and are given by (note that $\gamma > 0$)

$$\hat{y}_1 = -\frac{1}{\sqrt{\gamma}}, \quad \hat{y}_2 = 0, \quad \hat{y}_3 = \frac{1}{\sqrt{\gamma}} \tag{7.3.2a–c}$$

The linearised equations around these three fixed points are

$$\ddot{z}_1 + 2z_1 = 0, \quad \ddot{z}_2 - z_2 = 0, \quad \ddot{z}_3 + 2z_3 = 0 \tag{7.3.3a–c}$$

where $z_i(t) = y(t) - \hat{y}_i$. Seeking a solution of Equation (7.3.3a–c) in the form $e^{\lambda t}$, results in the eigenvalues $\lambda_{1;1,2} = \lambda_{3;1,2} = \pm j\sqrt{2}$ and $\lambda_{2;1,2} = \pm 1$. This shows that $\hat{y}_1$ and $\hat{y}_3$ are *centres*, with circular frequency $\omega_0 = \sqrt{2}$ and period $P = \pi\sqrt{2}$, while $\hat{y}_2$ is a (hilltop) *saddle*, with stable eigenvector $\mathbf{v}_s = [1, -1]^T$ (associated with the eigenvalue $\lambda_{2;2} = -1$) and unstable eigenvector $\mathbf{v}_u = [1, 1]^T$ (associated with the eigenvalue $\lambda_{2;1} = +1$). The presence of the saddle in the unperturbed case strongly characterises the system nonlinear dynamics. Note that the nonlinearity parameter γ influences only the position of the fixed points, since it does not appear in Equation (7.3.3a–c).

By considering the kinetic, the potential and the total energy, respectively given by

$$T = \frac{\dot{y}^2}{2}, \quad V = -\frac{y^2}{2} + \gamma\frac{y^4}{4}, \quad E = T + V \tag{7.3.4a–c}$$

it can be seen that

$$\dot{E} = (\ddot{y} - y + \gamma y^3)\dot{y} = 0 \tag{7.3.5}$$

i.e., the total energy is constant along the orbits, $E = E_0$, $E_0 = \dot{y}_0^2/2 - y_0^2/2 + \gamma y_0^4/4$ being determined by the initial position y_0 and the initial velocity $\dot{y}_0$ only. Thus, E is conserved along the orbits.

The symmetric twin-well potential energy $V(y)$, which is a characteristic of this system, is drawn in Figure 7.1 for different values of γ. The three fixed points are the stationary points of $V(y)$; note that $V(\hat{y}_1) = V(\hat{y}_3) = -1/(4\gamma)$ and $V(\hat{y}_2) = 0$, so that the depth of the wells is $1/(4\gamma)$. The centres are the minima of $V(y)$, while the saddle is the local maximum, and it separates the two symmetric potential wells.

Solving equation $E = E_0$ with respect to $\dot{y}$ gives

$$\dot{y} = \pm\sqrt{2(E_0 - V(y))} = \pm\sqrt{2E_0 + y^2 - \gamma\frac{y^4}{2}} \tag{7.3.6}$$

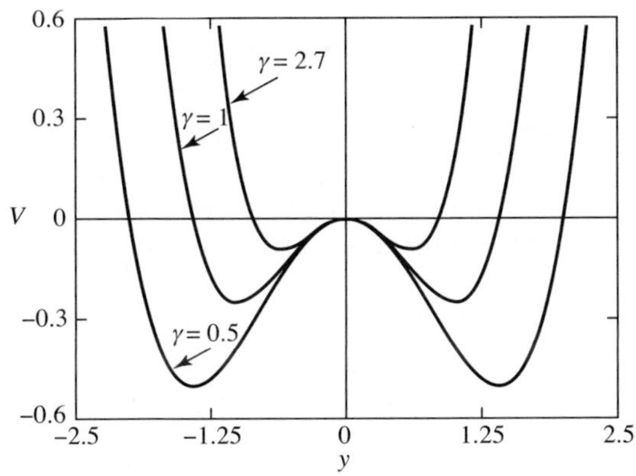

Figure 7.1 Potential energy V(y) for various values of γ.

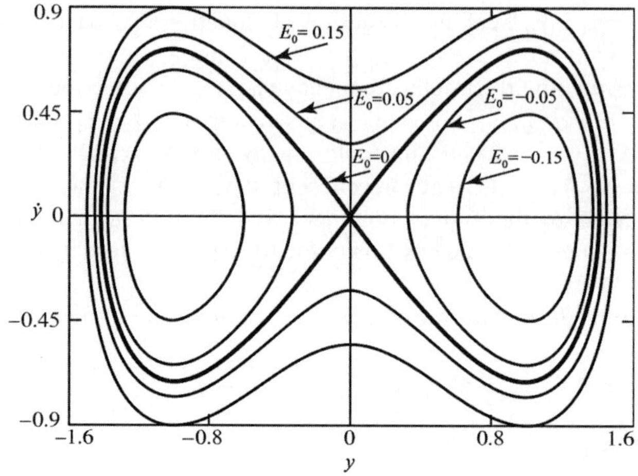

Figure 7.2 The phase portrait for $\gamma = 1$ and for various total energy values. The thick lines correspond to the two symmetric homoclinic orbits.

which: (i) reduces the order of the equation, and (ii) allows the orbits to be drawn in *phase space* without explicitly solving the equation, as shown in Figure 7.2.

From the phase portrait of Figure 7.2 it can be seen that there are three different kinds of motion. The first involves in-well, low-amplitude oscillations, which turn around the fixed points with a period depending on the amplitude. Those on the right part of the phase plane (the others by symmetry) range from

$$y_{\min} = \sqrt{\frac{1 - \sqrt{4\gamma E_0 + 1}}{\gamma}} \quad \text{to} \quad y_{\max} = \sqrt{\frac{1 + \sqrt{4\gamma E_0 + 1}}{\gamma}} \tag{7.3.7}$$

(this is obtained by solving $V(y) = E_0$), while the maximum velocity, which is achieved for $y = \hat{y}_3$, is given by

$$\dot{y}_{\max} = \sqrt{2E_0 + \frac{1}{2\gamma}} \tag{7.3.8}$$

These solutions exist for $-1/(4\gamma) = V_{\min} < E_0 < 0$, while for $E_0 > 0$ there are out-of-well large-amplitude swaying oscillations, which range from $y = -y_{\max}$ to $y = y_{\max}$, where $y_{\max}$ is given in Equation (7.3.7), with the same expression (7.3.8) for the maximum velocity. They turn around the two potential wells, also with an amplitude-dependent period.

For both in-well and out-of-well oscillations it is possible to write explicitly the solution $y(t)$ by means of elliptic functions ([31], pp. 154), but this is not done in this chapter. Instead, the period of the oscillations is computed explicitly. Equation (7.3.6) results in

$$\pm dt = \frac{dy}{\sqrt{2E_0 + y^2 - \gamma \frac{y^4}{2}}} = \frac{dx}{\sqrt{\gamma E_0 + x^2 - x^4}} \tag{7.3.9}$$

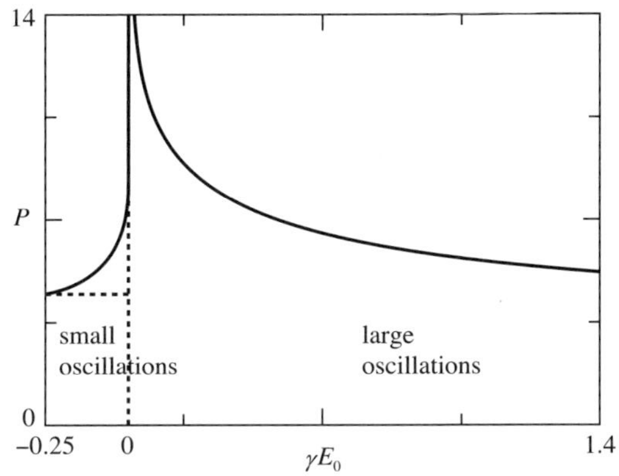

Figure 7.3 The period of the oscillations P as a function of γE_0.

where $x = y\sqrt{\gamma/2}$. The last expression in Equation (7.3.9) shows that solutions of the unperturbed equations, written in term of x instead of y, depend of the parameter γE_0 only. Integrating the right and the left right side gives

$$P = 2\int_{x_{\min}}^{x_{\max}} \frac{dx}{\sqrt{\gamma E_0 + x^2 - x^4}} \tag{7.3.10}$$

for the in-well oscillations (note that $x_{\max,\min} = \sqrt{1/2 \pm \sqrt{\gamma E_0 + 1/4}}$), and

$$P = 4\int_{0}^{x_{\max}} \frac{dx}{\sqrt{\gamma E_0 + x^2 - x^4}} \tag{7.3.11}$$

for the out-of-well oscillations. The period as a function of γE_0 is shown in Figure 7.3.

Equations (7.3.10) and (7.3.11) have the asymptotic expansion $P = \pi\sqrt{2} + (3\pi\sqrt{2}/4)(\gamma E_0 + 1/4) + \ldots$, which shows that, along with Figure 7.3, for $\gamma E_0 \to -1/4$ the period tends to that of small oscillations around the centres, $P = \pi\sqrt{2} \cong 4.443$. Moreover, the figure shows that (i) for $\gamma E_0 \to 0$ the period goes to infinity; and (ii) for $\gamma E_0 \to +\infty$ the period (slowly) goes to 0 as

$$\frac{4}{(\gamma E_0)^{1/4}}\int_{0}^{1} \frac{dx}{\sqrt{1 - x^4}} \cong \frac{5.244}{(\gamma\, E_0)^{1/4}} \tag{7.3.12}$$

Figure 7.3 further underlines that two families of oscillations are divided by the case $E_0 = 0$, which corresponds to a couple of (unperturbed) symmetric homoclinic orbits of the hilltop saddle $\hat{y}_2$, which are also called (with a clear meaning) separatrices. These orbits are the most important dynamical characteristic of Equation (7.3.1), and of Equation (7.1.2) as well. Although they are unstable solutions, they play a very important role in governing the system dynamics. This is a general feature of the nonlinear dynamics, where it is known that 'hidden' (i.e., unstable) solutions (saddles, homoclinic orbits, etc.) play a major role in the dynamical behaviour of the system.

Setting $E_0 = 0$ in Equation (7.3.9) and integrating the right and the left sides yields

$$\pm(t - t_0) = -\operatorname{arctanh}\left(\frac{2}{\sqrt{4 - 2\gamma y^2}}\right) = -\operatorname{arctanh}\left(\frac{1}{\sqrt{1 - x^2}}\right) \tag{7.3.13}$$

Solving the previous expression with respect to y gives

$$y_{\text{hom}}(t) = \pm\sqrt{\frac{2}{\gamma}}\frac{1}{\cosh(t - t_0)}, \quad x_{\text{hom}}(t) = \pm\frac{1}{\cosh(t - t_0)} \tag{7.3.14a,b}$$

which agrees with the limits given in Equation (7.3.7) for $E_0 = 0$. The t_0 in Equation (7.3.14a,b) reflects the fact that Equation (7.3.1) is *autonomous*, so that its solution is defined up to an arbitrary shift of time; the $\pm$ means that there are two (symmetric) unperturbed homoclinic orbits. Equations (7.3.14a,b) clearly show that they asymptotically tend to zero for $t \to \pm\infty$, which is equivalent to say that the period of these particular solutions is infinity, in agreement with Figure 7.3.

From Equations (7.3.14a,b) it can be seen that

$$\lim_{t \to +\infty} \frac{\dot{y}_{\text{hom}}(t)}{y_{\text{hom}}(t)} = -1, \quad \lim_{t \to -\infty} \frac{\dot{y}_{\text{hom}}(t)}{y_{\text{hom}}(t)} = 1 \tag{7.3.15a,b}$$

which shows that the orbits are tangent to the stable and unstable eigendirections. Indeed, the homoclinic orbits represent the coinciding stable and unstable manifold of the hilltop saddle [27,30,31].

7.3.2 The effect of damping

When damping $2\zeta\dot{y}$ is added to Equation (7.3.1), the system is no longer conservative, and

$$\dot{E} = (\ddot{y} - y + \gamma y^3)\dot{y} = -2\zeta\dot{y}^2 < 0 \tag{7.3.16}$$

i.e., the total energy decreases in time along any trajectory, which therefore approaches those solutions with minimum potential energy. As Equation (7.3.16) shows, these 'attracting' solutions must have $\dot{y} = 0$, and thus they are fixed points. They are the same as in the conservative case since the new term $2\zeta\dot{y}$ vanishes on them, and the dynamics in their neighbourhood are governed by the equations

$$\ddot{z}_1 + 2\zeta\dot{z}_1 + 2z_1 = 0, \quad \ddot{z}_2 + 2\zeta\dot{z}_2 - z_2 = 0, \quad \ddot{z}_3 + 2\zeta\dot{z}_3 + 2z_3 = 0 \tag{7.3.17a–c}$$

which are extensions of Equation (7.3.3). The eigenvalues are now

$$\lambda_{1;1,2} = \lambda_{3;1,2} = -\zeta \pm i\sqrt{2 - \zeta^2}, \quad \lambda_{2;1,2} = -\zeta \pm \sqrt{1 + \zeta^2} \tag{7.3.18a,b}$$

Supposing that the damping is small ($\zeta < \sqrt{2}$ is sufficient), it can be seen that there is a major modification with respect to the conservative case: while the saddle remains a saddle (it continues to have a positive and a negative real eigenvector, although they now differ from each other), the centres become foci, or attracting centres, or attractors, as their eigenvalues now have a negative real part (recall that $\zeta > 0$).

The attractors are those solutions having an open neighbourhood of initial conditions whose ensuing trajectory asymptotically tends to them (see [27,30,31] for a more precise definition of attractor). The appearance of attractor is the main element of novelty, and plays a major role in the dynamic response; furthermore, associated with the attractor there is also another basic concept of nonlinear dynamics, i.e., the basin of attraction, which is the largest set of initial conditions leading to the particular attractor. The basins of attraction of the two attractor $\hat{y}_1$ and $\hat{y}_3$ of Equation (7.1.2) for $F = 0$, $\gamma = 1$ and $\zeta = 0.025$ are given in Figure 7.4(a); the grey

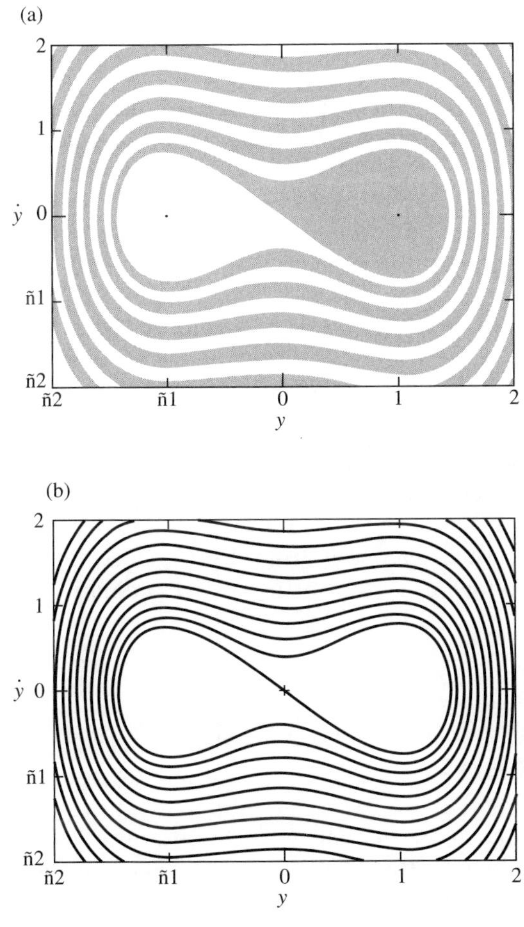

Figure 7.4 (a) The basins of attraction and (b) the stable manifold of the hilltop saddle of Equation (7.1.2) for $F = 0$, $\gamma = 1$ and $\zeta = 0.025$.

points are the initial conditions that asymptotically tend to the fixed point $\hat{y}_3$ in the right potential well, while the white points are the initial conditions that asymptotically tend to the fixed point $\hat{y}_1$ in the left potential well.

In Figure 7.4(b), on the other hand, the stable manifolds of the hilltop saddle $\hat{y}_2$ can be seen, which clearly constitute the boundary of the basins of attraction.

Figure 7.4(a) shows a very basic concept of nonlinear dynamics, which does not occur in linear systems – multistability: depending on the initial conditions, the trajectory may approach one or another coexisting attractor, so that the final outcome cannot be predicted on the base of the knowledge of the system only.

The basins of attraction of Figure 7.4(a) consist of a compact part around the attractor, from which a 'tongue' exits. This tongue first turns around the other basin of attraction, and then spirals around the 'core' of the basins.

7.3.3 The effect of the excitation

When the excitation $F\cos\Omega t$, is added to Equation (7.3.1), it is the same as Equation (7.1.2), and the energetic arguments used above no longer provide useful information. In fact, in this case

$$\dot{E} = (\ddot{y} - y + \gamma y^3)\dot{y} = -2\zeta\dot{y}^2 + \dot{y}F\cos\Omega t \tag{7.3.19}$$

no longer has a fixed sign but in general it oscillates in time between positive and negative values, so that E cannot be used as a Lyapunov function. $\dot{E}$ still vanishes on equilibrium points, which, however, no longer exist, as can be immediately checked by assuming $\ddot{y} = \dot{y} = 0$ in Equation (7.1.2) and noting that it has no time-independent solutions.

In this case the Duffing equation (7.1.2) has its most general dynamical behaviour, including chaotic dynamics. The key point for the onset of the complex dynamics is the fact that when the excitation is introduced, the (mathematical) dimension of the system increases from 2 to 3, since Equation (7.1.2) can be rewritten as the following system of first-order ordinary differential equations (ODEs)

$$\begin{cases} \dot{y}_1 = y_2, \\ \dot{y}_2 = -2\zeta y_2 + y_1 - \gamma y_1^3 + F\cos y_3, \\ \dot{y}_3 = \Omega \end{cases} \tag{7.3.20a–c}$$

It is the third dimension that allows for chaos; up to dimension 2 the system dynamics are much simpler (for example, it cannot have chaotic attractor), as shown by the Poincaré–Bendixon theorem [27].

The various aspects of the nonlinear dynamics of the forced system will be illustrated subsequently.

7.4 Nonlinear periodic oscillations

Although they are not the most important dynamical phenomena of this equation, the nonlinear periodic oscillations are a fundamental response feature and therefore deserve some attention.

Since the nonlinear periodic oscillations ensue from the linear harmonic oscillations by increasing the excitation amplitude, and since the latter occur around the fixed points of the conservative case, it is useful for the forthcoming analytical treatment to change the variable so that $y(t) = \hat{y}_{1,3} + z(t)$; Equation (7.1.2) then becomes

$$\ddot{z} + 2\zeta\,\dot{z} + 2z \pm 3\sqrt{\gamma}z^2 + \gamma\,z^3 = F\cos\Omega t \tag{7.4.1}$$

which also has a quadratic nonlinearity and a positive linear stiffness. In Equation (7.4.1) the minus sign is for the solution $\hat{y}_1$ and the plus sign to $\hat{y}_3$. In the following, only the latter case is considered without loss of generality (due to the symmetry of the system).

An approximate analytical treatment of the nonlinear oscillations of Equation (7.4.1) is first considered, since the exact solution is not known. The multiple scales method is applied by following the procedure given in the book of Nayfeh and Mook ([93], Section 4.2), which is referred to for further details and extensions.

Since there are quadratic nonlinear terms, an appropriate rescaling of Equation (7.4.1) is $\zeta = \varepsilon^2\bar{\zeta}$, $F = \varepsilon^3\bar{F}$, where ε is a small bookkeeping parameter. Assuming small but finite motion amplitudes, i.e., $z = \varepsilon\,\bar{z}$, results in

$$\ddot{\bar{z}} + 2\varepsilon^2\bar{\zeta}\dot{\bar{z}} + 2\bar{z} + 3\varepsilon\sqrt{\gamma}\bar{z}^2 + \varepsilon^2\gamma\bar{z}^3 = \varepsilon^2\bar{F}\cos\Omega t \tag{7.4.2}$$

A solution is sought in the form

$$\bar{z}(t) = z_0(T_0, T_1, T_2) + \varepsilon\,z_1(T_0, T_1, T_2) + \varepsilon^2 z_2(T_0, T_1, T_2) + \ldots \tag{7.4.3}$$

where $T_0 = t$ is the regular time and $T_1 = \varepsilon t$ and $T_2 = \varepsilon^2 t$ are *slow times*. Note that

$$\frac{dz_i}{dt} = \frac{\partial z_i}{\partial T_0} + \varepsilon\frac{\partial z_i}{\partial T_1} + \varepsilon^2\frac{\partial z_i}{\partial T_2}, \quad \frac{d^2 z_i}{dt^2} = \frac{\partial^2 z_i}{\partial T_0^2} + 2\varepsilon\frac{\partial^2 z_i}{\partial T_1 \partial T_0} + \varepsilon^2\left(\frac{\partial^2 z_i}{\partial T_1^2} + 2\frac{\partial^2 z_i}{\partial T_2 \partial T_0}\right) + \ldots \tag{7.4.4}$$

so that Equation (7.4.2) becomes

$$\begin{aligned} &\left(\frac{\partial^2 z_0}{\partial T_0^2} + 2z_0\right) + \varepsilon\left(\frac{\partial^2 z_1}{\partial T_0^2} + 2z_1 + 3\sqrt{\gamma}z_0^2 + 2\frac{\partial^2 z_0}{\partial T_1 \partial T_0}\right) \\ &\varepsilon^2\left(\frac{\partial^2 z_2}{\partial T_0^2} + 2z_2 + 2\frac{\partial^2 z_1}{\partial T_1 \partial T_0} + 6\sqrt{\gamma}z_1 z_0 + 2\bar{\zeta}\frac{\partial z_0}{\partial T_0} + 2\frac{\partial^2 z_0}{\partial T_2 \partial T_0} + \frac{\partial^2 z_0}{\partial {T_1}^2}\right. \\ &\left. + \gamma z_0^3 - \bar{F}\cos\Omega t\right) + \ldots = 0 \end{aligned} \tag{7.4.5}$$

Equating the coefficient of each power of ε to zero, gives a system of ordinary differential equations that can be solved to determine the unknown quantities z_0, z_1, z_2, ... in Equation (7.4.3).

To illustrate the main ideas, the case of the main resonance $\Omega \approx \sqrt{2}$ only is considered. For the subharmonic and superharmonic resonance cases the reader is referred to [93]. The *detuning parameter* σ is introduced that is a measure of the difference between the nonlinear resonance and the linear resonance, so that [93],

$$\Omega = \sqrt{2} + \varepsilon^2\sigma \tag{7.4.6}$$

Thus $\Omega t = \sqrt{2}t + \sigma\varepsilon^2 t = \sqrt{2}t + \sigma T_2$. The solution of the first problem is

$$z_0 = c_1(T_1, T_2)\sin(\sqrt{2}t) + c_2(T_1, T_2)\cos(\sqrt{2}t) \tag{7.4.7}$$

where c_1 and c_2 are constants. Note that in many papers on the subject, the solution is sought in the complex form $z_0 = A(T_1, T_2)e^{-j\sqrt{2}t} + \text{c.c.}$, where c.c. means complex conjugate. Substituting Equation (7.4.7) into the second equation ensuing from Equation (7.4.5) yields

$$\begin{aligned} &\frac{\partial^2 z_1}{\partial T_0^2} + 2z_1 + \cos(2\sqrt{2}t)\frac{3\sqrt{\gamma}}{2}(c_2^2 - c_1^2) + \sin(2\sqrt{2}t)3\sqrt{\gamma}c_1c_2 \\ &+ \cos(\sqrt{2}t)2\sqrt{2}\frac{\partial c_1}{\partial T_1} - \sin(\sqrt{2}t)2\sqrt{2}\frac{\partial c_2}{\partial T_1} + \frac{3\sqrt{\gamma}}{2}(c_1^2 + c_2^2) = 0 \end{aligned} \tag{7.4.8}$$

which permits the computation of z_1.

The terms $\cos(\sqrt{2}t)$ and $\sin(\sqrt{2}t)$ provide a solution of the form $z_1 \approx t\sin(\sqrt{2}t)$, which is unbounded in time and is thus not acceptable as it is not valid for all time. These are called *secular terms*, and must be eliminated from Equation (7.4.8). This requires

$$\frac{\partial c_1}{\partial T_1} = \frac{\partial c_2}{\partial T_1} = 0 \tag{7.4.9}$$

which means that c_1 and c_2 depend only on T_2. The particular integral of Equation (7.4.8) is

$$z_1 = \frac{\sqrt{\gamma}}{4}\Big(\cos(2\sqrt{2}t)(c_2^2 - c_1^2) + 2\sin(2\sqrt{2}t)c_1c_2 - 3(c_2^2 + c_1^2)\Big) \tag{7.4.10}$$

while the complementary function is not required in the present analysis.

Substituting Equations (7.4.10) and (7.4.7) into the third term in Equation (7.4.5) for Equation (7.4.9), gives

$$\begin{aligned} &\frac{\partial^2 z_2}{\partial T_0^2} + 2z_2 + \cos(3\sqrt{2}t)(\ldots) + \sin(3\sqrt{2}t)(\ldots) + \cos(2\sqrt{2}t)(\ldots) + \sin(2\sqrt{2}t)(\ldots) \\ &+ \cos(\sqrt{2}t)\left(2\sqrt{2}\frac{\partial c_1}{\partial T_2} + 2\sqrt{2}\zeta c_1 - 3\gamma c_2(c_1^2 + c_2^2) - \bar{F}\cos(\sigma T_2)\right) \\ &+ \sin(\sqrt{2}t)\left(-2\sqrt{2}\frac{\partial c_2}{\partial T_2} - 2\sqrt{2}\zeta c_2 - 3\gamma c_1(c_1^2 + c_2^2) + \bar{F}\sin(\sigma T_2)\right) + (\ldots) = 0 \end{aligned} \tag{7.4.11}$$

where only the secular producing terms have been included. They can be eliminated by setting equal to zero the expressions in square brackets in Equation (7.4.11):

$$2\sqrt{2}\frac{\partial c_1}{\partial T_2}+2\sqrt{2}\bar{\zeta}c_1-3\gamma c_2(c_1^2+c_2^2)=\bar{F}\cos(\sigma T_2) \tag{7.4.12}$$

$$2\sqrt{2}\frac{\partial c_2}{\partial T_2}+2\sqrt{2}\bar{\zeta}c_2+3\gamma c_1(c_1^2+c_2^2)=\bar{F}\sin(\sigma T_2) \tag{7.4.13}$$

These equations can be simplified assuming, without loss of generality, that

$$c_1(T_2)=a(T_2)\cos(b(T_2)),\quad c_2(T_2)=a(T_2)\sin(b(T_2)) \tag{7.4.14a,b}$$

Multiplying Equation (7.4.12) by $\sin b$ and Equation (7.4.13) by $\cos b$, and subtracting them, results in

$$2\sqrt{2}a\frac{\partial b}{\partial T_2}=-3\gamma a^3+\bar{F}\sin(\sigma T_2-b) \tag{7.4.15}$$

Multiplying Equation (7.4.12) with $\cos b$ and Equation (7.4.13) with $\sin b$, and summing them, yields

$$2\sqrt{2}\frac{\partial a}{\partial T_2}=-2\sqrt{2}\bar{\zeta}a+\bar{F}\cos(\sigma T_2-b) \tag{7.4.16}$$

The variable a is the slowly varying amplitude of oscillations, since Equations (7.4.14a,b) give

$$z_0=a(T_2)\sin(\sqrt{2}t+b(T_2)),\quad z_1=-a(T_2)^2\frac{\sqrt{\gamma}}{4}\left(3+\cos\left(2(\sqrt{2}t+b(T_2))\right)\right) \tag{7.4.17a,b}$$

For a periodic solution, bounded nonvanishing values of a are required. Constant values of the amplitude a, which fulfill the previous requirement, are considered. Setting $a(T_2)=a_0$ in Equation (7.4.16) gives

$$2\sqrt{2}\bar{\zeta}a_0=\bar{F}\cos(\sigma T_2-b(T_2)) \tag{7.4.18}$$

from which it can be seen that this solution is possible, provided that $b(T_2)=\sigma T_2-\beta$, where β is a constant. This results in the two algebraic equations

$$2\sqrt{2}a_0\sigma+3\gamma a_0^3=\bar{F}\sin\beta,\quad 2\sqrt{2}\bar{\zeta}a_0=\bar{F}\cos\beta \tag{7.4.19a,b}$$

Squaring and adding gives

$$a_0^2\left(2\sqrt{2}\sigma+3\gamma a_0^2\right)^2+8\bar{\zeta}^2a_0^2=\bar{F}^2 \tag{7.4.20}$$

which is the equation for a_0, while β can be computed by

$$\tan\beta=\frac{2\sqrt{2}\sigma+3\gamma a_0^2}{2\sqrt{2}\bar{\zeta}} \tag{7.4.21}$$

Equation (7.4.20) is cubic in $a_0^2 = u$,

$$9\gamma^2 u^3 + 12\sqrt{2}\gamma\sigma u^2 + 8(\bar{\zeta}^2 + \sigma^2)u - \bar{F}^2 = 0 \qquad (7.4.22)$$

and thus $a_0(\bar{F}, \sigma, \gamma, \bar{\zeta})$ can be obtained analytically. For $\gamma = 1$ and $\bar{\zeta} = 0.025$ the function $a_0(\sigma)$ is depicted in Figure 7.5 for different values of $\bar{F}$. From this graph a softening response of the system can be seen as the curve bends to the left. The maximum values are given by the *backbone curve* representing the amplitude–frequency relationship of the nonlinear *free oscillations*. This can be obtained by determining the maximum value of the function $a_0(\sigma)$, i.e., solving $da_0/d\sigma = 0$. By the implicit function theorem, $da_0/d\sigma = 0$ is equivalent to $\partial f/\partial \sigma = 0$, where f is

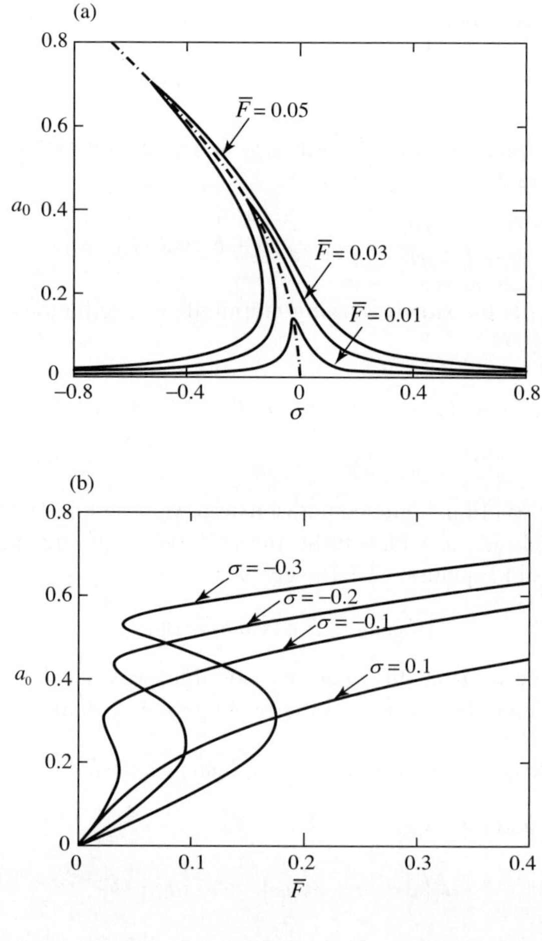

Figure 7.5 (a) Frequency-response curve $a_0(\sigma)$ for various excitation amplitudes, where the dashed-dotted line is the backbone curve, given by Equation (7.4.23); (b) forcing-response curves $a_0(\bar{F})$ for various detuning parameters and for $\gamma = 1$, $\bar{\zeta} = 0.025$.

the function on the left-hand side of Equation (7.4.22). This provides $u = -(2\sqrt{2}/3)(\sigma/\gamma)$, from which is obtained

$$a_{0,\max} = 0.9710\sqrt{\frac{-\sigma}{\gamma}} \tag{7.4.23}$$

which is dashed in Figure 7.5(a).

If instead of $\bar{F}$ the detuning parameter σ is fixed, the result is $a_0(\bar{F})$, which is depicted in Figure 7.5(b).

Figure 7.5(a) shows that for small values of $\bar{F}$ there is only one periodic solution for each value of the excitation frequency (represented by the detuning parameter σ), while for large values of $\bar{F}$ there is a region with three different periodic solution for each σ. This is the well-known nonlinear resonance hysteresis phenomenon [93], which is also demonstrated and discussed in Chapters 5, 6 as well as in Chapter 8. Figure 7.5(b) shows that the hysteresis occurs for negative values of σ because of the softening behaviour of the system.

There are three real solutions of Equation (7.4.22) if the discriminant

$$\Delta = -2187\gamma^4\bar{F}^4 - 1728\sqrt{2}\gamma^3\sigma(\sigma^2 + 9\bar{\zeta}^2)\bar{F}^2 - 18432\bar{\zeta}^2\gamma^2(\bar{\zeta}^2 + \sigma^2)^2 \tag{7.4.24}$$

is positive, and only one real solution if it is negative. Therefore, the inequality $\Delta > 0$ describes the parameter space (approximate) region where three solutions exist; outside, only one solution exists, and the boundary $\Delta = 0$ is then the locus of points where saddle-node bifurcations [27,30,31] occur. Solving $\Delta = 0$ with respect to $\bar{F}$ gives

$$\bar{F} = \frac{4}{9}\sqrt{\frac{2\sqrt{2}}{\gamma}}\sqrt{-9\bar{\zeta}^2\sigma - \sigma^3 \pm \sqrt{(\sigma^2 - 3\bar{\zeta}^2)^3}} \tag{7.4.25}$$

The functions $\bar{F}\sqrt{\gamma}$ are depicted in Figure 7.6 for $\bar{\zeta} = 0.025$ as a function of σ. Three solutions exist in the region inside the two curves. Note that Figure 7.6 is an

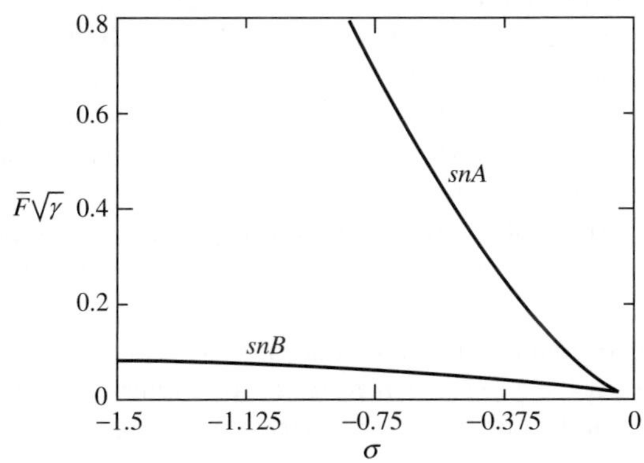

Figure 7.6 The saddle-node bifurcation curves in the $(\sigma, \bar{F}\sqrt{\gamma})$ plane for $\bar{\zeta} = 0.025$.

analytical approximation of the lower-right apex of Figure 7.11(a) (see later) and of Figure 1 of [58] (which is reproduced in Figure 7.11(b)).

The two curves coalesce at $\sigma = -\sqrt{3}\bar{\zeta}$ and $\bar{F} = 3.4077\sqrt{\bar{\zeta}^3/\gamma}$, which correspond to $F = \varepsilon^3\bar{F} = 3.4077\sqrt{\varepsilon^6\bar{\zeta}^3/\gamma} = 3.4077\sqrt{\zeta^3/\gamma}$ and $\Omega = \sqrt{2} + \varepsilon^2\sigma = \sqrt{2} - \sqrt{3}\varepsilon^2\bar{\zeta} = \sqrt{2} - \sqrt{3}\zeta$ (remember that $\zeta = \varepsilon^2\bar{\zeta}$). For $\gamma = 1$ and $\zeta = 0.025$ (the values used in the following numerical simulations) the vertex is at $\Omega \cong 1.371$ and $F \cong 0.0134$, in very good agreement with that determined numerically in Figure 7.11.

In addition to the method of multiple scales summarised in this section, there are several others approximate analytical methods for the study of the nonlinear resonance behaviour of twin-well Duffing equation. Among them, are the averaging method [11,12] and the harmonic balance method, which in [16] was applied together with continuation techniques.

7.5 Transition to complex response

7.5.1 Bifurcation diagrams, behaviour chart and basins of attraction

In the previous section the nonlinear resonance behaviour was investigated by determining analytically the large amplitude periodic oscillations that occur close to the main resonance. It was shown that a hysteresis phenomenon occurs (Figure 7.5), namely, a small-amplitude (nonresonant) periodic oscillation and a large-amplitude (resonant) periodic oscillation coexist in a certain region of the parameter space (Figure 7.6). However, even limiting the analysis to the case of the *principal resonance*, the scenario is more complex, and can be fully understood only by means of systematic numerical simulations.

As described in Section 7.2, numerical analysis of the twin-well Duffing equations has been addressed by many authors and with different purposes (see, for example, [33,48,53,55]). However, the most extended analysis was made by Szemplinska-Stupnicka and coworkers in a series of papers [46,49,57–60] where almost all aspects of the nonlinear dynamics have been investigated in depth. Since the main focus is on the dynamical behaviour around the principal resonance, reference [58] is probably the most important work as this is where the main phenomena investigated in other papers are summarised. However, so as not to duplicate exactly the numerical simulations conducted by Szemplinska-Stupnicka, a different damping ratio $\zeta = 0.025$ is chosen instead of $\zeta = 0.05$, which further sheds some light on the effects of damping. As will be seen, the main features are maintained, but some differences are observed.

The nonlinear resonance phenomenon is governed by the four competing attractor illustrated in the *bifurcation diagram* of Figure 7.7. They are (the labels in the figure correspond to those of [58], with S meaning *sink*):

- Attractor S_n, the nonresonant, small amplitude, in-well period-1 (i.e., the same period $2\pi/\Omega$ of the excitation) oscillations studied in Section 7.4. Note that

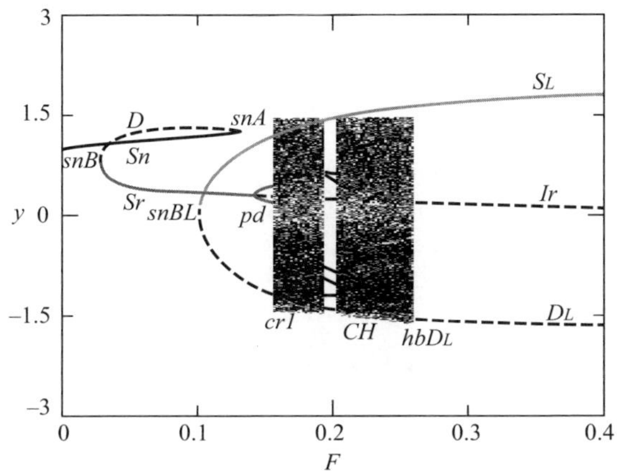

Figure 7.7 The bifurcation diagram of Equation (7.1.2) for $\zeta = 0.025$, $\gamma = 1$ *and* $\Omega = 1.15$. *The meaning of the various paths is described in the text.*

the S_n solution belonging to the other potential well is not shown, to simplify the figure.

- Attractor S_r, the resonant, large-amplitude, in-well period-1 oscillations studied in Section 7.4. Again, the S_r solution belonging to the other potential well is not shown.
- Attractor S_L, a cross-well, large-amplitude, period-1 solution, which is the unique attractor for large excitation values. It has no symmetric counterpart (contrary to S_n and S_r), as it is self-symmetric.
- Attractor *CH*, a cross-well (scattered) chaotic attractor.

Note that the following are not considered, (i) the 'weak' in-well chaotic attractor ensuing at the end of the *period-doubling cascade* (however this is shown in Figure 7.8(a)), since it is merged into the 'strong' *cross-well chaotic attractor* soon after being created (so that it is both practically and numerically irrelevant), and (ii) the periodic window within the chaotic attractor band.

Also, three principal saddles play an important role in the dynamical behaviour (Figure 7.7):

- Saddle *D*, the direct saddle joining resonant and nonresonant nonlinear oscillations through saddle-node bifurcation;
- Saddle I_r, the inverse saddle ensuing from a *period-doubling bifurcation* where the resonant oscillation loses stability;
- Saddle D_L, the direct saddle associated to the birth of the cross-well periodic oscillation S_L.

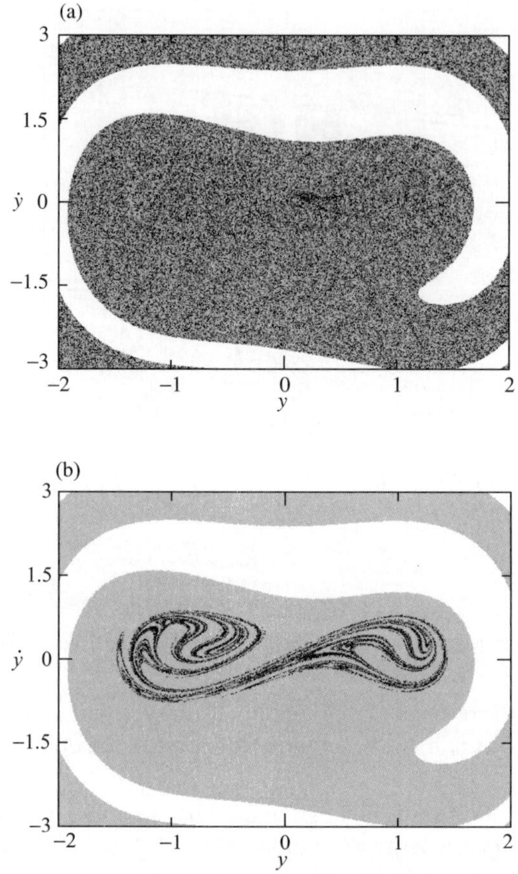

Figure 7.8 Overall attractor-basin phase portraits (a) before, $F = 0.155$, *and (b) after,* $F = 0.157$, *the global bifurcation cr1 determining the onset of the cross-well chaotic attractor (max Lyapunov exponent is* 0.138 *and capacity (fractal) dimension is* 1.518*) for* $\zeta = 0.025$, $\gamma = 1$ *and* $\Omega = 1.15$. *The white basin corresponds to the period-1 cross-well attractor.*

The paths of resonant (up to the period-doubling bifurcation *pd*) and nonresonant periodic attractor and the direct saddle correspond to those reported in Figure 7.5(b). The qualitative difference is simply due to the fact that in Figure 7.5(b) the amplitudes of the periodic motions are shown, while in Figure 7.7 the points corresponding to the stroboscopic Poincarè section are shown.

The range of existence of the previous attractor is delimited by the following local and *global bifurcations* (again, the labels are those of Figure 7.7 and of [58]):

- Bifurcation *snA* ($F \cong 0.131$): this is the (upper) saddle-node bifurcation where the nonresonant oscillation S_n and the direct saddle D coalesce and disappear for increasing F. It has been studied analytically in Section 7.4. Here, it is just noted that the numerical occurrence of *snA* is in very good agreement with that determined analytically.

- Bifurcation *snB* ($F \cong 0.028$): this is the (lower) saddle-node bifurcation where the resonant oscillation S_r and the direct saddle D jointly appear for increasing excitation amplitude. As with *snA*, it has been studied in Section 7.4 and the numerical result is in good agreement with the analytical prediction.
- Bifurcation *pd* ($F \cong 0.145$): this is the period-doubling bifurcation where the resonant periodic attractor loses stability with increasing excitation amplitude. It is followed by a classical period-doubling cascade which ends up with the boundary crisis *cr1* described later. Theoretical prediction of this event is not presented here; it can be found in [44].
- Bifurcation snB_L ($F \cong 0.101$): this is the saddle-node bifurcation at which the large-amplitude period-1 attractor S_L and its associated saddle D_L jointly appear as F is increased. An analytical approximation of this threshold by the subharmonic Melnikov method is provided in Section 7.5.2. The main difference with respect to [58] is that snB_L is below *cr1*, so that the crisis occurs (see next point) with the presence of a large amplitude periodic attractor, which however does not seem to directly modify the global bifurcation *cr1*.
- Bifurcation *cr1* ($F \cong 0.1567$): this is the first global bifurcation playing a major role in the system dynamics. It is responsible for the onset of the robust, cross-well chaotic attractor (Figure 7.8(b)). The attractor-basins phase portraits just before and after the bifurcation are reported in Figures 7.8(a) and (b), respectively, where also the basin of the coexisting cross-well period-1 attractor S_L is reported (in white).

  Apart from S_L, before the bifurcation there are two 'weak' in-well chaotic attractors, one per potential well. Their basins of attractions are fractal and strongly tangled (Figure 7.8(a), where the in-well attractors are barely visible). This (i) determines the strong sensitivity to initial conditions and practical unpredictability of the final outcome, thereby transforming the deterministic system into one with practically stochastic features, and (ii) is a consequence of the intersection of the hilltop saddle stable and unstable manifolds.

  The very strong fractality of Figure 7.8(a) comes from the fact that the homoclinic bifurcation of the hilltop saddle, which is the event triggering intertangling of the in-well attractor and that is studied analytically in Section 7.5.2, occurs at $F \cong 0.0408$, a value much lower than the present one, $F \cong 0.155$. Thus, the tangling, initially occurring around the unperturbed homoclinic orbits, had 'resources' to fully develop and to fill the 'whole' phase space.

  Contrary to the grey-grey basins boundaries, the white-grey boundary is regular. This is a consequence of the fact that the stable (which constitutes the boundary) and unstable manifolds of D_L are still disjoint. They undergo a homoclinic bifurcation later, determining the end of the scattered chaotic attractor (see the next point).

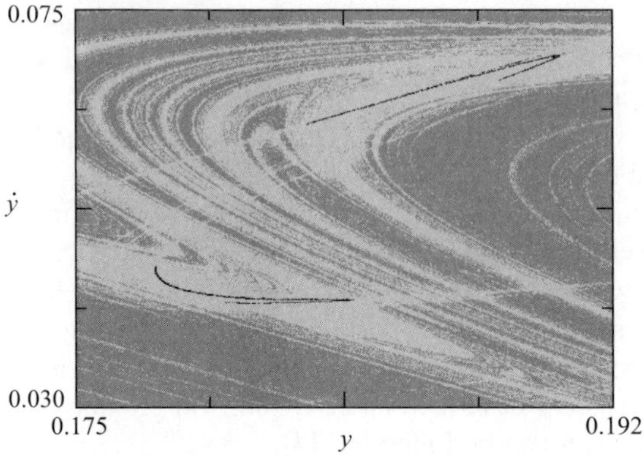

Figure 7.9 Enlargement of the attractor-basin phase portraits for $F = 0.1565$, *just before the boundary crisis cr1 for* $\zeta = 0.025$, $\gamma = 1$ *and* $\Omega = 1.15$.

The *cr1* bifurcation corresponds to a classical symmetric boundary crisis in which the in-well chaotic attractor simultaneously (by symmetry) touch their basin boundaries (as illustrated in the zoomed attractor-basin portrait of Figure 7.9) and suddenly disappear, with their former basins now belonging to the newly appeared scattered chaotic attractor (compare the grey colors in Figures 7.8(a) and (b)).

It is possible to show that the dynamical event that can be seen in the background is the homoclinic bifurcation of a period-3 saddle [58,63].

- Bifurcation hbD_L ($F \cong 0.2575$): this is the global bifurcation where the scattered chaotic attractor *CH* disappears, leaving S_L as the unique attractor. Just after being born at *cr1* (Figure 7.8(b)), *CH* starts to increase in size, until at $F \cong 0.2575$ it touches (Figure 7.10(a)) the boundary of its basin, and suddenly disappears through a classical boundary crisis. This event corresponds to the homoclinic bifurcation of D_L, as shown in Figure 7.10(b), and so it is clearly explained in terms of invariant manifolds. Comparing Figures 7.10(a) and (b) it is possible to see that the basin boundary of *CH* is just the stable manifold of D_L.

The dynamical behaviour described so far refers to the single value $\Omega = 1.15$ of the excitation frequency. To obtain the complete response around the principal resonance, the damping and nonlinearity are set so that $\zeta = 0.025$, $\gamma = 1$ and bifurcation diagrams are constructed for different excitation frequencies. The behaviour chart obtained in this way is shown in Figure 7.11(a), which corresponds to Figure 1 of [58]. This last picture is reproduced in Figure 7.11(b) to facilitate the comparison.

Figure 7.11(a) illustrates how the various bifurcation thresholds identified in Figure 7.7 vary with Ω. In addition, it can be seen that on segment *AB* there is another global bifurcation, not visible in Figure 7.7. This is a *subduction* [58], which corresponds to the sudden appearance (for increasing *F*) of the scattered chaotic attractor as a consequence of the saddle-node bifurcation *snA* where the nonresonant oscillation S_n disappears. *CH* 'inherits' the basin of attraction of S_n. Note that the

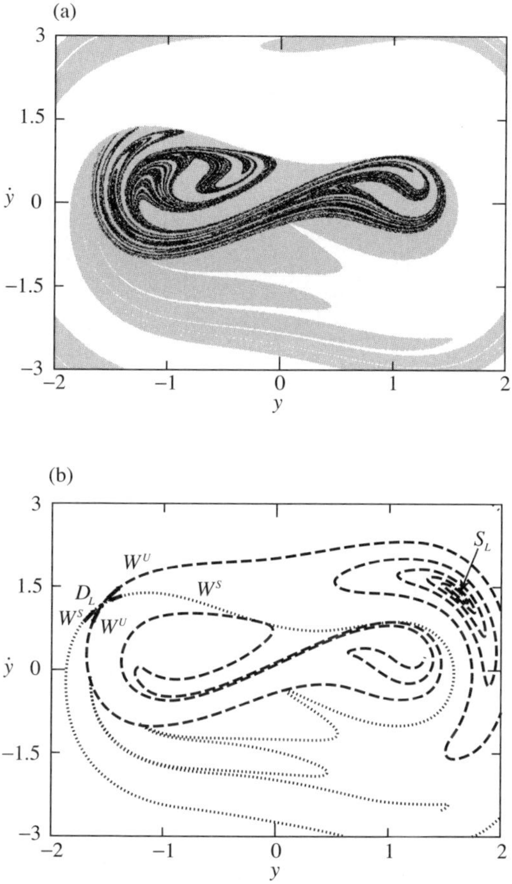

Figure 7.10 (a) Attractor-basin phase portraits (the chaotic attractor has max Lyapunov exponent equal to 0.181 *and capacity (fractal) dimension equal to* 1.587*) and (b) stable* W^S *and unstable* W^U *manifolds of* D_L *just before the homoclinic bifurcation of* D_L *corresponding to the boundary crisis for* $F = 0.257$, $\zeta = 0.025$, $\gamma = 1$ *and* $\Omega = 1.15$. *The white basin corresponds to the period-1 cross-well attractor.*

resonant oscillation S_r does not play a role since it had already disappeared because of the period-doubling cascade and the successive boundary crisis *cr1*.

The homoclinic bifurcation threshold hbD_H of the hilltop saddle D_H is shown, too, as it plays an important role in the topology of the boundaries of the basins of attraction. It is approximated analytically by using the Melnikov method in Section 7.5.2 (see Equation (7.5.15)).

Figure 7.11(a) shows the very good qualitative and quantitative agreement of the numerical curves *snA* and *snB* with the analytical curves shown in Figure 7.6. This confirms the effectiveness of the multiple scales method in detecting this aspect of the nonlinear resonance behaviour.

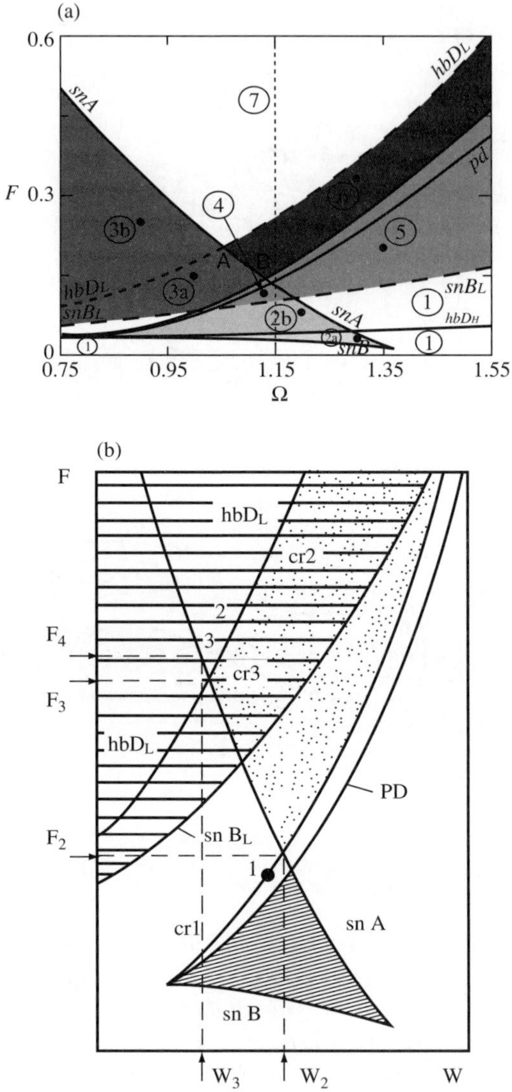

Figure 7.11 (a) The behaviour chart for $\zeta = 0.025$ and $\gamma = 1$. The dotted vertical line corresponds to the bifurcation diagram of Figure 7.7. The points correspond to the basins of attraction reported in the following. See Plate 2 for the coloured version of this figure. (b) Reprinted from [58], Copyright 1999, with permission from Springer Science + Business Media (Figure 1 therein is the behaviour chart for $\zeta = 0.05$).

The main difference between Figures 7.11(a) and (b), which highlights the role of damping, is that for $\zeta = 0.025$ the curve snB_L is lower, and partially below *pd*, while for $\zeta = 0.05$ is it always above *pd*. As will be seen, among other effects, this implies that while in [58] there is a parameter region in which *CH* is the unique attractor, here it always coexists with S_L.

The complex multistability behaviour is highlighted by the seven regions with different arrangement of the four attractor S_n, S_r, S_L and CH individuated before. They are:

Region 1: the unique attractor is S_n and so there are no special features of the basin of attraction.

Region 2: S_n and S_r coexist and compete with each other. Therefore there are four different attractors (two per potential well), as shown in Figure 7.12. This region is divided by hbD_H into two parts, 2a and 2b, below and above the hilltop homoclinic bifurcation threshold, respectively. The number of attractors does not change, but below hbD_H the left/right basin boundary is smooth

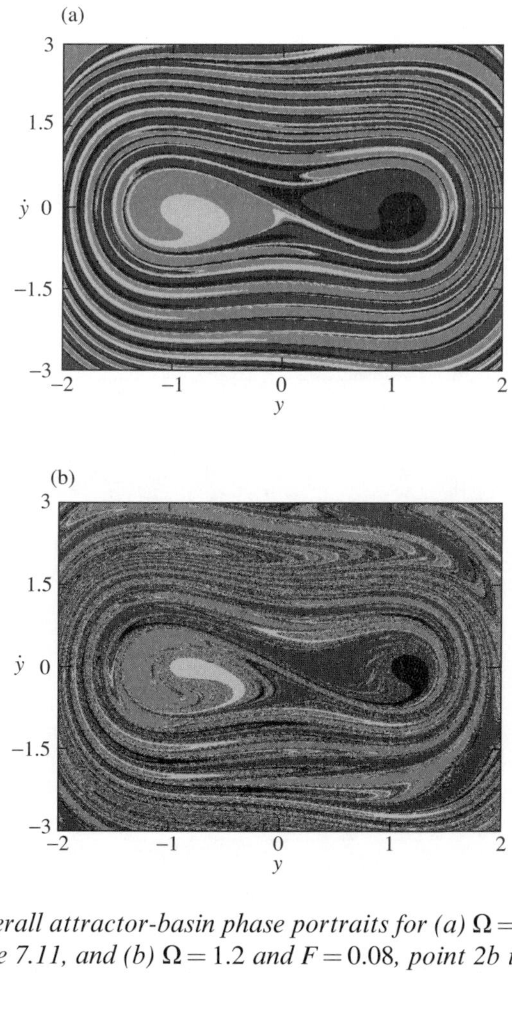

Figure 7.12 Overall attractor-basin phase portraits for (a) $\Omega = 1.3$ and $F = 0.03$, point 2a in Figure 7.11, and (b) $\Omega = 1.2$ and $F = 0.08$, point 2b in Figure 7.11, for $\zeta = 0.025$, $\gamma = 1$.

(Figure 7.12(a)), while above $hbD_{\rm H}$ the left/right boundary is fractal (Figure 7.12(b)), because the left/right basin boundary is the stable manifold of the hilltop saddle $D_{\rm H}$. The in-well resonant/nonresonant boundary, which is the stable manifold of D, is instead smooth in both cases, because D does not bifurcate homoclinically.

Region 3: $S_{\rm n}$ and $S_{\rm L}$ coexist and compete. There are two in-well nonresonant periodic attractor and one cross-well periodic attractor, as shown in Figure 7.13. Also this region is divided in two parts, now by $hbD_{\rm L}$. This explains why the white/grey boundaries of Figure 7.13(a) (below $hbD_{\rm L}$) are smooth, while the white/grey boundaries of Figure 7.13(b) (above $hbD_{\rm L}$) are fractal, since the stable manifold of $D_{\rm L}$ is the boundary of the basin of attraction of $S_{\rm L}$.

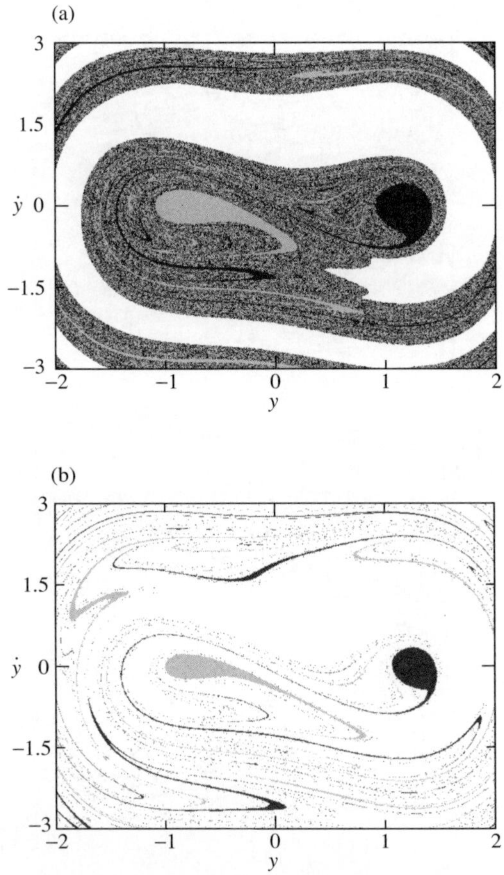

Figure 7.13 Overall attractor-basin phase portraits for (a) $\Omega = 1.0$ and $F = 0.15$, point 3a in Figure 7.11, and (b) $\Omega = 0.9$ and $F = 0.25$, point 3b in Figure 7.11, for $\zeta = 0.025$, $\gamma = 1$.

By comparing Figures 7.13(a) and (b) it can be seen that the extent of the basins of S_ns get smaller, because this point is closer to the saddle-node bifurcation *snA* where S_ns disappears. However, the extent of the 'compact' part of the basin is 'small' even for low excitation amplitudes, in spite of the fact that the whole basin covers a 'large' area. This is a consequence of the extended fractality, and means that the dynamic integrity of the solutions in both cases of Figure 7.13 is almost comparable, and small. This is an issue that has important practical consequences and is discussed further in Section 7.6.2.

Region 4: all the periodic attractor S_n, S_r and S_L coexist and compete. There are thus four in-well and one cross-well period-1 attractors, as shown in Figure 7.14. This is quite a small region, but it has the largest number of attractors, so that it is the most 'complex.' This region does not exist in Figure 7.11(b) since there snB_L is above the point B.

The white/grey basin boundary is smooth because the point is below hbD_L, but the grey/grey boundaries are fractal because it is above hbD_H.

Region 5: S_r and S_L coexist and compete. There are two in-well resonant periodic attractors and one cross-well periodic attractor, as shown in Figure 7.15. Although involving the resonant instead of nonresonant oscillations, this region is similar to 3a (compare Figures 7.13(a) and 7.15), with different shapes of the compact parts of the S_r grey basins (compare also with Figure 7.12).

Region 6: this is the unique region where the robust cross-well *chaotic attractor* exists, together with S_L (Figure 7.16; see also Figures 7.8(b) and 7.1(a)). Note that in [58] this region is divided into two parts by snB_L, the upper one similar to the present and the lower one with *CH* as the unique attractor.

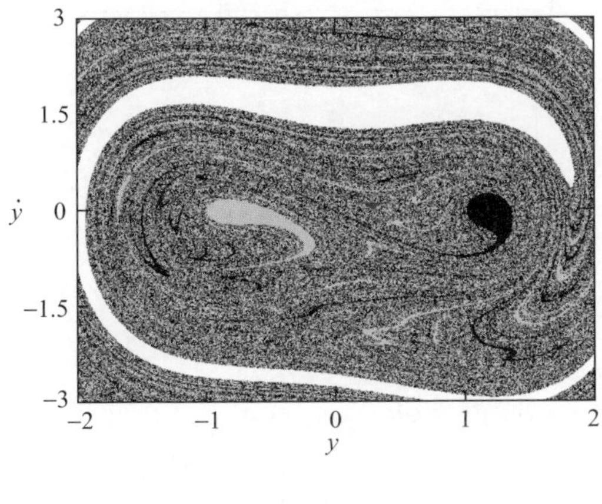

Figure 7.14 Overall attractor-basin phase portraits for $\Omega = 1.13$ and $F = 0.115$, point 4 in Figure 7.11, for $\zeta = 0.025$, $\gamma = 1$. See Plate 3 for the coloured version of this figure.

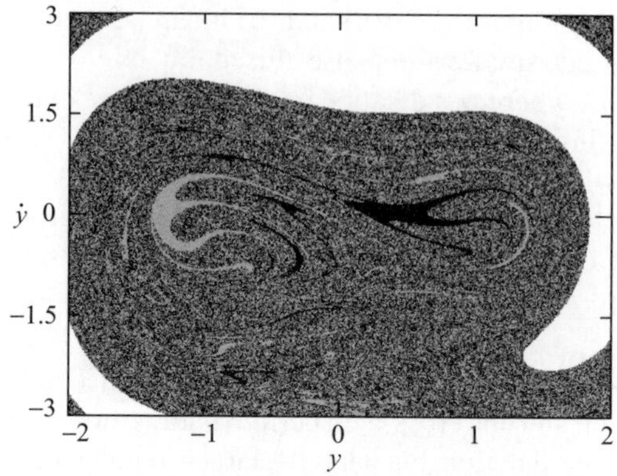

Figure 7.15 Overall attractor-basin phase portraits for $\Omega = 1.35$ *and* $F = 0.2$*, point 5 in Figure 7.11, for* $\zeta = 0.025$, $\gamma = 1$.

Region 7: the unique attractor is S_L and so, as in Region 1, there are no special features of the basin of attraction.

To end this section it should be noted that only the main attractor and bifurcational events have been illustrated to capture the key features of the nonlinear resonance. There are other bifurcations occurring in narrow sets of parameters space. Although of theoretical interest, they are of minor practical importance because they influence the dynamics only locally.

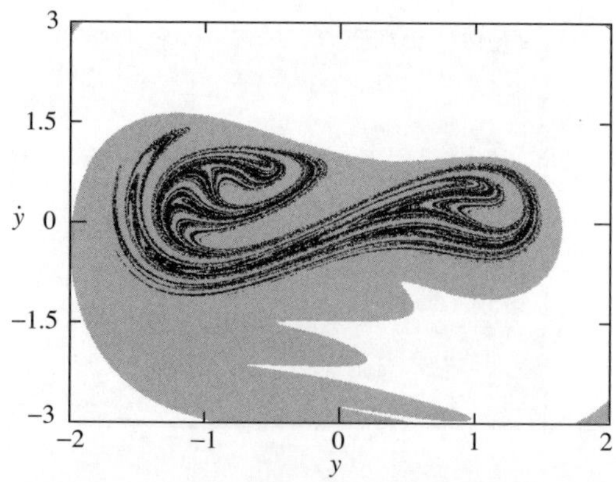

Figure 7.16 Overall attractor-basin phase portraits for $\Omega = 1.3$ *and* $F = 0.33$*, point 6 in Figure 7.11 for* $\zeta = 0.025$, $\gamma = 1$*. The chaotic attractor has max Lyapunov exponent equal to* 0.155 *and capacity (fractal) dimension equal to* 1.623.

7.5.2 Analytical prediction via the Melnikov method

In Region 2 of Figure 7.11 (see Figure 7.12) the practical relevance of the homoclinic bifurcation of the hilltop saddle has been seen. This threshold, as well as other relevant saddle-node bifurcations, can be analytically detected by the Melnikov method, which is an approximate method aimed at studying the persistence of periodic and homoclinic orbits of the unperturbed system (Section 7.3.1) under the perturbations (Sections 7.3.2 and 7.3.3).

In classical textbooks [27,30,31] and pioneering papers [2,13] a geometrical approach is followed. To complement this, an energetic approach [94,95] is considered in this section. This is less well known, but, in the authors' opinion, it is easier to understand.

The starting point for the present approach is the energy balance given by Equation (7.3.19). If $y_p(t)$ is a P-period orbit of the perturbed system (whose existence is herein initially assumed), then

$$0 = \int_0^P \dot{E} dt = \int_0^P (-2\zeta \dot{y}_p^2(t) + F\cos(\Omega t)\dot{y}_p(t))dt \tag{7.5.1}$$

is a necessary condition for its existence. If, on the contrary, $y_p(t)$ is a homoclinic orbit of the perturbed system, then the necessary condition (7.5.1) becomes

$$0 = \int_{-\infty}^{\infty} \dot{E} dt = \int_{-\infty}^{\infty} (-2\zeta \dot{y}_p^2(t) + F\cos(\Omega t)\dot{y}_p(t))dt \tag{7.5.2}$$

Equations (7.5.1) and (7.5.2) can be rewritten as

$$2\zeta \int_I \dot{y}_p^2(t)dt = F\int_I \cos(\Omega t)\dot{y}_p(t)dt \tag{7.5.3}$$

where $I = [0,P]$ or $]-\infty,\infty[$. Because of their practical relevance, in the following the analysis is restricted to subharmonic periodic orbits only, i.e., $P = n(2\pi/\Omega)$, where n is an integer and $2\pi/\Omega$ being the period of the external excitation. For the Melnikov method of ultrasubharmonic orbits the reader is referred to [27,30,31].

To stress that the perturbations are small with respect to the conservative case, it is assumed that $\zeta = \varepsilon\bar{\zeta}$, $F = \varepsilon\bar{F}$ (note the different scaling compared with Section 7.4). This nearly conservative assumption suggests that the perturbed solution $y_p(t)$ is close to a solution $y_0(t - t_0)$ of the unperturbed system (7.3.1) having the same characteristics (the same period for periodic orbits, *homoclinicity* for *homoclinic solutions*):

$$y_p(t) = y_0(t - t_0) + \varepsilon y_1(t) + \varepsilon^2 y_2(t) + \dots \tag{7.5.4}$$

where t_0 is introduced to emphasise that Equation (7.3.1) is autonomous, so that its solutions are defined up to an arbitrary time shifting. Substituting Equation (7.5.4) into Equation (7.5.3) yields

$$
\begin{aligned}
&2\zeta\left(\int_I \dot{y}_0^2(t-t_0)dt + 2\varepsilon\int_I \dot{y}_0(t-t_0)\dot{y}_1(t)dt + \ldots\right) \\
&= F\left(\int_I \cos(\Omega t)\dot{y}_0(t-t_0)dt + \varepsilon\int_I \cos(\Omega t)\dot{y}_1(t)dt + \ldots\right)
\end{aligned}
\tag{7.5.5}
$$

If ε is sufficiently small, the first-order terms are sufficient to determine the persistence of the solution under perturbations,

$$
2\zeta\int_I \dot{y}_0^2(t-t_0)dt = F\int_I \cos(\Omega t)\dot{y}_0(t-t_0)dt \tag{7.5.6}
$$

Only in special cases, for example double homoclinic tangencies, are the second-order terms required for a reliable analysis [96].

The great advantage of Equation (7.5.6) is that it allows information to be gained on the perturbed system by the sole knowledge of the solution $y_0(t-t_0)$ of the unperturbed system, and this makes the method very straightforward. This property is lost if higher-order terms are needed in Equation (7.5.5), because the determination of $y_1(t)$ requires solving the perturbed problem [96] (and this is why $y_1(t)$ is not invariant under time shifting).

The change of variable $\tau = t - t_0$ and trigonometric formulae permit the rewriting of Equation (7.5.6) in the simpler form (the extremes of integration do not change due to periodicity and homoclinicity)

$$
2\zeta\int_I \dot{y}_0^2(\tau)d\tau = F\left(\cos(\Omega t_0)\int_I \cos(\Omega\tau)\dot{y}_0(\tau)d\tau - \sin(\Omega t_0)\int_I \sin(\Omega\tau)\dot{y}_0(\tau)d\tau\right) \tag{7.5.7}
$$

By noting that for the conservative system Equation (7.3.6) holds, the left-hand side integral in Equation (7.5.7) becomes

$$
\int_I \dot{y}_0^2(\tau)d\tau = 2\int_{y_{0,\min}}^{y_{0,\max}} \sqrt{2E_0 + y_0^2 - \gamma\frac{y_0^4}{2}}dy_0 = \frac{4}{\gamma}\int_{x_{0,\min}}^{x_{0,\max}} \sqrt{\gamma E_0 + x_0^2 - x_0^4}dx = \frac{4}{\gamma}f_1(\gamma E_0) \tag{7.5.8}
$$

where $y_{0,\max}$ is given by Equation (7.3.7) in any case, while $y_{0,\min}$ is given by Equation (7.3.7) for in-well oscillations ($-1/4 < \gamma E_0 < 0$), $y_{0,\min} = 0$ for homoclinic *orbits* ($E_0 = 0$), and $y_{0,\min} = -y_{0,\max}$ for out-of-well oscillations ($E_0 > 0$).

Equation (7.5.8) shows that the integral is just the area inside the loop in the phase space (see Figure 7.2), and this explains why it has a discontinuity at $E_0 = 0$, i.e., across the homoclinic orbit, where it doubles, as clearly shown in Figure 7.17. Note that $y_{0,\max}$ ($x_{0,\max}$) and $y_{0,\min}$ ($x_{0,\min}$) are the points where the argument of the square root vanishes (the latter holds only for $-1/4 < \gamma E_0 < 0$).

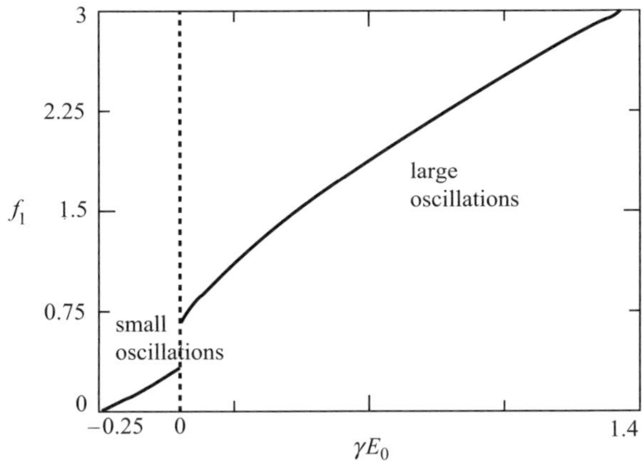

Figure 7.17 The function $f_1(\gamma E_0)$.

The function $f_1(\gamma E_0)$ can be written down in closed form by elliptic functions [27,30,31], but here it is shown in graphical form in Figure 7.17 and note that $f_1(\xi) = (\pi\sqrt{2}/4)(\xi + 1/4) + \ldots$ for $\xi \to -1/4$, $f_1(0^-) = 1/3$ (this value corresponds to homoclinic orbits) and $\lim_{\xi\to\infty} f_1(\xi) = \infty$.

In Equation (7.5.7) $y_0(\tau)$ is an arbitrary representation of the unperturbed solution, which can be assumed to be symmetric with respect to the point $\tau = 0$, i.e., $y_0(\tau) = y_0(-\tau)$. Thus, $\cos(\Omega\tau)\dot{y}_0(\tau)$ is an odd function, and its integral over the period vanishes. The nonvanishing integral on the right-hand side of Equation (7.5.7) becomes

$$\int_I \sin(\Omega\tau)\dot{y}_0(\tau)d\tau = -\Omega\int_I \cos(\Omega\tau)y_0(\tau)d\tau = -\Omega\sqrt{\frac{2}{\gamma}}\int_I \cos(\Omega\tau)x_0(\tau)d\tau \quad (7.5.9)$$

The following alternative expression is more useful in calculations of periodic orbits because it does not require explicit knowledge of $y_0(\tau)$:

$$\int_I \sin(\Omega\tau)\dot{y}_0(\tau)d\tau = 2\sqrt{\frac{2}{\gamma}}\int_{x_{0,\min}}^{x_{0,\max}} \sin\left(\Omega\int_{x_{0,\min}}^{x_0}\frac{d\xi}{\sqrt{\gamma E_0 + \xi^2 - \xi^4}}\right)dx_0 = 2\sqrt{\frac{2}{\gamma}}f_2(\gamma E_0, \Omega) \quad (7.5.10)$$

where Equation (7.3.9) has been used to obtain

$$\tau = \tau(x) = \int_{x_{0,\min}}^{x_0}\frac{d\xi}{\sqrt{\gamma E_0 + \xi^2 - \xi^4}} \quad (7.5.11)$$

Equation (7.5.11) shows that a symmetric representation $x_0(\tau)$ of the solution has been used such that $x_0(0) = x_{0,\min}$.

In the case of homoclinic orbits, i.e., $E_0 = 0$ and $x_0(\tau) = x_{\mathrm{hom}}(\tau)$ given by Equation (7.3.14b), there is, from Equation (7.5.9),

$$f_2(0,\Omega) = \Omega \int_0^\infty \frac{\cos(\Omega\tau)}{\cosh(\tau)}\, d\tau = \frac{\Omega\pi/2}{\cosh(\Omega\pi/2)} \tag{7.5.12}$$

while in the case of subharmonic orbits the function $f_2(\gamma E_0, \Omega)$ can be written in term of elliptic functions [27,30,31]. Since only subharmonic periodic orbits with period $P = n(2\pi/\Omega)$ are considered, by Equation (7.3.11) (see also Figure 7.3) there is a one-to-one relation between n and γE_0 (both for in-well and out-of-well oscillations), so that $f_2(\gamma E_0, \Omega)$ is actually $f_2(n, \Omega)$. Note that homoclinic orbits correspond to $n \to \infty$.

From the previous calculations and considerations, Equation (7.5.7) becomes

$$2\zeta \frac{4}{\gamma} f_1(n) = -F\sin(\Omega t_0) 2\sqrt{\frac{2}{\gamma}} f_2(n,\Omega) \tag{7.5.13}$$

that shows that the necessary condition for the existence of the perturbed subharmonic and homoclinic solution is

$$F > F^n_{\mathrm{cr}} = 2\zeta \sqrt{\frac{2}{\gamma}} \frac{f_1(n)}{|f_2(n,\Omega)|} \tag{7.5.14}$$

which is obtained by solving Equation (7.5.13) with respect to $\sin(\Omega t_0)$ and by noting that $-1 < \sin(\Omega t_0) < 1$. The function F^n_{cr} for in-well oscillation is plotted in Figure 7.18a for $n = 1, 2, 3$, together with F^h_{cr}, which is given below. This graph shows the rapid convergence of F^n_{cr} to F^h_{cr} for $n \to \infty$, and it is illustrated in Figure 7.18a with F^3_{cr} and F^h_{cr} overlapping.

Since for in-well oscillations the maximum frequency is $\sqrt{2}$ (see, for example, Figure 7.3 that shows that the minimum period is $\pi\sqrt{2}$), $f_2(n,\Omega)$ is defined only for $\Omega \in (0, n\sqrt{2})$, a fact that is confirmed by Figure 7.18(a). In particular, when $\Omega \to \sqrt{2}$, $x_0(\tau) \to (2/\sqrt{3})\sqrt{\sqrt{2}-\Omega}\cos(\Omega\tau + \varphi)$, so that $f_2(1,\Omega) \to (\pi/\sqrt{3})\sqrt{\sqrt{2}-\Omega}$. Since $f_1(1) \to (2\pi/3)(\sqrt{2}-\Omega)$ for $\Omega \to \sqrt{2}$, it has been shown that $F^1_{\mathrm{cr}} \to 0$ proportionally to $\sqrt{\sqrt{2}-\Omega}$ for $\Omega \to \sqrt{2}$, as is shown in Figure 7.18(a).

For out-of-well oscillations there are similar curves, $\bar{F}^n_{\mathrm{cr}}$, which are plotted in Figure 7.18(b) and that again rapidly converge to F^h_{cr} for $n \to \infty$. The only difference is that, since for out-of-well oscillations there is no upper bound for the frequency of the unperturbed oscillations (see Figure 7.3), $\bar{F}^n_{\mathrm{cr}}(\Omega)$ is now defined for all Ω. Furthermore, by symmetry, it is possible to see that the functions $\bar{F}^n_{\mathrm{cr}}$ exist only for odd values of n (Figure 7.18(b)).

So far, it has been shown that Equation (7.5.14) is a necessary condition. The fact that it is also a sufficient condition for the existence of perturbed solution is a more delicate mathematical issue, and the reader is referred to mathematically oriented textbooks [27,30,31] for a rigorous proof. Here, it is simply noted that the thresholds F^n_{cr} and $\bar{F}^n_{\mathrm{cr}}$ correspond to a saddle-node bifurcation where one stable and one unstable

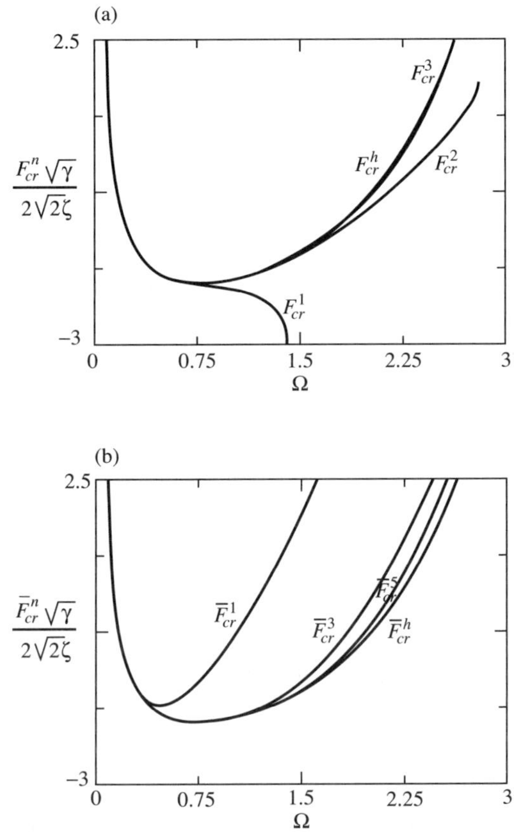

Figure 7.18 Functions (a) $F^n_{\rm cr}\sqrt{\gamma}/(2\sqrt{2}\zeta)$ and (b) $\bar{F}^n_{\rm cr}\sqrt{\gamma}/(2\sqrt{2}\zeta)$. The function $F^h_{\rm cr}\sqrt{\gamma}/(2\sqrt{2}\zeta)$ is also reported in both cases.

period n cycles are created. Thus, it is a sufficient condition for the existence of a periodic attractor.

The attractor born at $F^n_{\rm cr}$ and $\bar{F}^n_{\rm cr}$ may soon lose stability, typically by a period-doubling bifurcation, and they are possibly not so important in practice, but their existence is theoretically established just above $F^n_{\rm cr}$ and $\bar{F}^n_{\rm cr}$. The lowest index curves are very important, since it has been verified numerically that $F^1_{\rm cr}$ coincides with *snB* up to the vertex of the apex of Figure 7.6 and Figure 7.11. This gives another analytical expression of this local bifurcation threshold, while $\bar{F}^1_{\rm cr}$ coincides with $snB_{\rm L}$, and then provides an analytical expression of this curve. This, in particular, explains why $snB_{\rm L}$ is lower for $\zeta = 0.025$ compared with the case $\zeta = 0.05$ of [58]; more precisely, since $\bar{F}^1_{\rm cr}$ is linear in ζ, $snB_{\rm L}$ practically halves.

The previous considerations show the important role played by $F^1_{\rm cr}$ and $\bar{F}^1_{\rm cr}$ in the neighbourhood of the principal resonance, while the other curves do not seem to be important. This is possibly because they are too close to $F^h_{\rm cr}$ for this range of Ω.

For homoclinic orbits, the situation is slightly more involved. In fact, above the threshold

$$F_{\mathrm{cr}}^{h} = \frac{2\zeta}{3}\sqrt{\frac{2}{\gamma}\frac{\cosh(\Omega\pi/2)}{\Omega\pi/2}} \tag{7.5.15}$$

which is shown in Figure 7.18, a homoclinic orbit of the perturbed system exists (and this is not trivial since the perturbed system is not conservative). However, a homoclinic orbit belongs, by definition, to the stable and unstable manifolds (the loci of initial conditions that tend to the saddle forward and backward, respectively, in time) of the saddle, which is maintained in the perturbed case. It can then be concluded that for $F > F_{\mathrm{cr}}^{h}$ the stable and the unstable manifolds intersect, while they keep disjoint for $F < F_{\mathrm{cr}}^{h}$. Evidently, for $F = F_{\mathrm{cr}}^{h}$ the manifolds are tangent between each other, and this transition from detachment to intersection through a tangency of manifolds is called a homoclinic bifurcation. Though not pertaining to the hilltop saddle D_{H} considered here, a numerical example of manifolds tangency has been shown in Figure 7.10 for the D_{L} saddle.

The *homoclinic tangency*, in addition to triggering the *homoclinic transversal intersection* discussed above, is a very complex phenomenon. For example, Newhouse [97] theoretically showed that for certain parameter values near those at which homoclinic tangency occurs, infinitely many coexisting periodic attractors (called Newhouse sinks) exist. However, their basins of attraction could be so small as to be unobservable in physical and numerical experiments.

The stable and unstable manifold intersection is a situation that is very important from a theoretical point of view. In fact, by the Smale–Birkhoff theorem [27,30,31], it guarantees the existence of a *Smale horseshoe* in the system dynamics. This in turn implies that there are chaotic dynamics in the sense that the following complex phenomena occur, although to a varying extent:

1) Fractal basin boundaries. Various examples have been reported in Sections 7.5.1 and 7.6.2.
2) Sensitivity to initial conditions, i.e., certain solutions starting from arbitrarily close initial conditions diverge in time.
3) Chaotic transient, i.e., solutions may behave chaotically, even for a long time, before approaching a nonchaotic attractor.
4) Presence of the so-called chaotic saddles, i.e., of infinitely many saddles of any period, which of course affect the system dynamics.

The presence of a chaotic attractor is, however, not guaranteed, although a homoclinic intersection is a necessary (but not sufficient, indeed) condition for its existence, and although in some cases the chaotic transient could be so long to be confused or identified, at least from an engineering point of view, with a truly chaotic attractor.

To end this section it is noted that the Melnikov method is one of very few analytical methods that can be used to predict the occurrence of chaos (in the sense specified above) in dynamical systems.

7.6 Nonclassical analyses

In the previous sections some 'classical' analyses of nonlinear dynamics of the twin-well Duffing equation have been summarised. However, Equation (7.1.1) has been used also for 'nonclassical' analyses, applications and developments, which further underlines the archetypal nature of the equation. In this section the control of chaos through elimination (or shift in parameters space) of homoclinic bifurcation (Section 7.6.1) is considered, and in Section 7.6.2 dynamical integrity is considered.

7.6.1 Control of homoclinic bifurcation

Controlling a dynamical system means acting on and/or modifying the system to obtain desired goals in term of dynamical response. The control action and the expected results can be very general and can vary significantly from case to case; the reader is referred to Section 7.2 for a survey of control methods applied to the twin-well Duffing equation. Here, attention is restricted to the following cases:

- The control is open-loop (no feedback), i.e., it does not depend on the system state, and so sensors are not required.
- The control action is by external and parametric excitation added to the primary external excitation $F\cos\Omega t$. Since open-loop control is considered and the excitation is periodic, it is assumed that the control excitations are also periodic with the same period $2\pi/\Omega$.
- The goal is the elimination, or better the shift in parameter space, of a homoclinic bifurcation responsible for unwanted dynamical events (see Section 7.5.1 to see the effects of homoclinic bifurcation). In particular, the homoclinic bifurcation of the hilltop saddle is considered, which can be detected analytically [88], and the homoclinic bifurcation of D_{L}, which is detected numerically [89].

According to the previous points, for the purpose of control the following equation is considered

$$\ddot{y} + 2\zeta\dot{y} - (1 + f_{p1}^c(t))y + (\gamma + f_{p2}^c(t))y^3 = F\cos\Omega t + f_{\mathrm{e}}^c(t) \tag{7.6.1}$$

where $f_{p1}^c(t)$, $f_{p2}^c(t)$ are parametric excitations and $f_{\mathrm{e}}^c(t)$ is the external control excitation. They are periodic with period $2\pi/\Omega$, and have zero mean, so the following Fourier representations hold:

$$f_{p1}^c(t) = \sum_{\mathrm{n}=1}^{N} f_{p1,n}\cos(n\Omega t + \phi_n) \tag{7.6.2}$$

$$f_{p2}^{c}(t) = \sum_{m=1}^{M} f_{p2,m} \cos(m\Omega t + \varphi_m) \tag{7.6.3}$$

$$f_{e}^{c}(t) = \sum_{j=2}^{J} f_{e,j} \cos(j\Omega t + \psi_j) \tag{7.6.4}$$

Note that in Equation (7.6.4) $f_{e,1} = 0$, otherwise the trivial choice $f_{e,1} = -F$ and $\psi_1 = 0$ shows that the control problem is trivial because in this case the excitation is simply eliminated.

Control of the homoclinic bifurcation of the hilltop saddle D_H is considered. As shown in Section 7.5.2, this threshold can be determined by the Melnikov method, which means that the results of Section 7.5.2 can be extended. The energy balance (7.3.19) becomes

$$\dot{E} = -2\zeta\dot{y}^2 + F\cos(\Omega t)\dot{y} + f_{p1}^{c}(t)y\dot{y} - f_{p2}^{c}(t)y^3\dot{y} + f_{e}^{c}(t)\dot{y} \tag{7.6.5}$$

Integrate from $t \to -\infty$ to $t \to +\infty$ and suppose that the right-hand side of Equation (7.6.5) is small, i.e., only first order terms in the equivalent of Equation (7.5.5) are considered. Recall that $x = y\sqrt{\gamma/2}$, so that (see Section 7.5.2):

$$\begin{aligned}
\frac{4\zeta}{\gamma}\int_{-\infty}^{\infty} \dot{x}_{\text{hom}}^2(\tau)d\tau &= -F\sqrt{\frac{2}{\gamma}}\sin(\Omega t_0)\int_{-\infty}^{\infty} \sin(\Omega\tau)\dot{x}_{\text{hom}}(\tau)d\tau \\
&- \frac{2}{\gamma}\sum_{n=1}^{N} f_{p1,n}\sin(n\Omega t_0 + \phi_n)\int_{-\infty}^{\infty} \sin(n\Omega\tau)x_{\text{hom}}(\tau)\dot{x}_{\text{hom}}(\tau)d\tau \\
&+ \frac{4}{\gamma^2}\sum_{m=1}^{M} f_{p2,m}\sin(m\Omega t_0 + \varphi_m)\int_{-\infty}^{\infty} \sin(m\Omega\tau)x_{\text{hom}}^3(\tau)\dot{x}_{\text{hom}}(\tau)d\tau \\
&- \sqrt{\frac{2}{\gamma}}\sum_{j=2}^{J} f_{e,j}\sin(j\Omega t_0 + \psi_j)\int_{-\infty}^{\infty} \sin(j\Omega\tau)\dot{x}_{\text{hom}}(\tau)d\tau
\end{aligned} \tag{7.6.6}$$

Note that $x_{\text{hom}}(\tau) = 1/\cosh(\tau)$, see Equation (7.3.14b); so (by taking the minus sign the results are obtained by symmetry)

$$\int_{-\infty}^{\infty} \dot{x}_{\text{hom}}^2(\tau)d\tau = \frac{2}{3},$$

$$\int_{-\infty}^{\infty} \sin(j\Omega\tau)\dot{x}_{\text{hom}}(\tau)d\tau = -j\Omega\int_{-\infty}^{\infty} \frac{\cos(j\Omega\tau)}{\cosh(\tau)}d\tau = -\frac{\pi j\Omega}{\cosh\left(\frac{\pi j\Omega}{2}\right)},$$

$$\int_{-\infty}^{\infty}\sin(n\Omega\tau)x_{\text{hom}}(\tau)\dot{x}_{\text{hom}}(\tau)d\tau = -\frac{n\Omega}{2}\int_{-\infty}^{\infty}\frac{\cos(n\Omega\tau)}{\cosh^2(\tau)}d\tau = -\frac{\pi(n\Omega)^2}{2}\frac{1}{\sinh\left(\frac{\pi n\Omega}{2}\right)},$$

$$\int_{-\infty}^{\infty}\sin(m\Omega\tau)x_{\text{hom}}^3(\tau)\dot{x}_{\text{hom}}(\tau)d\tau = -\frac{m\Omega}{4}\int_{-\infty}^{\infty}\frac{\cos(m\Omega\tau)}{\cosh^4(\tau)} = -\frac{\pi(m\Omega)^2}{26}\frac{((m\Omega)^4+4)}{\sinh\left(\frac{\pi m\Omega}{2}\right)} \tag{7.6.7}$$

Thus, Equation (7.6.6) becomes

$$\begin{aligned}\frac{8\zeta}{3\gamma} = {} & F\sqrt{\frac{2}{\gamma}}\sin(\Omega t_0)\frac{\pi\Omega}{\cosh\left(\frac{\pi\Omega}{2}\right)} + \frac{2}{\gamma}\sum_{\text{n=1}}^{N} f_{p1,n}\sin(n\Omega t_0+\phi_n)\frac{\pi(n\Omega)^2}{2}\frac{1}{\sinh\left(\frac{\pi n\Omega}{2}\right)} \\ & - \frac{4}{\gamma^2}\sum_{\text{m=1}}^{M} f_{p2,m}\sin(m\Omega t_0+\varphi_m)\frac{\pi(m\Omega)^2}{26}\frac{((m\Omega)^4+4)}{\sinh\left(\frac{\pi m\Omega}{2}\right)} \\ & + \sqrt{\frac{2}{\gamma}}\sum_{\text{j=2}}^{J} f_{e,j}\sin(j\Omega t_0+\psi_j)\frac{\pi j\Omega}{\cosh\left(\frac{\pi j\Omega}{2}\right)} = def = \frac{8\chi(\Omega t_0)}{3\gamma}\end{aligned} \tag{7.6.8}$$

The function $\chi(\Omega t_0)$is simpler than it may appear, since it is just a function with period 2π, and by simple trigonometric formulae it can be rewritten in the form

$$\chi(\Omega t_0) = \sum_{\text{l=1}}^{L}(\alpha_l\sin(l\Omega t_0)+\beta_l\cos(l\Omega t_0)) = \sum_{\text{l=1}}^{L}\mu_l\sin(l\Omega t_0+v_l) \tag{7.6.9}$$

where $L = \max\{J,N,M\}$, $\mu_l = \sqrt{\alpha_l^2+\beta_l^2}$, $\tan v_l = \beta_l/\alpha_l$ and where

$$\alpha_1 = F\frac{3}{4}\sqrt{\frac{\gamma}{2}}\frac{\pi\Omega}{\cosh\left(\frac{\pi\Omega}{2}\right)} + f_{p1,1}\frac{3\pi\Omega^2}{8}\frac{\cos(\phi_1)}{\sinh\left(\frac{\pi\Omega}{2}\right)} - f_{p2,1}\frac{3\pi\Omega^2}{52\gamma}\frac{(\Omega^4+4)\cos(\varphi_1)}{\sinh\left(\frac{\pi\Omega}{2}\right)},$$

$$\beta_1 = f_{p1,1}\frac{3\pi\Omega^2}{8}\frac{\sin(\phi_1)}{\sinh\left(\frac{\pi\Omega}{2}\right)} - f_{p2,1}\frac{3\pi\Omega^2}{52\gamma}\frac{(\Omega^4+4)\sin(\varphi_1)}{\sinh\left(\frac{\pi\Omega}{2}\right)},$$

$$\begin{aligned}\alpha_l = {} & f_{p1,l}\frac{3\pi(l\Omega)^2}{8}\frac{\cos(\phi_l)}{\sinh\left(\frac{\pi l\Omega}{2}\right)} - f_{p2,l}\frac{3\pi(l\Omega)^2}{52\gamma}\frac{((l\Omega)^4+4)\cos(\varphi_l)}{\sinh\left(\frac{\pi l\Omega}{2}\right)} \\ & + \frac{3}{4}\sqrt{\frac{\gamma}{2}}f_{e,l}\frac{\pi l\Omega\cos(\psi_l)}{\cosh\left(\frac{\pi l\Omega}{2}\right)},\ l>1,\end{aligned}$$

$$\beta_l = f_{p1,l}\frac{3\pi(l\Omega)^2}{8}\frac{\sin(\phi_l)}{\sinh\left(\frac{\pi l\Omega}{2}\right)} - f_{p2,l}\frac{3\pi(l\Omega)^2}{52\gamma}\frac{((l\Omega)^4+4)\sin(\varphi_l)}{\sinh\left(\frac{\pi l\Omega}{2}\right)} + \frac{3}{4}\sqrt{\frac{\gamma}{2}}f_{e,l}\frac{\pi l\Omega\sin(\psi_l)}{\cosh\left(\frac{\pi l\Omega}{2}\right)},\ l > 1 \tag{7.6.10a–d}$$

Note that $\chi(\Omega t_0)$ takes into account all the control excitations $f^c_{p1}(t)$, $f^c_{p2}(t)$ and $f^c_e(t)$, in a compact way, so it can be considered as an equivalent excitation [98].

From Equation (7.6.8) the condition for the existence of the perturbed homoclinic orbit (and related chaotic behaviour) is

$$\zeta < \max\{\chi(\Omega t_0)\} = \bar{M} \tag{7.6.11}$$

where the single number $\bar{M}$ summarises the effects of the control excitation on the system hilltop saddle homoclinic behaviour. Note that Equation (7.6.11) agrees with the notion that for the persistence of homoclinic orbits in the perturbed system a sufficiently small amount of damping is needed, since it is the damping that tends to keep the stable and unstable manifold disjoint. It is the excitation that triggers the intersection of the manifolds [95].

In the simplest case $L = 1$ (only parametric controlling excitations, see for example [99]), Equation (7.6.11) yields

$$\zeta < \mu_1 = \sqrt{\alpha_1^2 + \beta_1^2} \tag{7.6.12}$$

which in the space of control parameters $f_{p1,1}$, φ_1, $f_{p2,1}$ and ϕ_1 defines the chaotic region. Thus, it is sufficient to choose $f_{p1,1}$, φ_1, $f_{p2,1}$ and ϕ_1 from outside this region to eliminate chaos. To show this by an example, consider $f_{p2,1} = 0$, which in turn implies that ϕ_1 disappears and the only two parameters are $f_{p1,1}$ and φ_1. The inequality given in Equation (7.6.12) becomes

$$\left(\frac{f_{p1,1}}{F}\right)^2 + a\left(\frac{f_{p1,1}}{F}\right)\cos(\phi_1) + b > 0, \quad a = 4\sqrt{\frac{\gamma}{2}}\frac{\tanh\left(\frac{\pi\Omega}{2}\right)}{\Omega} > 0,$$
$$b = \frac{a^2}{4} - 64\frac{\sinh^2\left(\frac{\pi\Omega}{2}\right)}{9\pi^2\Omega^4}\left(\frac{\zeta}{F}\right)^2 \tag{7.6.13a–c}$$

which in the parameter space $(\varphi_1, f_{p1,1}/F)$ gives a region whose boundary is the closed curve C depicted in Figure 7.19 for various values of b.

Note the following:

1) To have useful control excitation, there must be a homoclinic intersection for $f_{p1,1} = 0$ (otherwise the uncontrolled system is nonchaotic and control is

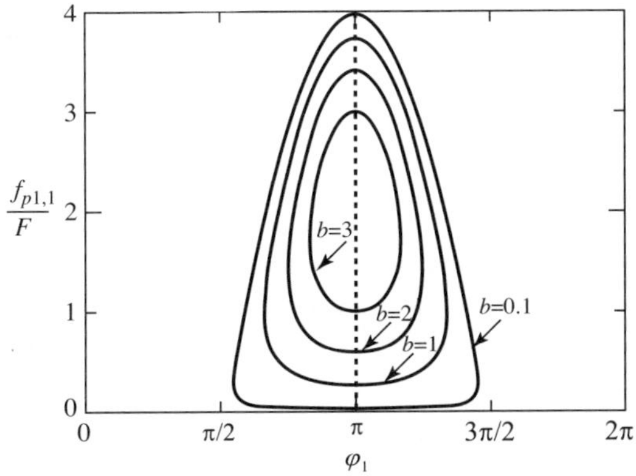

Figure 7.19 The curves C for a = 4 and for various values of b.

useless), namely, it is necessary that $b \geq 0$, which in turn gives Equation (7.5.15);

2) Control is achieved inside the region delimited by C, where it is possible to eliminate chaos from the system;
3) The curve C exists provided $a^2 > 4b$; this always occurs in the present case because

$$a^2 - 4b = 256 \frac{\sinh^2\left(\frac{\pi\Omega}{2}\right)}{9\pi^2\Omega^4} \left(\frac{\zeta}{F}\right)^2 > 0;$$

4) For a fixed a, i.e., for a fixed excitation frequency, the smaller the value of b (down to 0) the larger is the interior part of C, i.e., the larger is the region where the control is theoretically effective. Note that b is small when either the damping ζ is large or the excitation amplitude F is small;
5) The maximum horizontal extension of C is achieved for $f_{p1,1}/F = \sqrt{b}$;
6) The maximum vertical extension of C is obtained for $\varphi_1 = \pi$;
7) For $\varphi_1 = \pi$ control exists for the lowest value of $f_{p1,1}$ (more precisely for $f_{p1,1}/F = (a - \sqrt{a^2 - 4b})/2$), i.e., chaos is eliminated for the lowest parametric control effort on the linear term. This shows why $\varphi_1 = \pi$ is the optimal choice for the phase φ_1.

In the case when $L > 1$, the situation is more involved and more appealing. In this case considered here the value of $\bar{M}$ in Equation (7.6.11) can be modified. More precisely, $\bar{M}$ can be reduced to shrink the region in parameter space where 'chaotic' behaviour exists, or equivalently enlarge the region where chaotic behaviour is eliminated.

Actually, much more can be done. The control excitations can be chosen to reduce $\bar{M}$ as much as possible, in the pursuit of optimal control. This issue has been addressed in several papers by the authors [72,88–90,98,100–106]. Due to the lack of space, however, only the simplest situation called 'one-side' control is summarised.

From a mathematical point of view the optimisation problem is

$$\text{minimise } \bar{M} = \max\{\chi(\Omega t_0)\} \text{ by varying the Fourier coefficients } \alpha_l \text{ and } \beta_l \text{ (or alternatively } \mu_l \text{ and } \nu_l\text{) of } \chi(\Omega t_0). \tag{7.6.14}$$

Before proceeding, it is necessary to stress that if both the parametric excitations are present, the problem can become trivial: if $\chi(\Omega t_0)=0$ (i.e., no chaos) by the assumption $f_{p1,l}=f_{p2,l}=f_{e,l}=0$ for $l>1$ (which provides $\alpha_l=\beta_l=0$ for $l>1$), $f_{p1,1}$ and $f_{p2,1}$ can be chosen so that $\alpha_1=\beta_1=0$ (this is straightforward because Equation (7.6.10a) and Equation (7.6.10b) form a linear system with respect to $f_{p1,1}$ and $f_{p2,1}$).

Thus, the meaningful cases are those when the previous situation does not occur, for example when only superharmonic control terms are applied, namely $f_{p1,1}=f_{p2,1}=0$. In this case

$$\alpha_1 = F\frac{3}{4}\sqrt{\frac{\gamma}{2}}\frac{\pi\Omega}{\cosh\left(\dfrac{\pi\Omega}{2}\right)}, \quad \beta_1=0 \tag{7.6.15a,b}$$

cannot be modified by the controlling excitation, so that the optimisation problem (7.6.14) is needed. It has been solved in [100] for $L=\infty$ and in [88] for finite values of L, which is the case that applies to the present situation. The optimal solutions are reported in Table 7.1 for increasing values of L, and the related optimal functions $\chi(\Omega t_0)$ are depicted in Figure 7.20. In particular, this figure allows the effect of decreasing $\bar{M}$ (the maximum of $\chi(\Omega t_0)$) to be seen in the controlled case. This means that the size of the nonchaotic region by Equation (7.6.11) is increased.

The enlargement of the nonchaotic zone can be further seen by comparing $\bar{M}_L$ for the optimal solutions with $\bar{M}_h = 1$ of the reference case for harmonic excitation. This gives the so-called 'gain' $G_L = \bar{M}_h/\bar{M}_L$, which is also given in Table 7.1. It provides a measure of the enlargement of the nonchaotic zone, i.e., of the effectiveness of the control excitations [72,90,100].

Once the optimal parameters α_l and β_l have been determined by solving the optimal problem (7.6.14) (see Table 7.1), the optimal controlling excitation can be determined by solving Equations (7.6.10c) and (7.6.10d) with respect to $f_{p1,l}$, φ_l, $f_{p2,l}$, ϕ_l, $f_{e,l}$ and ψ_l. Apart from the case when there is only one controlling excitation, this is

Table 7.1 The solutions of the optimisation problem defined by Equation (7.6.14) with increasing finite number of controlling terms. The coefficients not reported vanish.

L	$\bar{M}_L$	G_L	β_2/α_1	α_3/α_1	β_4/α_1	α_5/α_1
2	0.7071	1.4142	0.353553			
3	0.6180	1.6180	0.552756	−0.170789		
4	0.5773	1.7321	0.673525	−0.333274	−0.096175	
5	0.5550	1.8019	0.751654	−0.462136	−0.215156	0.059632

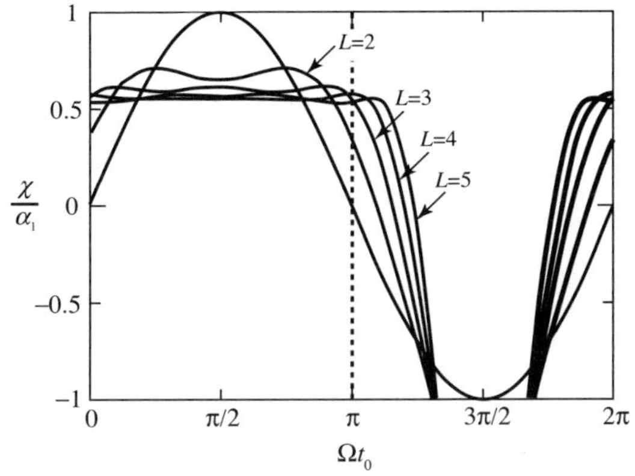

Figure 7.20 The optimal functions curves $\chi(\Omega t_0)/\alpha_1$. The harmonic function $\sin(\Omega t_0)/\alpha_1$ corresponding to the uncontrolled case $L = 1$, is also shown for reference.

an under determined system. Thus, while the solution of the mathematical optimisation problem (7.6.14) is unique, the optimal solution is not unique in terms of the physical excitation.

An example of the effectiveness of the proposed method in eliminating the homoclinic intersection of the hilltop saddle is shown in Figure 7.21. Here only the external control excitation with one superharmonic is considered (i.e., $f_{p1}^{c}(t) = f_{p2}^{c}(t) = 0$ and $f_{e}^{c}(t) = f_{e,2}\cos(2\Omega t + \psi_2)$). In this case the optimal solution is obtained for $\psi_2 = \pi/2$ and

$$f_{e,2} = 0.3535F \frac{\cosh(\pi\Omega)}{2\cosh(\pi\Omega/2)} \tag{7.6.16}$$

which in the case $\Omega = 1$ provides $f_{e}^{c}(t) = -0.8168F\sin(2t)$ and $F_{cr}^{h} = 1.506\,\zeta/\sqrt{\gamma}$.

The method for eliminating homoclinic intersections illustrated so far can be applied to any saddle whose intersection of manifolds is responsible for an unwanted dynamical event. The only difference compared to the hilltop case is that the Melnikov method cannot be applied, and so a numerical approach is necessary. This includes the optimisation problem, but it does not change the main idea of the control method. This issue has been developed in [89], which is referred to for further details.

Here, in Figure 7.22 it is shown how, by the simple (nonoptimal) addition of the external controlling excitation $f_{e}^{c}(t) = -0.2F\cos(3t)$, it is possible to eliminate the homoclinic tangency of Figure 7.10(b). This permits anticipation (as excitation amplitudes are decreased) of the boundary crisis hbD_L and the associated sudden destruction of the scattered chaotic attractor *CH* (see Section 7.5.1 and in particular Figure 7.7), thus reducing the chaotic region of the system.

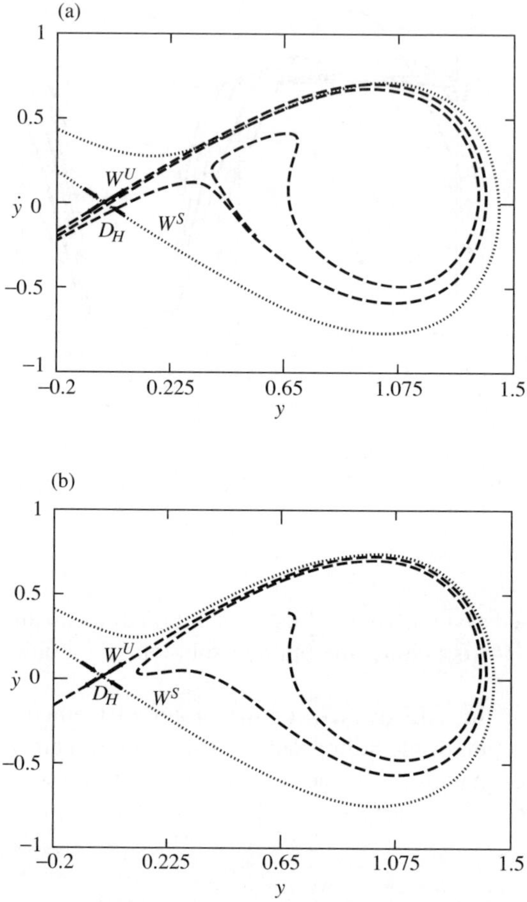

Figure 7.21 Stable manifold W^S and unstable manifold W^U of the hilltop saddle D_H for (a) uncontrolled and (b) controlled case for $F = 0.045$, $\zeta = 0.025$, $\gamma = 1$ and $\Omega = 1$. Note that in this case $F^h_{\rm cr} = 0.0376$, while with the controlling superharmonic the critical threshold is $F_{\rm cr} = 0.0532$.

7.6.2 Dynamical integrity

Following the pioneering work of Thompson and coworkers [53,107–109], it was realised that the stability of an attractor is not enough for practical applications. The compact part of the basin of attraction surrounding the attractor could be so small because of fractality to make the theoretically stable oscillation practically useless. This is because small perturbations can lead to the basin of another attractor and thus to practical instability and unpredictability. An example of this situation is given in Figure 7.8(a), where two stable periodic attractors, one per potential well, exist, but

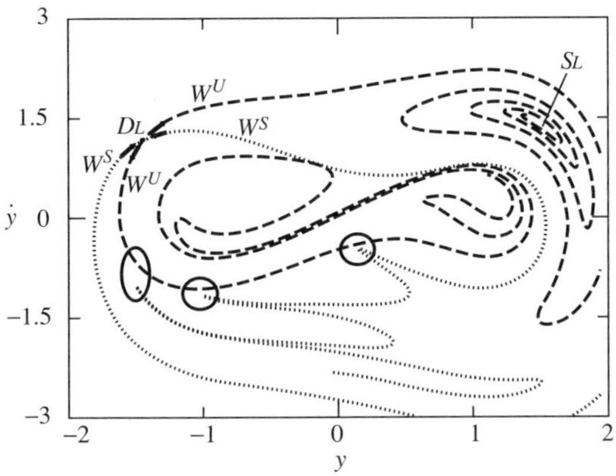

Figure 7.22 Stable manifold W^S and unstable manifold W^U of the saddle D_L for the controlled case $f_e^c(t) = -0.2F\cos(3t)$ and for $F = 0.257$, $\zeta = 0.025$, $\gamma = 1$ and $\gamma = 1.15$. Compare with Figure 7.10(b) to appreciate how the control detaches the manifolds (circles are points of previous tangency).

their basins of attractions are so intertwined that this clearly represents an unsafe situation.

The *integrity of dynamical systems* has been recently the subject of renewed interest in view of different applications, see for example [64,65,110] and references quoted therein. In this section some considerations and results are summarised with specific reference to Equation (7.1.2).

The fundamental issues necessary for an accurate investigation of the dynamical integrity are (i) the appropriate definition of safe basin, along with a quantitative measure capable of detecting the relevant aspects, and (ii) the analysis of how the integrity changes as the system parameters change.

7.6.2.1 Safe basins

The safe basin is the union, in phase space, of all initial conditions guaranteeing some desired dynamical response. It can be the convergence in time toward one (or more) attractor(s), in this case coinciding with a (union of) classical basin(s) of attraction, or it can be the nonescape from a given potential well. Also, the opposite can be considered in principle, for example, the safe basin can be the union of initial conditions guaranteeing the escape from a potential well.

The safe basin can be determined either by considering only the steady-state dynamics or by paying attention also to the transient dynamics. This may be in cases where the short-term dynamics are the most important part of the response or some unsafe phenomena, such as a temporary escape from the potential well, may occur during the transient. In several applications, the differences are actually minor and this distinction remains mainly a theoretical issue.

7.6.2.2 Integrity measures

Various parameters have been proposed to measure the magnitude of the safe basins [64,108]. The differences reflect the two underlying competing requirements of being computationally simple and of capturing the main features of the integrity. Only two of them, whose combined use seems to provide enough information without excessive computational efforts, are considered.

The *global integrity measure* (*GIM*) is the normalised hypervolume (area in 2D cases) of the safe basin. In this case normalisation means that the hyper-volume of the safe basin corresponding to the actual value of a varying parameter is divided by that corresponding to a reference value, so that *GIM* is a dimensionless number. It is the most intuitive and easy integrity measure to calculate, but it is not satisfactory in all cases in which the safe basin is strongly fractal, as shown in Figure 7.8(a): for both attractors the area of the basin of attraction is large, and thus they have a large GIM, in spite of the fact that the integrity of these attractors is clearly very low.

The *integrity factor* (*IF*), on the other hand, is the normalised radius of the largest hypersphere (circle in 2D cases) entirely belonging to the safe basin. Examples are given in Figure 7.23. The *IF* is as computationally easy as *GIM*, but it succeeds in eliminating the unsafe fractal tongues from the integrity measure. It is, in fact, a measure of the compact part of the safe basin, which is the largest convex set entirely belonging to the basin that guarantees the system dynamical integrity.

7.6.2.3 Reduction of integrity and erosion profiles

Integrity measures are useful for studying how the safety of the system changes when the parameters vary. Plotting an integrity measure as a function of a varying parameter provides integrity profiles that are also named 'erosion profiles', since the attention is usually focused on the reduction of dynamical integrity.

Indeed, in practical applications, a very important parameter is the excitation amplitude, whose increase usually means a reduction of integrity, namely, a loss of structural safety, which justifies the name erosion. An example of this behaviour is represented by the nonresonant oscillation herein investigated (Figure 7.24). However, also presented is the case of resonant oscillation in which the integrity initially increases up to a maximum, and only later undergoes an overall reduction (Figure 7.24).

The reduction of integrity of dynamical systems is commonly due to the occurrence of topological mechanisms – usually a local or global bifurcation – which sometimes remain hidden if the outcomes are given only in terms of erosion profiles. These phenomena are central for understanding of system dynamics and also for their control.

7.6.2.4 Integrity of the twin-well Duffing oscillator

To illustrate the main features of *dynamical integrity* of the system (7.1.2), the safe basins are chosen to be the classical basins of attraction, and only the *IF* measure is considered. Furthermore, the parameters are fixed such that $\zeta = 0.025$, $\gamma = 1$ and

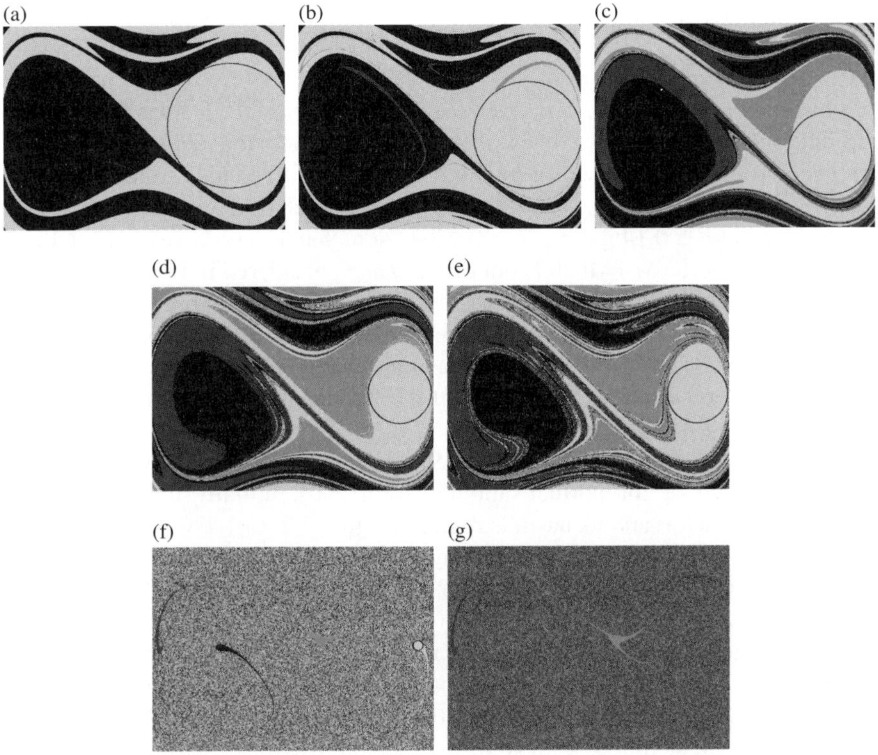

Figure 7.23 Basins of attractions for $\zeta = 0.025$, $\gamma = 1$ and $\Omega = 1.15$ and for increasing values of the excitation amplitude F: (a) $F = 0.027$; (b) $F = 0.029$; (c) $F = 0.041$; (d) $F = 0.060$; (f) $F = 0.065$; (g) $F = 0.130$; (h) $F = 0.132$. The circle is that involved in the definition of the IF for the right nonresonant oscillation S_n. The windows are defined by $-1.4 \leq x \leq 1.4$ and $-1 \leq \dot{x} \leq 1$.

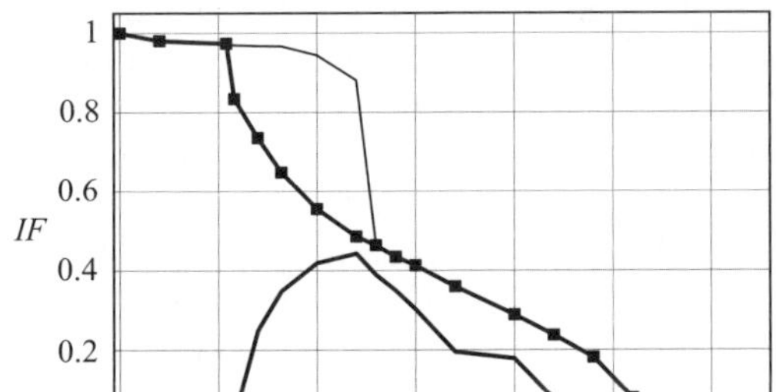

Figure 7.24 Erosion profiles of the right resonant S_r (thick), nonresonant S_n (squares) and right-well (thin) attractors considered jointly for $\zeta = 0.025$, $\gamma = 1$ and $\Omega = 1.15$.

$\Omega = 1.15$ and the excitation amplitude F is increased. This corresponds to Figure 7.7, which is implicitly referred to in the following analysis for understanding the underlying dynamical behaviour.

Illustrated in Figure 7.23 are seven basins of attraction for increasing excitation amplitude. The figures have been selected to illustrate the effects of the main bifurcational events discussed in Section 7.5.1 on the dynamical integrity: *snB* at $F \cong 0.028$; hbD_H at $F^h_{cr} \cong 0.0408$ (this has been computed in Section 7.5.2); *snA* at $F \cong 0.131$; *pd* at $F \cong 0.145$; *cr1* at $F \cong 0.1567$. Note that the cross-well period-1 large amplitude oscillation S_L (which is out of the range considered in Figure 7.23) is not considered, since it has a regular basin of attraction for low and medium values of excitation amplitude (see for example Figure 7.8). In fact, there are only resonant S_r and nonresonant S_n attractors. It should be noted that the integrity analysis naturally ends at $F \cong 0.1567$ with the appearance of the 'unique' cross-well chaotic attractor *CH* (see Figure 7.7).

The erosion profiles of the right S_r and S_n are shown in Figure 7.24 with thick lines (the reference circle for normalisation is for $F = 0$). Initially there is only the nonresonant attractor, and its basin is regular (Figure 7.23(a)). For increasing F, the basin modifies only slightly, and accordingly *IF* is almost constant.

When $F \cong 0.028$ the resonant attractor appears through *snB*. It is born inside the former basin of S_n (compare Figures 7.23(a) and (b)), so that the compact part of its basin instantaneously decreases in size, a fact that is clearly highlighted by the sudden reduction of the erosion curve of S_n in correspondence of the new born curve of S_r (Figure 7.24). Note that when $F \cong 0.028$ the size of the basin of S_n does not reduce significantly, so that the *GIM* only decreases slightly, thus missing this integrity reduction phenomenon. This is an example of the general superiority of the *IF* with respect to *GIM* in measuring dynamical integrity.

The next relevant dynamical phenomenon is the homoclinic bifurcation of the hilltop saddle, occurring at $F \cong 0.0408$. It entails initial fractalisation of the basin boundaries of left/right well attractors (see Figure 7.23(c)). Contrary to other cases [64,65,110], here the fractalisation does not rapidly affect the compact part of the basins (this occurs for larger values of F, even larger than $F = 0.060$ – see Figure 7.23(d)), and so it has no immediate effects on the dynamical integrity. This is why nothing special occurs for $F \cong 0.0408$ in the erosion profiles of Figure 7.24.

After the dramatic reduction at *snB*, the *IF* of S_n reduces continuously due to the parallel growth of the basin of S_r, which reaches a maximum at $F \cong 0.060$, corresponding to the case of Figure 7.23(d). It then decreases (Figure 7.23(e)) due to development of homoclinic tangling inside its basin (the penetration of the fractal tongues is already visible in the upper right part of Figure 7.23(d) and is confirmed in Figure 7.23(e)). After this point, both erosion profiles 'smoothly' decrease up to $F \cong 0.131$, where S_n disappears through *snA* (compare Figures 7.23(f) and (g)). At this level of excitation, however, the dynamical integrity is very small, not only because of the disappearing S_n, but also for the surviving S_r, as shown in Figure 7.23(f) and confirmed by the fact that $IF \cong 10\%$.

In the final path S_r undergoes a classical period-doubling cascade route to chaos (Figure 7.7). This, however, does not modify the dynamical integrity of periodic in-well solutions, and the erosion profile remains approximately constant up to the final crisis *cr1* that determines the disappearance of the in-well attractor.

In the previous analysis, the resonant and nonresonant attractors S_r and S_n were considered separately. However, in some cases, it is not important to consider specific attractors, but rather all in-well attractors, so that the safe basin is just the union of the basins of attraction of all attractors belonging to a given potential well. This is the case when scattered cross-well dynamics are the unwanted event.

In this case the mutual erosion of in-well basins of attraction is not of interest, instead the overall reduction of in-well integrity due to the mutual erosion between different wells is considered. To highlight this, in Figure 7.24 also the erosion profile of the cumulative right-well attractors (thin line) is plotted. Outside the interval $F \in (0.028, 0.131)$, i.e., where only one attractor per well exists, the overall erosion profile coincides with that of S_n and S_r, respectively.

The major difference with the two profiles is just in the initial part. In fact, the appearance of S_r does not produce any effect, since the basins of S_n and of S_r are indistinguishable. So, the profile remains flat up to $F \cong 0.0408$, where the homoclinic bifurcation triggers the fractality and entails a (moderate but visible) change of slope in the curve. The fractal tongues then begin entering the compact part of the safe basins, where they suddenly penetrate at about $F \cong 0.065$ (see Figure 7.23(e)) corresponding to the maximum extent of the safe basin of S_r. This phenomenon is clearly tracked by the overall erosion profile, which undergoes a sudden jump.

After the jump, the erosion profile of right-well attractors practically coincide with that of S_n, in spite of the fact that now $F < 0.131$. This is because the penetrated fractal tongues mostly affect the basin of attraction of S_r from which in-well and then cross-well chaos arises. This keeps the basins of S_n and S_r disjoint so that their union has a compact part substantially coinciding with the one of S_n (see Figures 7.23(e)–(g)).

To conclude this section it is noted that the erosion profile of right-well attractors has the same qualitative behaviour of those observed in other dynamical systems (see for example [65,88,98,110]).

7.7 Summary

In this chapter some of the dynamical features of the twin-well Duffing oscillator that correspond to the oscillator with a negative linear-positive cubic stiffness have been reviewed. Different techniques have been used to investigate various dynamical phenomena and to highlight the relevant performance and relationships. In reformulating approaches from the literature some elements of novelty have been incorporated. The chapter was basically divided into four parts.

1) **General dynamical behaviour.** The conservative case was first considered. This gives the skeleton of the main dynamical behaviour and permits analytical computation because of the energy integral. Competing effects of damping (dissipating energy from the system) and excitation (pumping energy into the

system) were then studied. The aim of this part was to provide the necessary background for the subsequent studies.

2) **Nonlinear periodic oscillations.** This issue was initially investigated theoretically by a standard application of the classical multiple scale method. This enables the detection of saddle-node bifurcation occurring in the neighbourhood of the nonlinear resonance. Also the Melnikov method for periodic oscillations was applied to predict some relevant bifurcation.
3) **Complex behaviour.** The transition to complex dynamical behaviour was initially studied by a purely numerical approach using the combined use of bifurcation diagrams, attractor-basins phase portraits, and stable and unstable manifold detection. This complemented the numerical simulation for periodic nonlinear oscillations. The occurrence of a scattered, robust, chaotic attractor was highlighted, together with the global bifurcations (crises) leading to its sudden appearance/disappearance. The whole dynamical picture was summarised in a comprehensive behaviour chart. The application of the classical Melnikov method allowed the detection of the homoclinic bifurcation of the hilltop saddle, which is an important dynamical event triggering the basins erosion. However, this was not directly responsible for any event in the case studied.
4) **'Nonclassical' analysis.** To complement the 'classical' analyses, two 'nonclassical' developments were also considered. The first was the control of homoclinic bifurcation, which was obtained by suitably modifying the shape of the excitation or, equivalently, by adding controlling excitations to the primary excitation. The main features were illustrated, and it was shown how this method, when appropriately applied, is able to eliminate any kind of homoclinic bifurcation responsible for unwanted dynamical events. The second issue addressed was the study of dynamical integrity and of its reduction (erosion) by varying parameters. In the authors' opinion this is an important issue, and was more fundamental than the stability of attractors in real applications.

References

[1] W.-Y. Tseng, J. Dugundji, Nonlinear vibrations of a buckled beam under harmonic excitation. *ASME Journal of Applied Mechanics*, 38, 467–476, 1971.

[2] P.J. Holmes, A nonlinear oscillator with a strange attractor. *Philosophical Transactions of the Royal Society of London A*, 292 (1394), 419–448, 1979.

[3] W.W. Bolotin, *Dynamic Stability of Elastic Systems*, Holden Day, San Francisco, 1964.

[4] F.C. Moon, P.J. Holmes, A magnetoelastic strange attractor. *Journal of Sound and Vibration*, 65, 275–296, 1979.

[5] F.C. Moon, Experiments on chaotic motion of a forced nonlinear oscillator – Strange attractors. *ASME Journal of Applied Mechanics*, 47, 638–644, 1980.

[6] P. Holmes, J. Marsden, A partial differential equation with infinitely many periodic orbits: chaotic oscillations of a forced beam. *Archive of Rational Mechanics and Analysis*, 76, 135–165, 1981.

[7] B.D. Greenspan, P.J. Holmes, Homoclinic orbits, subharmonics and global bifurcations in forced oscillations. G.I. Barenblatt, G. Iooss, D. D. Joseph eds. Nonlinear Dynamics and Turbulence, Pitman, London, pp. 172–214, 1983.

[8] P.J. Holmes, F.C. Moon, Strange attractors and chaos in nonlinear mechanics. *Journal of Applied Mechanics*, 50, 1021–1032, 1983.

[9] P.J. Holmes, D. Whitley, On the attracting set for Duffing's equation. *Physica D*, 7, 111–123, 1983.

[10] R.A. Mahaffey, Anharmonic oscillator description of plasma oscillations. *Physics of Fluids*, 19, 1387–1391, 1976.

[11] C. Holmes, P. Holmes, Second order averaging and bifurcations to subharmonics in Duffing's equation. *Journal of Sound and Vibration*, 78, 161–174, 1981.

[12] K. Yagasaki, Second-order averaging and Melnikov analyses for forced nonlinear oscillators. *Journal of Sound and Vibration*, 190, 587–609, 1996.

[13] V.K. Melnikov, On the stability of the center for time periodic perturbations. *Transactions of the Moscow Mathematical Society*, 12, 1–57, 1963.

[14] E.H. Dowell, C. Pezeshki, On the understanding of chaos in Duffing's equation including a comparison with experiments. *Journal of Applied Mechanics*, 53, 5–9, 1986.

[15] C. Pezeshki, E.H. Dowell, An examination of initial condition maps for the sinusoidally excited buckled beam modeled by the Duffing's equation. *Journal of Sound and Vibration*, 117, 219–232, 1987.

[16] K.B. Blair, C.M. Krousgrill, T.N. Farris, Harmonic balance and continuation techniques in the dynamic analysis of Duffing's equation. *Journal of Sound and Vibration*, 202, 717–731, 1997.

[17] J.A. Gottwald, L.N. Virgin, E.H. Dowell, Experimental mimicry of Duffing equation. *Journal of Sound and Vibration*, 158, 447–467, 1992.

[18] S.W. Shaw, A.G. Haddow, On 'roller-coaster' experiments for nonlinear oscillators. *Nonlinear Dynamics*, 3, 375–384, 1992.

[19] J.P. Cusumano, B.W. Kimble, A stochastic interrogation method for experimental measurements of global dynamics and basin evolution: application to a two-well oscillator. *Nonlinear Dynamics*, 8, 213–235, 1995.

[20] M.D. Todd, L.N. Virgin, An experimental verification of basin metamorphoses in a nonlinear mechanical system. *International Journal of Bifurcation and Chaos*, 7, 1337–1357, 1997.

[21] L.N. Virgin, M.D. Todd, C.J. Begley, S.T. Trickey, E.H. Dowell, Transient global behavior in nonlinear experimental oscillators. F. C. Moon ed. *New Applications of Nonlinear and Chaotic Dynamics in Mechanics*, Kluwer, pp. 353–362, 1999.

[22] S.T. Trickey, L.N. Virgin, Bottlenecking phenomenon near a saddle-node remnant in a Duffing oscillator. *Physics Letters A*, 248, 185–190, 1998.

[23] W. Xu, Q. He, T. Fang, H. Rong, Global analysis of crisis in twin-well Duffing system under harmonic excitation in the presence of noise. *Chaos Solitons and Fractals*, 23, 141–150, 2005.

[24] Y. Tang, F. Yang, G. Chen, T. Zhou, Classification of homoclinic tangencies for periodically perturbed systems. *Chaos, Solitons & Fractals*, 28 (1), 76–89, 2006.

[25] S.L.T. de Souza, I.L. Caldas, R.L. Viana, J.M. Balthazar, R.M.L.R.F. Brasil, Basins of attraction changes by amplitude constraining of oscillators with limited power supply. *Chaos, Solitons & Fractals*, 26, 1211–1220, 2005.

[26] L. Cveticanin, M. Zukovic, Melnikov's criteria and chaos in the systems with fractional order deflection. *Journal of Sound and Vibration*, 326, 768–779, 2009.

[27] J. Guckenheimer, P.J. Holmes, *Nonlinear Oscillations, Dynamical Systems and Bifurcation of Vector Fields*, Springer-Verlag, New York, 1983.

[28] J.M.T. Thompson, H.B. Stewart, *Nonlinear Dynamics and Chaos*, John Wiley, Chicester, 1986.

[29] F. Moon, *Chaotic Vibrations*, Wiley & Sons, New York, 1987.

[30] S. Wiggins, *Global Bifurcation and Chaos-Analytical Methods*, Springer-Verlag, New York, 1988.

[31] S. Wiggins, *Introduction to Applied Nonlinear Dynamical Systems and Chaos*, Springer-Verlag, New York, 1990.

[32] E. Ott, *Chaos in Dynamical Systems*, Cambridge University Press, Cambridge, New York, 1993.

[33] F.C. Moon, G.X. Li, Fractal basin boundaries and homoclinic orbits for periodic motion in a two-well potential. *Physical Review Letters*, 55, 1439–1444, 1985.

[34] F.C. Moon, G.-X. Li, The fractal dimension of the two-well potential strange attractor. *Physica D*, 17, 99–108, 1985.

[35] B.H. Tongue, Existence of chaos in a one-degree-of-freedom system. *Journal of Sound and Vibration*, 110, 69–78, 1986.

[36] D.M. Tang, E.H. Dowell, On the threshold force for chaotic motions for a forced buckled beam. *Journal of Applied Mechanics*, 55, 190–196, 1988.

[37] G. Schmidt, Onset of chaos and global analytical solutions for Duffing's oscillator. *ZAMM*, 16, 129–140, 1986.

[38] E.H. Dowell, Chaotic oscillations in mechanical systems. *Computational Mechanics*, 199–216, 1988.

[39] E.H. Dowell, C. Pezeshki, On necessary and sufficient conditions for chaos to occur in Duffing's equation. An heuristic approach. *Journal of Sound and Vibration*, 121, 195–200, 1988.

[40] K. Higuchi, E.H. Dowell, Effect of constant transverse force on chaotic oscillations of sinusoidally excited buckled beam. W. Schiehlen ed. *Nonlinear Dynamics in Engineering Systems*, Springer-Verlag, Berlin, pp. 99–106, 1990.

[41] E.H. Dowell, A chaotic scenario. *Journal of Sound and Vibration*, 144, 179–180, 1991.

[42] W. Szemplinska-Stupnicka, A discussion on necessary and sufficient conditions for steady-state chaos. *Journal of Sound and Vibration*, 152, 369–372, 1992.

[43] J. Rudowski, W. Szemplinska-Stupnicka, On an approximate criterion for chaotic motion in a model of a buckled beam. *Ingenieur-Archiv*, 57, 243–255, 1987.

[44] W. Szemplinska-Stupnicka, The refined approximate criterion for chaos in a two-state mechanical oscillator. *Ingenieur-Archiv*, 58, 354–366, 1988.

[45] W. Szemplinska-Stupnicka, J. Rudowski, Local methods in predicting occurrence of chaos in twin-well potential systems: superharmonic frequency region. *Journal of Sound and Vibration*, 152, 57–72, 1992.

[46] W. Szemplinska-Stupnicka, J. Rudowski, Steady-states in the twin-well potential oscillator: Computer simulations and approximate analytical studies. *Chaos – International Journal of Nonlinear Science*, 3, 375–385, 1993.

[47] Y. Ueda, H. Nakajima, T. Hikihara, H.B. Stewart, Forced two-well potential Duffing's oscillator. F. M. A. Salam and M. L. Levi eds. *Dynamical Systems Approaches to Nonlinear Problems in Systems and Circuits*, SIAM, Philadelphia, pp. 128–137, 1988.

[48] Y. Ueda, S. Yoshida, H.B. Stewart, J.M.T. Thompson, Basin explosions and escape phenomena in the twin-well Duffing oscillator: Compound global bifurcations organizing behaviour.' *Philosophical Transactions of the Royal Society of London A*, 332, 169–186, 1990.

[49] K.L. Janicki, W. Szemplinska-Stupnicka, Subharmonic Resonances and Criteria for Escape and Chaos in a Driven Oscillator. *Journal of Sound and Vibration*, 180, 253–269, 1995.

[50] K.L. Janicki, W. Szemplinska-Stupnicka, Subharmonic resonances in a driven oscillator: bifurcation structures and transitions to chaos. *European Journal of Mechanics, A/Solids*, 16 (4), 671–694, 1997.

[51] W. Szemplinska-Stupnicka, Cross-well chaos and escape phenomena in driven oscillators. *Nonlinear Dynamics*, 3, 225–243, 1992.

[52] W. Szemplinska-Stupnicka, The analytical predictive criteria for chaos and escape in nonlinear oscillators: A survey. *Nonlinear Dynamics*, 7, 129–147, 1995.

[53] A.N. Lansbury, J.M.T. Thompson, H.B. Stewart, Basin erosion in the twin-well Duffing oscillator: Two distinct bifurcation scenarios. *International Journal of Bifurcation and Chaos*, 2, 505–532, 1992.

[54] H.B. Stewart, A chaotic saddle catastrophe in forced oscillators. F.M.A. Salam, M. L. LeviEds. *Dynamical Systems Approaches to Nonlinear Problems in Systems and Circuits*, SIAM, Philadelphia, pp. 138–149, 1988.

[55] A.L. Katz, E.H. Dowell, From single well chaos to cross-well chaos: A detailed explanations in terms of manifold intersections. *International Journal of Bifurcation and Chaos*, 4 (4), 933–941, 1994.

[56] M. Zakrzhevsky, Global stable oscillations near unstable equilibrium positions: The hilltop effect. F. C. Moon ed. *New Applications of Nonlinear and Chaotic Dynamics in Mechanics*, Kluwer, Eindhoven, pp. 117–124, 1999.

[57] W. Szemplinska-Stupnicka, K.L. Janicki, Basin boundary bifurcations and boundary crisis in the twin-well Duffing oscillator: Scenarios related to the saddle of the large resonant orbit. *International Journal of Bifurcation and Chaos*, 7, 129–146, 1997.

[58] W. Szemplinska-Stupnicka, A. Zubrzycki, E. Tyrkiel, Properties of Chaotic and Regular Boundary Crisis in Dissipative Driven Nonlinear Oscillators. *Nonlinear Dynamics*, 19, 19–36, 1999.

[59] W. Szemplinska-Stupnicka, E. Tyrkiel, Sequences of global bifurcations and the related outcomes after crisis of the resonant attractor in a nonlinear oscillator. *International Journal of Bifurcation and Chaos*, 7, 2437–2457, 1997.

[60] W. Szemplinska-Stupnicka, E. Tyrkiel, Common Features of the Onset of the Persistent Chaos in Nonlinear Oscillators: A Phenomenological Approach. *Nonlinear Dynamics*, 27, 271–293, 2002.

[61] E. Tyrkiel, On the role of chaotic saddles in generating chaotic dynamics in nonlinear driven oscillators. *International Journal of Bifurcation and Chaos*, 15, 1215–1238, 2005.

[62] M.S. Soliman, J.M.T. Thompson, Basin organization prior to a tangled saddle-node bifurcation. *International Journal of Bifurcation and Chaos*, 1, 107–118, 1990.

[63] H.B. Stewart, Y. Ueda, Catastrophes with indeterminate outcome. *Proceedings of the Royal Society of London A*, 432, 113–123, 1991.

[64] G. Rega, S. Lenci, Identifying, evaluating, and controlling dynamical integrity measures in nonlinear mechanical oscillators. *Nonlinear Analysis T. M. & A.*, 63, 902–914, 2005.

[65] G. Rega, S. Lenci, Dynamical integrity and control of nonlinear mechanical oscillators. *Journal on Vibration and Control*, 14, 159–179, 2008.

[66] A.L. Fradkov, A.Y. Pogromsky, *Introduction to Control of Oscillations and Chaos*, World Scientific, Singapore, 1998.

[67] G. Rega, S. Lenci, *Recent Advances in Control of Complex Dynamics in Mechanical and Structural Systems*, M. Sanjuan ed., World Scientific, Singapore, in press, 2009.

[68] M.K. Sifakis, S.J. Elliott, Strategies for the control of chaos in a Duffing–Holmes oscillator. *Mechanical Systems and Signal Processing*, 14, 987–1002, 2000.

[69] K. Pyragas, Continuous control of chaos by self-controlling feedback. *Physics Letters A*, 170, 421–428, 1992.

[70] E.R. Hunt, Stabilizing high-periodic orbits in a chaotic system: The diode resonator. *Physical Review Letters*, 67, 1953–1955, 1991.

[71] E. Ott, C. Grebogi and J.A. Yorke, Controlling chaos. *Physical Review Letters E*, 64, 1196–1199, 1990.

[72] S. Lenci, G. Rega, A unified control framework of the nonregular dynamics of mechanical oscillators. *Journal of Sound and Vibration*, 278, 1051–1080, 2004.

[73] R. Lima and M. Pettini, Suppression of chaos by resonant parametric perturbations. *Physical Review A*, 41, 726–733, 1990.

[74] L. Fronzoni, M. Giocondo, M. Pettini, Experimental evidence of suppression of chaos by resonant parametric perturbations. *Physical Review A*, 43, 6483–6487, 1991.

[75] B.R. Nana Nbendjo, R. Tchoukuegno, P. Woafo, Active control with delay of vibration and chaos in a double-well Duffing oscillator. *Chaos Solitons and Fractals*, 18, 345–353, 2003.

[76] B.R. Nana Nbendjo, P. Woafo, Active control with delay of horseshoes chaos using piezoelectric absorber on a buckled beam under parametric excitation. *Chaos Solitons and Fractals*, 32, 73–79, 2007.

[77] F.C. Moon, M.A. Johnson, W.T. Holmes, Controlling chaos in a two-well oscillator. *International Journal of Bifurcation and Chaos*, 6, 337–347, 1996.

[78] U. Dressler, T. Ritz, A. Schenk zu Schweinsberg, R. Doerner, B. Hübinger, W. Martienssen, Tracking unstable periodic orbits in a bronze ribbon experiment. *Physical Review E*, 51, 1845–1848, 1995.

[79] G. Chen, X. Dong, On feedback control of chaotic continuous-time systems. *IEEE Transactions on Circuits and Systems – I: Fundamental Theory and Applications*, 40, 591–601, 1993.

[80] H. Nijmeijer, Adaptive/robust control of chaotic systems. D. H. van Campen ed. *Interaction Between Dynamics and Control in Advanced Mechanical Systems*, Kluwer, Eindhoven, pp. 255–262, 1997.

[81] B. Ravindra, A.K. Mallik, Dissipative control of chaos in nonlinear vibrating systems. *Journal of Sound and Vibration*, 211, 709–715, 1998.

[82] G.M. Mahmoud, A.A. Mohamed, A.A. Shaban, Strange attractors and chaos control in periodically forced complex Duffing's oscillators. *Physica A: Statistical Mechanics and its Applications*, 292 (1–4), 193–206, 2001.

[83] S. Bowong, F.M.M. Kakmeni, Chaos control and duration time of a class of uncertain chaotic systems. *Physics Letters A*, 316, 206–217, 2003.

[84] K. Yamasue, T. Hikihara, Domain of attraction for stabilized orbits in time delayed feedback controlled Duffing systems. *Physical Review E*, 69, 056209, 2004.

[85] Y. Lei, W. Xu, Y. Xu, T. Fang, Chaos control by harmonic excitation with proper random phase. *Chaos Solitons and Fractals*, 21, 1175–1181, 2004.

[86] R. Aguilar-López, R. Martínez-Guerra, Chaos suppression via observer based active control scheme: Application to Duffing's oscillator. *Chaos Solitons and Fractals*, 32, 1887–1897, 2007.

[87] J. Alvarez-Ramirez, G. Espinosa-Paredes, Exploiting the attractor structure for chaos feedback control. The Duffing oscillator. *International Journal of Bifurcation and Chaos*, 14, 3661–3670, 2004.

[88] S. Lenci, G. Rega, Optimal control of nonregular dynamics in a Duffing oscillator. *Nonlinear Dynamics*, 33, 71–86, 2003.

[89] S. Lenci, G. Rega, Optimal numerical control of single-well to cross-well chaos transition in mechanical systems. *Chaos, Solitons & Fractals*, 15, 173–186, 2003.

[90] S. Lenci, G. Rega, Global optimal control and system-dependent solutions in the hardening Helmholtz-Duffing oscillator. *Chaos, Solitons & Fractals*, 21, 1031–1046, 2004.

[91] H.J. Cao, G.R. Chen, Global and local control of homoclinic and heteroclinic bifurcations. *International Journal of Bifurcation and Chaos*, 15, 2411–2432, 2005.

[92] H.J. Cao, G.R. Chen, A simplified optimal control method for homoclinic bifurcations. *Nonlinear Dynamics*, 42, 43–61, 2005.

[93] A.H. Nayfeh, D.T. Mook, *Nonlinear Oscillations*, Wiley, Blacksburg, 1979.

[94] Y. Ketema, A physical interpretation of Melnikov's method. *International Journal of Bifurcation and Chaos*, 2, 1–9, 1992.

[95] G. Rega, S. Lenci, J.M.T. Thompson, Controlling chaos: the OGY method, its use in mechanics, and an alternative unified framework for control of non-regular dynamics, in M. Thiel, J. Kurths, C. Romano, A. Moura and G. Károlyi (eds.), *Nonlinear Dynamics and Chaos: Advances and Perspectives*, Understanding Complex Systems, DOI 10.1007/978-3-642-04629-2_11, Springer-Verlag, Berlin, Heidelberg, 2010, pp. 211–269.

[96] S. Lenci, G. Rega, Higher-order Melnikov functions for single-d.o.f. oscillators: Theoretical treatment and applications. *Mathematical Problems in Engineering*, 2004, 145–168, 2004.

[97] S.E. Newhouse, The abundance of wild hyperbolic sets and non-smooth stable sets for diffeomorphisms. *Publications Mathématiques de l'IHÉS*, 50, 101–151, 1979.

[98] S. Lenci, G. Rega, Optimal control of homoclinic bifurcation: Theoretical treatment and practical reduction of safe basin erosion in the Helmholtz oscillator. *Journal of Vibrations and Control*, 9, 281–315, 2003.

[99] R. Chacon, *Control of Homoclinic Chaos by Weak Periodic Perturbations*, World Scientific, Singapore, 2005.

[100] S. Lenci, G. Rega, A procedure for reducing the chaotic response region in an impact mechanical system. *Nonlinear Dynamics*, 15, 391–409, 1998.

[101] S. Lenci, On the suppression of chaos by means of bounded excitations in an inverted pendulum. *SIAM Journal of Applied Mathematics*, 58, 1116–1127, 1998.

[102] S. Lenci, G. Rega, Controlling nonlinear dynamics in a two-well impact system. Parts I & II. *International Journal of Bifurcation and Chaos*, 8, 2387–2424, 1998.

[103] S. Lenci, G. Rega, Heteroclinic bifurcations and optimal control in the nonlinear rocking dynamics of generic and slender rigid blocks. *International Journal of Bifurcation and Chaos*, 15, 1901–1918, 2005.

[104] S. Lenci, G. Rega, Optimal control and anti-control of the nonlinear dynamics of a rigid block. *Philosophical Transactions of the Royal Society of London A*, 364, 2353–2381, 2006.

[105] S. Lenci, G. Rega, Control of pull-in dynamics in a nonlinear thermoelastic electrically actuated microbeam. *Journal of Micromechanics and Microengineering*, 16, 390–401, 2006.

[106] S. Lenci, G. Rega, Control of the homoclinic bifurcation in buckled beams: infinite dimensional vs. reduced order modeling. *International Journal of Nonlinear Mechanics*, 43, 474–489, 2008.

[107] M.S. Soliman, J.M.T. Thompson, Global dynamics underlying sharp basin erosion in nonlinear driven oscillators. *Physical Review A*, 45, 3425–3431, 1992.

[108] M.S. Soliman, J.M.T. Thompson, Integrity measures quantifying the erosion of smooth and fractal basins of attraction. *Journal of Sound and Vibration*, 135, 453–475, 1989.

[109] J.M.T. Thompson, Chaotic phenomena triggering the escape from a potential well. *Proceedings of the Royal Society of London A*, 421, 195–225, 1989.

[110] S. Lenci, G. Rega, Competing dynamic solutions in a parametrically excited pendulum: attractor robustness and basin integrity. *ASME Journal of Computational and Nonlinear Dynamics*, 3, 041010-1-9, 2008.

8

Forced harmonic vibration of an asymmetric Duffing oscillator

Ivana Kovacic[1] and Michael J. Brennan[2]
[1]*University of Novi Sad, Faculty of Technical Sciences, Serbia*
[2]*University of Southampton, Institute of Sound and Vibration Research, United Kingdom*

8.1 Introduction

In physics, symmetry is the property that some quantity does not change under a set of transformations. The Nobel laureate P.W. Anderson wrote in his article [1] that '... symmetry is of great importance in physics ... It is only slightly overstating the case to say that physics is the study of symmetry. The first demonstration of the power of this idea might have been by Newton, who might have asked himself the question: 'What if the matter here in my hand obeys the same laws as that up in the sky'...' However, asymmetry, which implies that the aforementioned requirement is not met, is also widely present, not only in physics, but in all branches of science as well as in engineering systems.

In oscillatory systems, asymmetry can be with respect to the dependence of the restoring force on the displacement as well as the type of excitation. The system is asymmetric, for instance, if the restoring force is an odd function and a constant force acts on it. One illustrative example of it is an isolation system with a *quasi-zero*

The Duffing Equation: Nonlinear Oscillators and their Behaviour, First Edition. Edited by Ivana Kovacic and Michael J. Brennan.

stiffness characteristic on which an additional constant load acts (see Chapter 2 and [2]). Besides this, another example of an asymmetric system is the one when the restoring force is not an odd function, i.e., when it comprises also a quadratic term, as shown in Chapter 2, in the case of some electric circuits [3] or cables [4–7]. Of interest here are asymmetric oscillators that have a single equilibrium position and a corresponding *single-well (one-well) potential.*

This chapter is devoted to the study of asymmetric forced oscillatory systems with the aim of showing some of the dynamic responses possible, and the variety of tools that can be used to investigate their dynamic behaviour.

8.2 Models of the systems under consideration

The equation of motion for the oscillator that has a pure cubic (PC) nonlinearity is:

$$\ddot{y}+2\zeta\dot{y}+\gamma y^3=F_0+F\cos\Omega t \tag{8.2.1}$$

where y, ζ, γ, F_0, F and Ω represent, respectively, the nondimensional variables of displacement, damping ratio, coefficient of the cubic nonlinearity, constant force, and the amplitude and frequency of the harmonic excitation force; overdots denote derivatives with respect to nondimensional time t.

By shifting the origin of the coordinate system so that

$$z=y-\sqrt[3]{\frac{F_0}{\gamma}} \tag{8.2.2}$$

the equivalent oscillator with both quadratic and cubic (EQC) nonlinearity is obtained

$$\ddot{z}+2\zeta\dot{z}+\omega_0^2 z+\beta z^2+\gamma z^3=F\cos\Omega t \tag{8.2.3}$$

where

$$\omega_0^2=3\sqrt[3]{\gamma F_0^2},\quad \beta=3\sqrt[3]{\gamma^2 F_0} \tag{8.2.4}$$

Note that this coordinate transformation affects the geometric parts only, i.e., the parts related to the restoring force, while the other mechanical properties such as the damping ratio and the external force remain the same. Thus, an increase in the constant force F_0 results in an increase in the natural frequency of the EQC oscillator and an increase in the coefficient of the quadratic term. Ordering of these coefficients and how they relate to each other quantitatively also depends on the coefficient of the cubic term, which is not affected by the coordinate transformation and remains the same. The restoring force of the PC oscillator is $F_{\mathrm{rPC}}=\gamma y^3-F_0$, while the corresponding restoring force of the EPC oscillator is thus given by $F_{\mathrm{rEQC}}=\omega_0^2 z+\beta z^2+\gamma z^3$. Both of these forces are shown in Figures 8.1(a) and (b), as functions of the corresponding deflection for $F_0=0.1$ and $\gamma=0.0783$ (this value is related to a particular configuration of the isolation system studied in [2]). In addition, their potential energies $V_{PC} = \gamma\frac{y^4}{4} - F_0 y + \frac{3}{4}\sqrt[3]{\frac{F_0^4}{\gamma}}$ and $V_{EQC} = \omega_0^2\frac{z^2}{2} + \beta\frac{z^3}{3} + \gamma\frac{z^4}{4}$ are plotted in Figures 8.1(c) and (d). It should be noted that the constant term has been added to V_{PC} in order to compare mutually the energy stored in these two systems with respect to the reference point $z = 0$, where the potential energy is

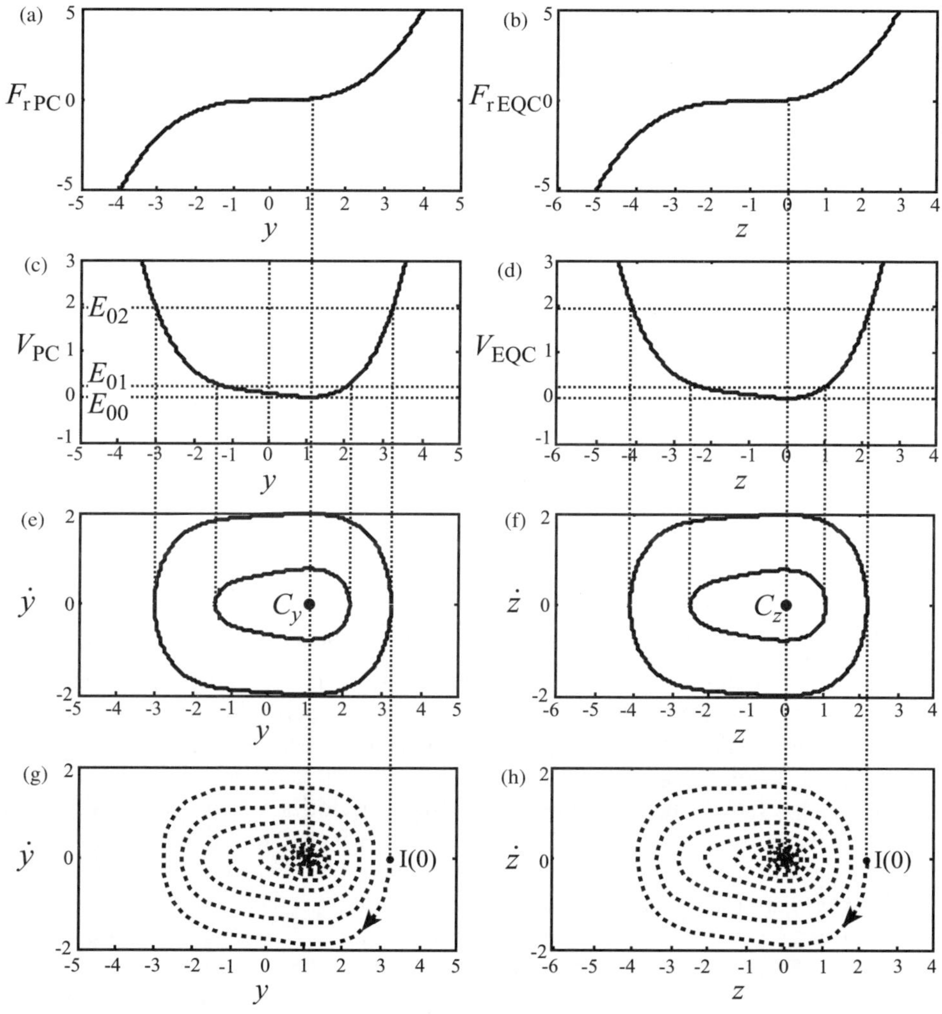

Figure 8.1 (a) Restoring force of the PC oscillator with respect to displacement y; (b) Restoring force of the EQC oscillator with respect to displacement z; (c) Potential energy of the PC oscillator; (d) Potential energy of the EQC oscillator; (e) Phase trajectories force of the conservative PC oscillator; (f) Phase trajectories of the conservative EQC oscillator; (g) Phase trajectory of the nonconservative PC oscillator; (h) Phase trajectory of the nonconservative EQC oscillator.

assumed to be zero. Thus, the potential energies have a minimum at $y = \sqrt[3]{\frac{F_0}{\gamma}}$ and $z = 0$, respectively, representing a single-well potential.

By considering the conservative freely vibrating systems, i.e., Equations (8.2.1) and (8.2.3) for $\zeta = 0$ and $F = 0$:

$$\ddot{y} + \gamma y^3 = F_0 \tag{8.2.5}$$

$$\ddot{z} + \omega_0^2 z + \beta z^2 + \gamma z^3 = 0 \tag{8.2.6}$$

the following expressions for the phase trajectories can be derived:

$$\dot{y} = \pm\sqrt{2\left(E_0 - \gamma\frac{y^4}{4} + F_0 y\right)} \tag{8.2.7}$$

$$\dot{z} = \pm\sqrt{2\left(E_0 - \omega_0^2\frac{z^2}{2} - \beta\frac{z^3}{3} - \gamma\frac{z^4}{4}\right)} \tag{8.2.8}$$

where E_0 represents a constant defined by the initial energy level $E_0 = \frac{1}{2}\dot{y}^2(0) + \gamma\frac{y^4(0)}{4} - F_0 y(0) + \frac{3}{4}\sqrt[3]{\frac{F_0^4}{\gamma}} \equiv \frac{1}{2}\dot{z}^2(0) + \omega_0^2\frac{z^2(0)}{2} + \beta\frac{z^3(0)}{3} + \gamma\frac{z^4(0)}{4}$. Equations (8.2.7) and (8.2.8) are used to plot the phase trajectories of both oscillators for $\dot{y}(0) = 0$ and $y(0) = 2\sqrt[3]{\frac{F_0}{\gamma}}$ and $y(0) = 3\sqrt[3]{\frac{F_0}{\gamma}}$. The corresponding energy levels are labelled by E_{01} and E_{02} in Figures 8.1(c) and (d). In addition, the energy level corresponding to the minima of potential energies is labelled by E_{00}. In this case, the level curves degenerate into the singular points C_y and C_z (Figures 8.1(e) and (f)), the so-called centres. When the energy levels are less than these values, there is no real solution. For higher values, such as E_{01} and E_{02} shown, the level curves are closed trajectories surrounding the *centre*. It should also be noted that Figure 8.1 as a whole, demonstrates the asymmetry of the system and helps to visualise and understand the coordinate transformation introduced.

If no harmonic excitation exists, but the systems considered are damped, i.e., $F = 0$ and $\zeta \neq 0$, the closed trajectories surrounding centres turn into *foci* (*spirals*), as shown in Figures 8.1(g) and (h). These phase trajectories were obtained by carrying out the direct numerical integration of the equations of motion for $\zeta = 0.025$ and for the initial conditions corresponding to the initial energy level E_{02} defined above, labelled by I(0) in Figures 8.1(g) and (h). The phase trajectories reflect the shrinking ellipse, spiralling in towards the points C_y and C_z.

The system described by Equation (8.2.3) can also be represented as an oscillator in which the natural frequency is equal to unity:

$$z'' + 2\hat{\zeta}z' + z + \hat{\beta}z^2 + \hat{\gamma}z^3 = \hat{F}\cos\hat{\Omega}\hat{t} \tag{8.2.9}$$

where a new nondimensional time and a frequency are introduced, as well as the following parameters

$$\hat{t} = \sqrt[6]{27\gamma F_0^2}\,t, \quad \hat{\Omega} = \frac{\Omega}{\sqrt[6]{27\gamma F_0^2}},$$
$$\hat{\zeta} = \frac{\zeta}{\sqrt[6]{27\gamma F_0^2}}, \quad \hat{\beta} = \sqrt[3]{\frac{\gamma}{F_0}}, \quad \hat{\gamma} = \sqrt[3]{\frac{\gamma^2}{27F_0^2}}, \quad \hat{F} = \sqrt[3]{\frac{F^3}{27\gamma F_0^2}} \tag{8.2.10}$$

The primes in Equation (8.2.9) denote differentiation with respect to $\hat{t}$. A change in F_0 now affects the coefficients of the nonlinear terms, the damping coefficient and the magnitude of the harmonic force. This implies that all the physical features of the systems are different.

There are systems in which the restoring force has a linear-plus-quadratic-plus-cubic term (the so-called Helmholtz–Duffing oscillators), but no such coupling between coefficients exists. Some of the illustrative examples are structural systems (elastic cables, as described in Chapter 2) with the initial curvature subjected to vertical forcing under planar excitation (see, for example, [5]), whose behaviour can be described by

$$\ddot{z} + 2\zeta\dot{z} + z + c_2 z^2 + c_3 z^3 = F\cos\Omega t \tag{8.2.11}$$

where the coefficients c_2 and c_3 are mutually dependent and are related to the cable properties and the mode-shape functions [5]. Note that the coefficients in front of the quadratic and cubic term are labelled by β and γ in Chapter 2, Equation (2.10.19), but are changed in Equation (8.2.11) to c_2 and c_3, respectively, to emphasise the difference with respect to the corresponding coefficients in Equation (8.2.9)). The equilibrium points associated with the oscillator given by Equation (8.2.11) are $z_1 = 0$ and $z_{2,3} = \frac{-c_2 \pm \sqrt{c_2^2 - 4c_3}}{2c_3}$. For a heavy suspended cable that can only resist tensile forces, the values of these coefficients are such that there is only one physically admissible stable equilibrium, i.e., $c_2^2 - 4c_3 < 0$, and the corresponding potential is asymmetric and has a single-well potential [4,7]. This type of Helmholtz–Duffing oscillator with a single-well potential is labelled subsequently as a SWHD oscillator and is considered in Section 8.4 and 8.6.

8.3 Regular response of the pure cubic oscillator

This section is mainly concerned with obtaining and analysing the response of the PC oscillator given by Equation (8.2.1) at the frequency of excitation, i.e., in the *primary resonance* regime. Subharmonic and superharmonic responses are briefly discussed at the end of this section.

8.3.1 Primary resonance: transient solution

When the response of the system is such that the fundamental component with period 2π dominates over higher harmonics, the solution of Equation (8.2.1) is given by [8]

$$y = a(t)\sin\Omega t + b(t)\cos\Omega t + c(t) \tag{8.3.1}$$

The amplitudes $a(t)$ and $b(t)$ are assumed to be slowly varying compared to $\sin\Omega t$ and $\cos\Omega t$ so that $\ddot{a}$ and $\ddot{b}$ can be neglected [8].

Then,

$$\dot{y}(t) = (\dot{a} - b\Omega)\sin\Omega t + (a\Omega + \dot{b})\cos\Omega t + \dot{c} \tag{8.3.2}$$

and

$$\ddot{y}(t) = (-a\Omega^2 - 2\dot{b}\Omega)\sin\Omega t + (2\dot{a}\Omega - b\Omega^2)\cos\Omega t + \ddot{c} \tag{8.3.3}$$

Also, the cubic term in y can be approximated by

$$y^3(t) = \left(\frac{3}{4}a^3 + \frac{3}{4}ab^2 + 3ac^2\right)\sin\Omega t + \left(\frac{3}{4}b^3 + \frac{3}{4}a^2 b + 3bc^2\right)\cos\Omega t + \frac{3}{2}a^2 c$$
$$+ \frac{3}{2}cb^2 + c^3 + \text{higher harmonics} \tag{8.3.4}$$

When Equations (8.3.1)–(8.3.4) are substituted into Equation (8.2.1), and the terms are re-arranged so that the coefficients of $\sin\Omega t$, $\cos\Omega t$ as well as the terms that are independent of frequency are grouped together, the following *autonomous system* of equations is derived

$$-a\Omega^2 - 2\dot{b}\Omega + 2\zeta(\dot{a} - b\Omega) + \gamma\left(\frac{3}{4}a^3 + \frac{3}{4}ab^2 + 3ac^2\right) = 0,$$
$$2\dot{a}\Omega - b\Omega^2 + 2\zeta(a\Omega + \dot{b}) + \gamma\left(\frac{3}{4}b^3 + \frac{3}{4}a^2 b + 3bc^2\right) = F, \tag{8.3.5a–c}$$
$$\ddot{c} + 2\zeta\dot{c} + \gamma\left(\frac{3}{2}a^2 c + \frac{3}{2}cb^2 + c^3\right) = F_0$$

which can be rewritten as

$$\dot{a} = \frac{\zeta}{2\Omega^2 + 2\zeta^2}(aG + 2\zeta\Omega b) + \frac{\Omega}{2\Omega^2 + 2\zeta^2}(bG + F - 2\zeta\Omega a) \equiv \bar{Y}_1(a, b, c, d),$$
$$\dot{b} = -\frac{\Omega}{2\Omega^2 + 2\zeta^2}(aG + 2\zeta\Omega b) + \frac{\zeta}{2\Omega^2 + 2\zeta^2}(bG + F - 2\zeta\Omega a) \equiv \bar{Y}_2(a, b, c, d),$$
$$\dot{d} = -2\zeta d - \frac{3}{2}\gamma c\left(a^2 + \frac{3}{2}b^2\right) - \gamma c^3 + F_0 \equiv \bar{Y}_3(a, b, c, d),$$
$$\dot{c} = d \equiv \bar{Y}_4(a, b, c, d), \tag{8.3.6a–d}$$

where

$$G = \Omega^2 - \frac{3}{4}\gamma(a^2 + b^2) - 3\gamma c^2 \tag{8.3.7}$$

The assumption that the solution to Equation (8.2.1) has the form of Equation (8.3.1), which has been used to derive the autonomous system (8.3.6), should be examined closely. It relates to the response in which the first harmonic is dominant compared to the higher harmonics, which should be checked numerically before general conclusions about the system behaviour are made. If necessary, higher harmonics can be included by following both the procedure given above and the one presented subsequently. However, in such a case, the equations derived and the calculations involved generally become cumbersome.

Equations (8.3.6) and (8.3.7) describe the *transient motion* of the oscillator given by Equation (8.2.1). They can also be used to study the steady-state response, which is considered in the following subsection.

8.3.2 Primary resonance: steady-state solution

The steady-state response occurs when $\dot{a} = \dot{b} = \dot{c} = \dot{d} = 0$. Equation (8.3.6d) implies $d = 0$, which is trivial. Thus, it is omitted subsequently. The steady-state response is defined by

$$\bar{Y}_i(a, b, c, 0) = 0, \quad i = 1, \ldots, 3 \tag{8.3.8}$$

If the following substitutions are introduced

$$a = -A_1 \sin\theta, \quad b = A_1 \cos\theta, \quad c \equiv A_0 \tag{8.3.9a–c}$$

the steady-state response can be expressed as

$$y = A_0 + A_1 \cos(\Omega t + \theta) \tag{8.3.10}$$

In this way, the constants A_0 and A_1 represent, respectively, the amplitude of a DC term and of the first harmonic, while θ is its phase.

By using Equations (8.3.9a–c), the system of equations (8.3.8) yields the following system of nonlinear algebraic equations in terms of A_0, A_1 and θ

$$\begin{aligned} Y_1(A_0, A_1, \theta) &\equiv -A_1\Omega^2 + 3\gamma A_0^2 A_1 + \frac{3}{4}\gamma A_1^3 - F\cos\theta = 0, \\ Y_2(A_0, A_1, \theta) &\equiv -2\zeta A_1 \Omega - F\sin\theta = 0, \\ Y_3(A_0, A_1, \theta) &\equiv \gamma A_0^3 + \frac{3}{2}\gamma A_0 A_1^2 - F_0 = 0 \end{aligned} \tag{8.3.11a–c}$$

Equations (8.3.11a–c) can be combined to give the implicit equation for the frequency response of A_0:

$$\begin{aligned} &25\gamma^3 A_0^9 - 20\gamma^2\Omega^2 A_0^7 - 15\gamma^2 F_0 A_0^6 + 4\gamma\Omega^2(\Omega^2 + 4\zeta^2)A_0^5 + 16\gamma F_0 \Omega^2 A_0^4 \\ &+ 3\gamma(2F^2 - 3F_0^2)A_0^3 - 4F_0\Omega^2(\Omega^2 + 4\zeta^2)A_0^2 + 4F_0^2\Omega^2 A_0 - F_0^3 = 0 \end{aligned} \tag{8.3.12}$$

According to Descartes's rule of signs [9], the number of positive roots of the polynomial (8.3.12) is either equal to the number of sign changes in the sequence of its coefficients, where vanishing terms are disregarded, or it is less than that number by a positive even integer. This implies that the system can have a maximum number of five, three or one steady-state values. For given values of ζ, γ, F_0 and F, Equation (8.3.12) is solved numerically to find the number of real roots. The way in which the maximum number of the steady-state values of A_0 and A_1 depends on F_0 and F is shown in Figure 8.2 for the case when $\zeta = 0.025$ and $\gamma = 0.0783$. It can be seen that for some small values of the magnitude of harmonic excitation F, which corresponds to the white region, there are no multivalued responses, and hence the corresponding *frequency–response curve* (FRC) is similar to that for a linear harmonically excited system. For the majority of combinations shown, which correspond to the light grey region in Figure 8.2, the system has a maximum of three steady-state values. There is also a region where the combinations of F_0 and F yield five steady-state amplitudes for some forcing frequencies; this corresponds to the dark grey region in Figure 8.2. The multivaluedness of the response raises the question of its *stability* and the occurrence of the *jump phenomenon*. The occurrence of the *jumps* in the system response is associated with a *saddle-node bifurcation* and vertical tangency of the FRC [9] and is discussed in the following subsections.

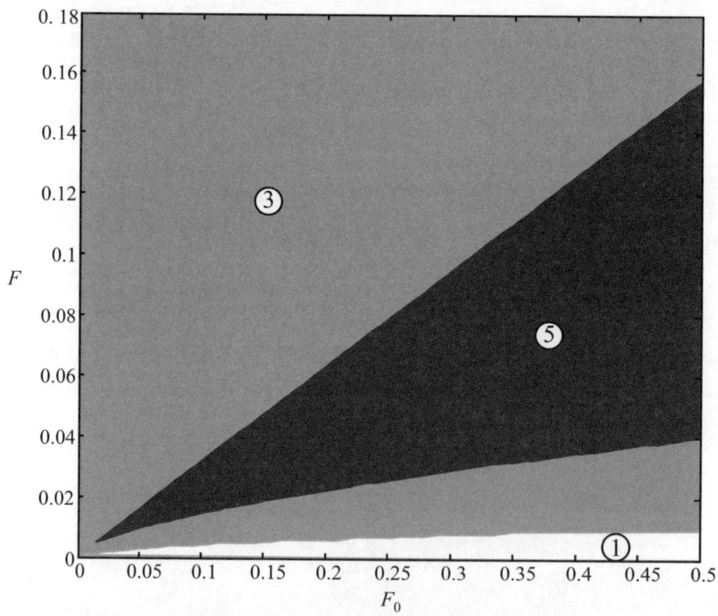

Figure 8.2 The maximum number of the steady-state amplitudes: one (white region), three (light grey region) or five (dark grey triangles), as a function of the nondimensional constant force F_0 and magnitude of the harmonic force F for $\zeta = 0.025$ and $\gamma = 0.0783$.

8.3.2.1 Stability of the steady-state solution

In the case when several steady-state solutions exist, not all of them will correspond to stable motion. Thus, stability analysis of the approximate solution given by Equation (8.3.10) is necessary. To carry out this analysis, a small perturbation $u(t)$ is introduced as follows

$$y + u(t) = A_0 + A_1\cos(\Omega t + \theta) + u(t) \tag{8.3.13}$$

Substituting Equation (8.3.13) into Equation (8.2.1), the following linearised variational equation is obtained

$$\ddot{u} + 2\zeta\dot{u} + 3\gamma(A_0 + A_1\cos(\Omega t + \theta))^2 u = 0 \tag{8.3.14}$$

Using the substitution $u(t) = e^{-\zeta t}v(t)$, Equation (8.3.14) transforms to *Hill's equation* [2,7,10]

$$\ddot{v} + \left(\sigma_0 + 2\sum_{n=1}^{2}\sigma_n\cos n(\Omega t + \theta)\right)v = 0 \tag{8.3.15}$$

where

$$\sigma_0 = 3\gamma A_0^2 + \frac{3\gamma}{2}A_1^2 - \zeta^2, \quad \sigma_1 = 3\gamma A_0 A_1, \quad \sigma_2 = \frac{3}{4}\gamma A_1^2 \tag{8.3.16}$$

According to *Floquet theory*, a system driven parametrically, such as the system modelled by Equation (8.3.15), can exhibit resonance whenever the driving frequency is equal to $2\sqrt{\sigma_0}/n$, where $\sqrt{\sigma_0}$ is the normalised natural frequency of the system and n is an integer [11]. For the stability analysis of oscillations having the same frequency as the approximate harmonic balance solution, the second unstable region is of interest, i.e., $n = 2$. By virtue of Floquet theory, the solution to Equation (8.3.15) can be assumed to be of the form

$$v(t) = e^{\mu\tau}(B + \sin(\Omega t + \varphi)) \tag{8.3.17}$$

where μ is the characteristic Floquet exponent. On the boundary between the stable and unstable region, the real part of the term $(-\zeta \pm \mu)$ is equal to zero. The stability of the approximate solution is determined by the condition that the real part of the term $(-\zeta \pm \mu)$ should be negative. Since the characteristic Floquet exponent can be either real or imaginary, this condition is equivalent to $\zeta > 0$ and $\zeta^2 > \mu^2$. Substituting Equation (8.3.17) into Equation (8.3.15) and applying the harmonic balance method gives

$$\begin{bmatrix} \mu^2 + \sigma_0 & \sigma_1\cos\theta & -\sigma_1\sin\theta \\ -2\sigma_1\sin\theta & -2\mu\Omega - \sigma_2\sin 2\theta & \mu^2 - \Omega^2 + \sigma_0 - \sigma_2\cos 2\theta \\ 2\sigma_1\cos\theta & \mu^2 - \Omega^2 + \sigma_0 + \sigma_2\cos 2\theta & 2\mu\Omega - \sigma_2\sin 2\theta \end{bmatrix} \begin{Bmatrix} B \\ \sin\varphi \\ \cos\varphi \end{Bmatrix} = 0 \tag{8.3.18}$$

Nontrivial solutions exist only when the determinant of the coefficient matrix

$$\Delta_1(\mu) \equiv \begin{vmatrix} \mu^2+\sigma_0 & \sigma_1\cos\theta & -\sigma_1\sin\theta \\ -2\sigma_1\sin\theta & -2\mu\Omega-\sigma_2\sin 2\theta & \mu^2-\Omega^2+\sigma_0-\sigma_2\cos 2\theta \\ 2\sigma_1\cos\theta & \mu^2-\Omega^2+\sigma_0+\sigma_2\cos 2\theta & 2\mu\Omega-\sigma_2\sin 2\theta \end{vmatrix} \tag{8.3.19}$$

vanishes. The previous stability conditions lead to $\Delta_1(\zeta)=0$ on the boundary between the stable and unstable regions and $\Delta_1(\zeta)>0$ in a stable region. Thus, it follows that

$$\sigma_0\sigma_2^2-2\sigma_1^2\sigma_2+2\sigma_1^2(\sigma_0-\Omega^2)-\sigma_0(\sigma_0-\Omega^2)^2>0 \tag{8.3.20}$$

Substituting the parameters as given by Equations (8.3.16) leads to the following stability limit:

$$\Delta_1=\left(\Omega^2-3\gamma A_0^2-\frac{3}{2}\gamma A_1^2\right)^2-\frac{9}{16}\gamma^2A_1^4+4\zeta^2\Omega^2-\frac{6\gamma A_0^2A_1^2\left(\Omega^2-3\gamma A_0^2-\frac{3}{4}\gamma A_1^2\right)}{A_0^2+\dfrac{A_1^2}{2}}=0 \tag{8.3.21}$$

To find stability limits directly on the FRCs, use can be made of the following procedure [3,12]. Finding the total derivatives of Y_1, Y_2 and Y_3 defined by Equations (8.3.11) with respect to Ω and solving the system obtained with respect to $\partial A_1/\partial\Omega$, $\partial A_0/\partial\Omega$ and $\partial\theta/\partial\Omega$, gives

$$\frac{\partial A_1}{\partial\Omega}=\frac{\Delta_{21}}{\Delta_2},\qquad \frac{\partial A_0}{\partial\Omega}=\frac{\Delta_{22}}{\Delta_2},\qquad \frac{\partial\theta}{\partial\Omega}=\frac{\Delta_{23}}{\Delta_2} \tag{8.3.22}$$

where

$$\Delta_2=\begin{vmatrix} \dfrac{\partial Y_1}{\partial A_0} & \dfrac{\partial Y_1}{\partial A_1} & \dfrac{\partial Y_1}{\partial\theta} \\ \dfrac{\partial Y_2}{\partial A_0} & \dfrac{\partial Y_2}{\partial A_1} & \dfrac{\partial Y_2}{\partial\theta} \\ \dfrac{\partial Y_3}{\partial A_0} & \dfrac{\partial Y_3}{\partial A_1} & \dfrac{\partial Y_3}{\partial\theta} \end{vmatrix} \tag{8.3.23}$$

and Δ_{2i} $(i=1,2,3)$ stands for the determinant obtained by replacing the ith column in Equation (8.3.23) with the terms $-\frac{\partial Y_1}{\partial\Omega}$, $-\frac{\partial Y_2}{\partial\Omega}$, $-\frac{\partial Y_3}{\partial\Omega}$. Expanding the determinant in Equation (8.3.23) leads to:

$$\Delta_2 = \left(3\gamma A_0^2 A_1 + \frac{3}{2}\gamma A_1^3\right) \cdot \left(-\left(\Omega^2 - 3\gamma A_0^2 - \frac{3}{2}\gamma A_1^2\right)^2 + \frac{9}{16}\gamma^2 A_1^4 - 4\zeta^2\Omega^2 - \frac{6\gamma A_0^2 A_1^2\left(\Omega^2 - 3\gamma A_0^2 - \frac{3}{4}\gamma A_1^2\right)}{A_0^2 + \frac{A_1^2}{2}}\right) \tag{8.3.24}$$

Comparing Equations (8.3.21) and (8.3.24), it is seen that when $\Delta_1 = 0$ then $\Delta_2 = 0$. This leads to the conclusion that at the stability limits one has

$$\frac{\partial A_1}{\partial \Omega} = \frac{\partial A_0}{\partial \Omega} = \frac{\partial \theta}{\partial \Omega} = \infty \tag{8.3.25}$$

Thus, the FRCs have vertical tangents at the stability limits. When these limits are known, the regions of FRCs between two vertical tangents corresponding to stable and unstable solutions can be determined. This is done using the stability condition for the second unstable region of the linearised variational equation given by Equation (8.3.20).

8.3.2.2 Saddle-node bifurcation

Saddle-node bifurcations are one of the most fundamental *bifurcations* in nonlinear dynamics [13]. They are also called tangent bifurcations because they are locations of vertical tangencies or *fold bifurcation* due to the geometry at such points. A saddle-node bifurcation corresponds to a collision and disappearance of two *fixed points* (*equilibria*) in dynamical systems.

The approximate saddle-node *bifurcation set* in terms of F_0 and Ω (the other parameters are fixed) is calculated from Equations (8.3.11a–c) by using the condition of vertical tangencies and is shown in Figure 8.3. It consists of two pairs of bifurcation curves: BA and AC, meeting at a cusp bifurcation point A, and QP and PR, meeting at the other cusp P. These bifurcation curves indicate how the fixed points are created or annihilated. The change in the number of the fixed points is also denoted in this figure. On the curves BA and QS (solid line) the saddle-node bifurcation occurs that corresponds to the coalescence of two fixed points (one stable and one unstable) when one fixed point exists. Their disappearance is related to the curve AS, TC, and RT (dashed line). The curve SP (dotted line) depicts the situation when there is a bifurcation from three fixed points (two stable and one unstable) to five (three stable and two unstable). Along ST and TP (dashed-dotted line) the opposite holds – one stable and one unstable fixed point disappear. At the cusps A and P, further degeneracy occurs and there is a multiply repeated root. Figure 8.3(b) shows saddle-node bifurcation curves computed numerically. They confirm that the behaviour of the

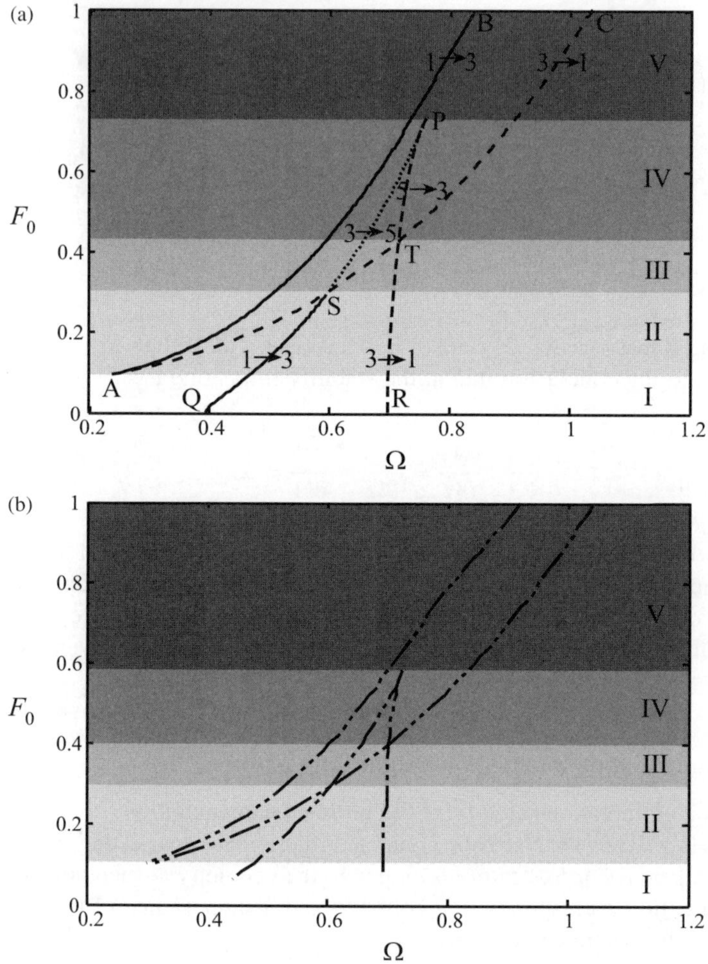

Figure 8.3 (a) An approximate saddle-node bifurcation set in the (Ω, F_0) plane for $\zeta = 0.025$, $\gamma = 0.0783$ and $F = 0.1$: bifurcation from one to three fixed points (solid line), bifurcation from three to one fixed point (dashed line), bifurcation from three to five fixed points (dotted line) and bifurcation from five to three fixed points (dashed-dotted line). Regions labelled by I–V correspond to the cases with distinguishable FRCs; (b) A numerically computed saddle-node bifurcation set ('-..-' line).

system is captured qualitatively well by the theoretical results given in Figure 8.3(a), although the first approximation is used only. Generally, Figure 8.3 indicates that, if the saddle-node bifurcations exist, there can be from two to four of them depending on the values of F_0. The (Ω, F_0) plane in Figure 8.3 is divided into five regions labelled by I–V based on the locations of cusps and the intersections of the bifurcation curves [14]. Characteristic FRCs relating to the regions are considered in the next subsection.

8.3.2.3 Jump phenomenon

Figures 8.4(a)–(e) show the FRCs corresponding to the amplitude of the first harmonic for all five distinguishable regions recognised and labelled in Figure 8.3. Analytical results, calculated on the basis of Equations (8.3.11a–c) are plotted, together with the results from the stability analysis, where the dashed parts of the FRCs represent the unstable regions. In addition, numerical results, which were obtained by integrating the equation of motion (8.2.1) directly, are also given in circles for confirmation. It should be noted that the values of F_0 and the frequency regions shown were chosen so that the first harmonic dominates with respect to other harmonics. This is also checked numerically because of the assumed form of the solution (8.3.10). The choice of parameters ensures that there is consistent accuracy of the assumed solution, which includes the first harmonic only. What is noticeable is the multivaluedness of the response in some frequency regions and the saddle-node bifurcation points, where the vertical tangents of the FRCs exist.

All of these five typical shapes are also shown in Figures 8.5(a)–(e) to explain qualitatively the phenomena associated with their shapes.

For a very small value of F_0, which corresponds to the Case I shown in Figure 8.5 (a), the FRC is bent to the right, which represents a typical hardening behaviour. Such behaviour is expected, because the constant force in the equation of motion (8.2.1) is very small so that the positive (hardening) cubic nonlinearity defines the response of the system (see Chapter 5). If the excitation frequency is slowly increased, i.e., varied quasistatically, starting from point *1*, the amplitude A_1 follows the path labelled by *1-2-3-4*. Point *2* is a jump-down point, where the amplitude decreases suddenly. If the frequency is slowly decreased starting from point *4*, the path *4-5-6-1* is followed, where point *5* is a jump-up point, associated with a sudden increase of the amplitude. Thus, the response depends on the frequency sweep, i.e., it exhibits ‘path dependence’, which is called the *hysteresis* phenomenon. Sudden changes in the amplitude are the reasons why the saddle-node bifurcation belongs to the class of discontinuous or catastrophic bifurcations [13].

Similar behaviour with only two jumps in these two sweeps of the frequency corresponds to very large values of F_0, i.e., to Case V, shown in Figure 8.5(e). However, the FRC is bent to the left here, which corresponds to softening behaviour. Point *2* is a jump-up point, while point *5* represents a jump-down point. Two extreme cases, Case I and V and the change from hardening to softening behaviour can be related to the transformation of the original model of the PC oscillator (8.2.1) to the transformed EQC oscillator (8.2.3), (8.2.4), presented in Section 8.2. Thus, the larger the value of F_0, the larger the value of the coefficient of the quadratic nonlinearity in Equation (8.2.3), which introduces a softening effect into the response [10], which is demonstrated below. The cases between these two extremes are characterised with multiple bends of the FRCs and the mixture of the hardening and softening behaviour.

When the constant force is increased so that Case II exists, the upper branch of the FRC of the harmonic response bends first to the left, experiencing a softening effect and then to the right experiencing a hardening effect (Figure 8.5(b)). Thus, the system experiences four jumps in its response: two jumps occur when increasing frequency

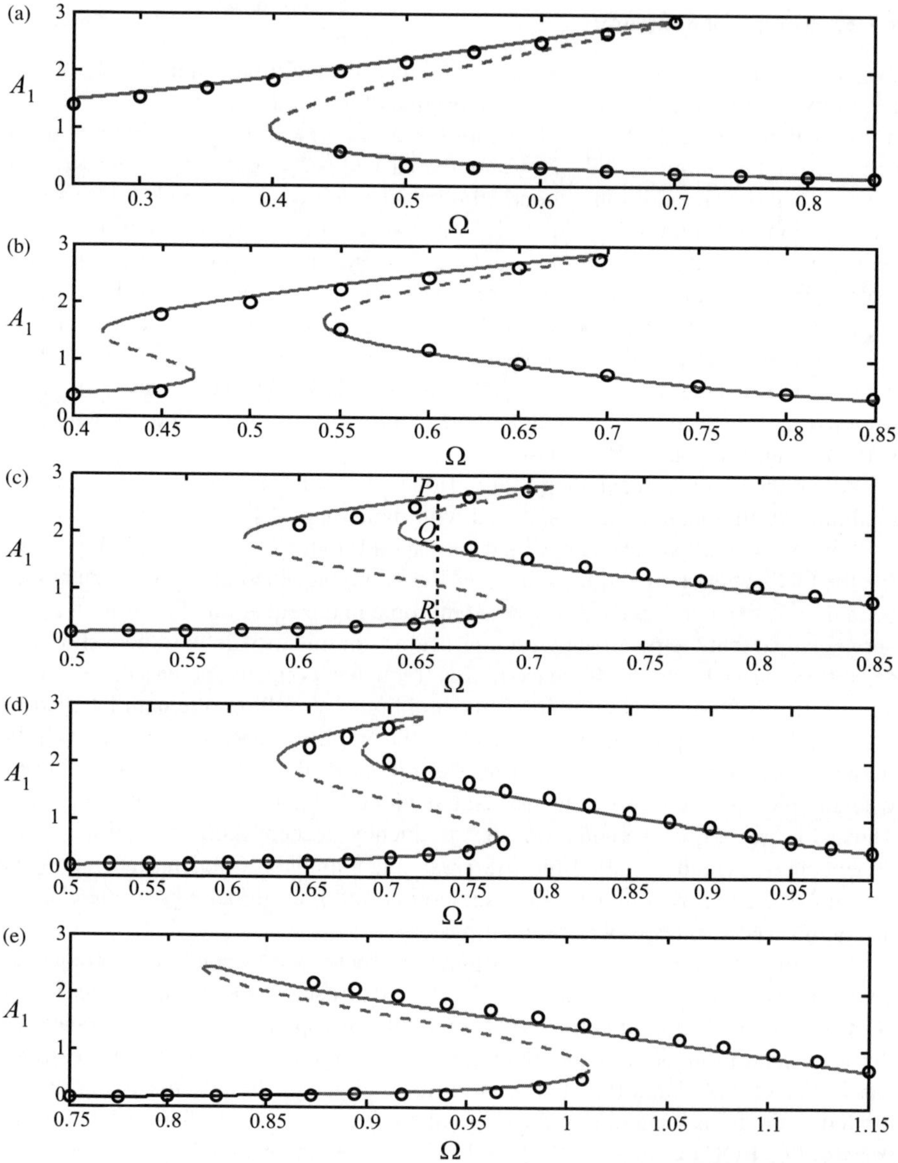

Figure 8.4 FRCs corresponding to the regions labelled in Figure 8.3; stable analytical solutions (solid line), unstable analytical solutions (dashes line), numerical solutions (circles): (a) Case I, $F_0 = 0.01$; (b) Case II, $F_0 = 0.2$; (c) Case III, $F_0 = 0.4$; Case IV, $F_0 = 0.5$; (d) Case V, $F_0 = 0.95$. Points P, Q, R refer to Figure 8.7.

on the path *1-2-3-4-5-6* and two jumps when decreasing frequency when the path *6-7-8-9-10-1* is followed. The overall time-history response of the system corresponding to this *hysteretic behaviour*, obtained by carrying out numerical integration of the equation of motion given by Equation (8.2.1), is illustrated in Figures 8.6(a) and (b) for slowly increasing and decreasing frequency, respectively, and illustrates the double-jump phenomenon [15].

The FRC corresponding to a further increase in F_0 and Case III, is given in Figure 8.5(c) with the characteristic paths and jumps labelled. It can be seen that the upper branch of the FRC of the harmonic response bends even more to the left. During

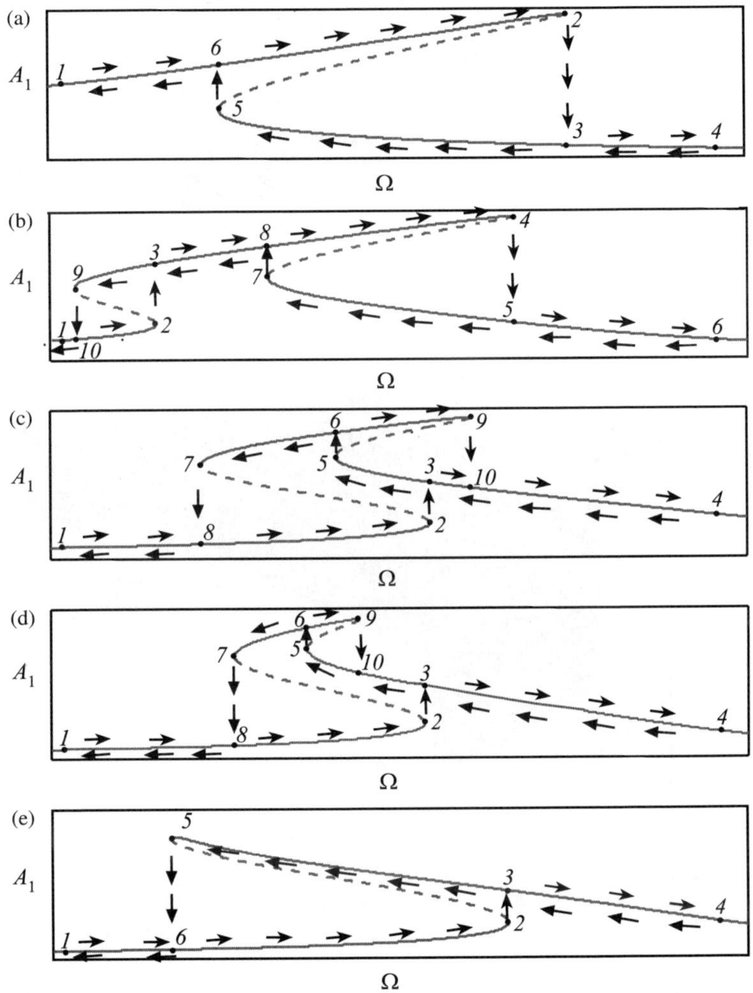

Figure 8.5 Typical FRCs corresponding to Case I-Case V with the paths corresponding to frequency sweep-up and sweep-down labelled.

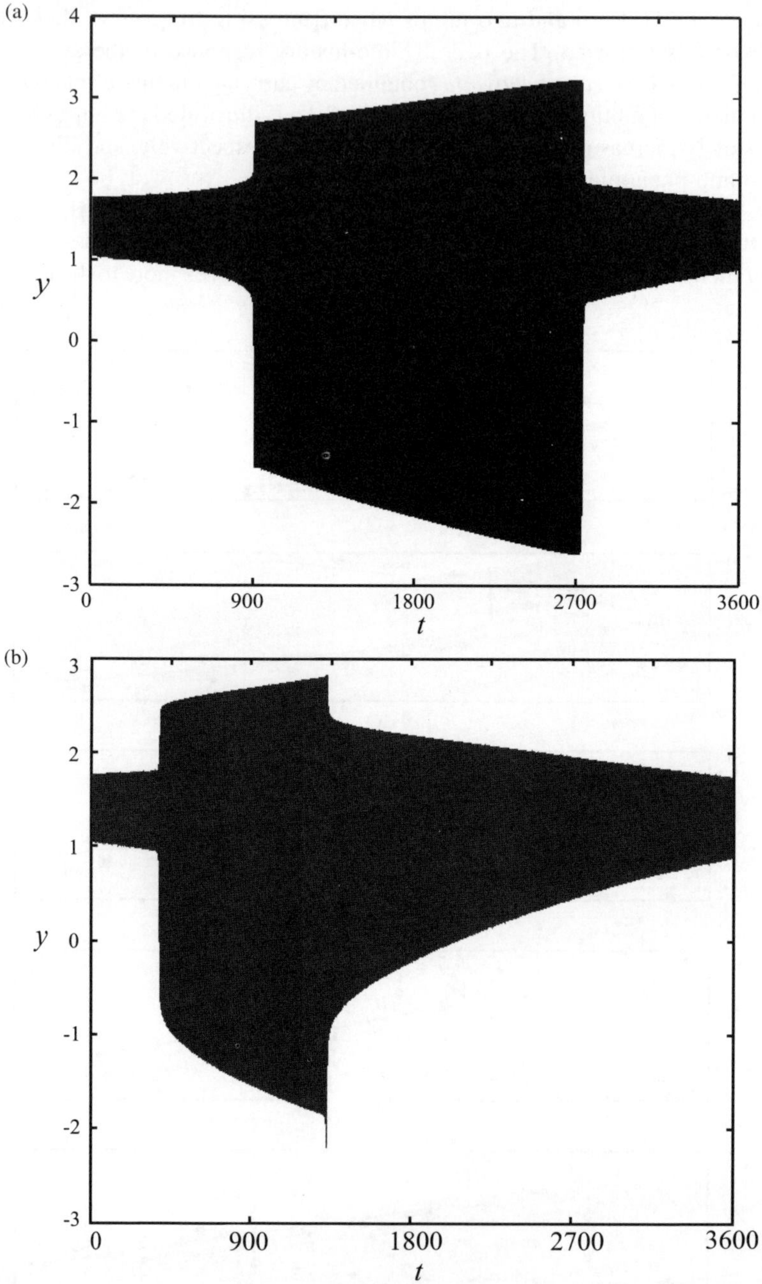

Figure 8.6 The overall time-history diagram for the case shown in Figure 8.4(b) for the frequency rate $d\Omega/dt = 1.3389 \times 10^{-4}$ *obtained by: (a) increasing the frequency from* $\Omega(t = 0) = 0.35$ *to* $\Omega(t = 3600) = 0.85$*; (b) decreasing the frequency from* $\Omega(t = 0) = 0.85$ *to* $\Omega(t = 3600) = 0.35$.

the sweep-up, the path *1-2-3-4* is followed, while the sweep-down results in the response numerated by *4-5-6-7-8-1*. The jump-down point *9* is not accessible in these one-direction-sweeps, but can be reached if the frequency is increased and decreased subsequently, following the path *3-5-6-9*.

The additional feature of this case is that there is a frequency region where five *steady states* exist – three stable and two unstable. This corresponds to the frequency range between the curves SP and ST in Figure 8.3(a). Which of these three stable steady states is achieved for a fixed value of frequency depends on the initial conditions. For example, if the frequency is adjusted to $\Omega = 0.66$, three steady states P, Q and R exist, as labelled in Figure 8.4(c). The *basins/domains of attraction* for these three coexisting periodic attractors are shown in Figure 8.7(a), indicating which initial conditions lead to each of them. In addition, the corresponding phase projections and Poincaré points are presented in Figures 8.7(b)–(d).

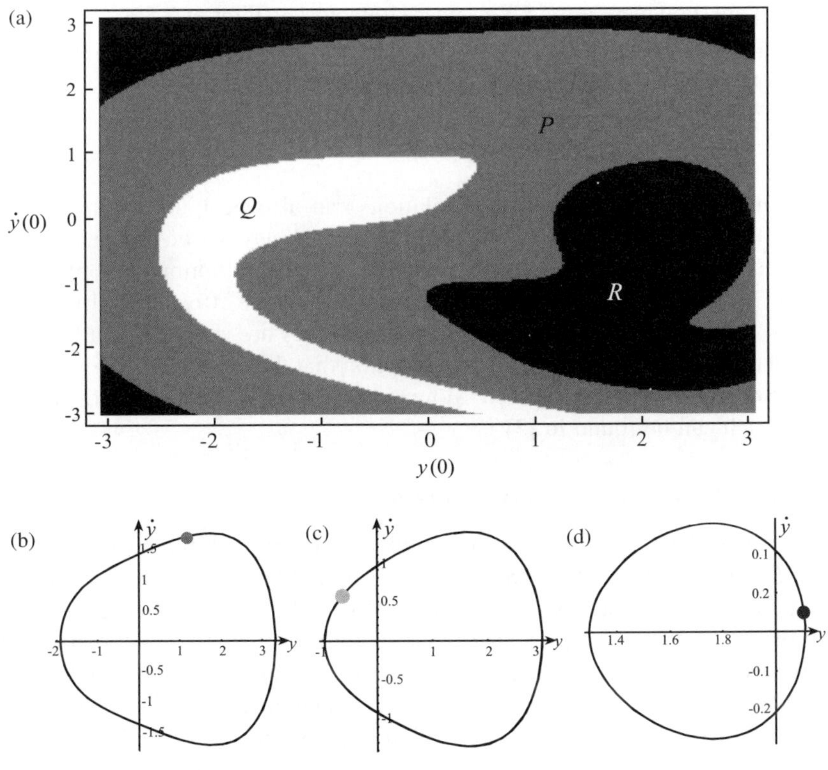

Figure 8.7 (a) Basins/domains of attraction P, Q and R from Figure 8.4(c) when $\Omega = 0.66$*; Poincaré points and phase projections obtained for: (b)* $P(0)$*:* $y(0) = 1$, $\dot{y}(0) = 0$*;* $Q(0)$*:* $y(0) = 1$, $\dot{y}(0) = -3$*;* $R(0)$*:* $y(0) = 1$, $\dot{y}(0) = -1$.

A typical FRC corresponding to Case IV is shown in Figure 8.5(d). Besides multiple bends, it is also characterised by the fact that there is a frequency region in which five steady states exist. During the sweep-up the path *1-2-3-4* is followed, and during the sweep-down, the path *4-5-6-7-8* is followed. As with Case VI, the peak can be reached by combining the increase and the decrease of frequency in the path *3-5-6-9*. What is different in comparison to Case III, is that the value of the jump-up frequency *2* is higher than the frequency at which the peak *9* occurs. The system can exhibit one jump when the frequency is increased and two jumps when it is decreased.

In Figures 8.4 and 8.5 only the FRCs corresponding to the amplitude of the first harmonic A_1 are shown. Those corresponding to the DC term A_0 and the phase θ are not given for brevity, but can be seen in [14]. They are also characterised by the jump and hysteresis phenomena. In all these cases, when the DC term experiences a jump-up, the oscillatory term experiences a jump-down and vice versa.

Transient motion obtained numerically from Equation (8.2.1) for the initial conditions $P(0)$: $y(0)=1$, $\dot{y}(0)=0$; $Q(0)$: $y(0)=1$, $\dot{y}(0)=-3$; and $R(0)$: $y(0)=1$, $\dot{y}(0)=-1$ are shown in Figure 8.8, as phase trajectories (Figures 8.8(a)–(c)) and time histories (Figures 8.8(d)–(f)). Unlike the phase trajectory of the nonconservative unforced system shown in Figure 8.1(g), which spirals towards the equilibrium point C_y with a zero velocity, the phase trajectories of the forced system spiral out or spiral in towards nontrivial steady states, whose phase projections are given in Figures 8.7(b)–(d).

It should be noted that transient solutions can also be found from Equations (8.3.6a–d) and (8.3.7). However, the results obtained may not be as accurate due to several reasons. Besides the limitation imposed by the assumption that only the first harmonic exists in the solution (8.3.1), the accuracy can be affected by the choice of the initial conditions. If they are prescribed so that they are very different from those corresponding to the steady state, the assumption that the amplitudes vary slowly, as given in Subsection 8.3.1, is violated. More reasons and details about the drawback of this approach can be found in [3].

8.3.2.4 Effects of damping on the bifurcation set

The influence of the change of damping on the shape of the bifurcation sets, can be seen in Figure 8.9. All parts of this figure are plotted for $\gamma = 0.0783$, $F = 0.1$, and contain the bifurcation sets obtained analytically, with the legend being the same as that in Figure 8.3. It should be noted that the bifurcation sets were also obtained numerically. Being qualitatively very similar to the analytical ones, they are not shown here for clarity. The bifurcation sets corresponding to smaller damping than that used to produce Figure 8.3 are plotted in Figures 8.9(a) and (b), for $\zeta = 0.0125$ and $\zeta = 0.00625$, respectively.

It can be seen that the structure of the bifurcation sets shown is similar to that given in Figure 8.3, with two pairs of bifurcation curves and two cusps. Because of this, all five cases labelled in Figure 8.3 can also be identified here, and the reasons for the

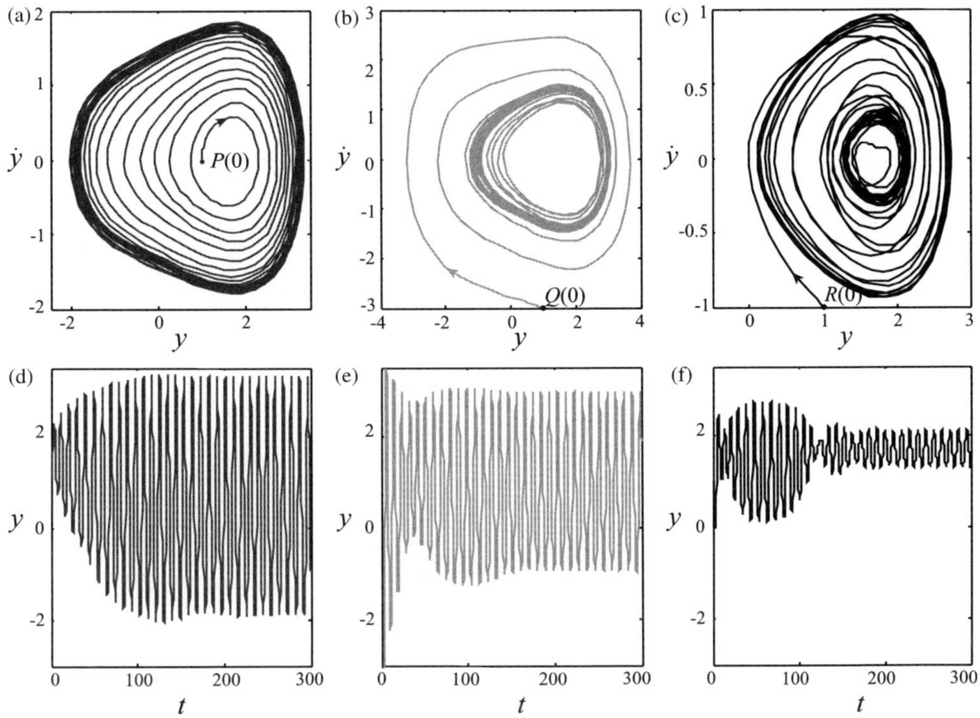

Figure 8.8 Phase plots of the transient motion for the initial conditions defined in Figure 8.7 (a) $P(0)$; (b) $Q(0)$; (c) $R(0)$; (d)–(f) corresponding time histories.

appearance of jumps are similar to those described above relating to Figure 8.5. It can also be noted that the left pair of curves and the corresponding cusp only change slightly with a decrease in damping, while the right pair is affected more. The left curve of this pair comes closer to the left branch of the other pair, or, in other words, the curve QP from Figure 8.3 becomes closer to the curve BA from the same figure. This indicates that the frequencies at which the jump-down and jump-up occur are closer to each other. The right curve from the right pair is almost vertical, which means that the corresponding jump occurs at approximately the same frequency regardless of the values the constant force: at $\Omega \approx 1$ for $\zeta = 0.0125$ (see Figure 8.9(a)) and at $\Omega \approx 1.4$ for $\zeta = 0.00625$ (see Figure 8.9(b)). With a reduction in damping its cusp moves toward higher frequencies and higher values of the magnitude of the constant force. In the special case, when damping is zero, it goes to infinity, since two branches of the FRC do not meet.

The bifurcation sets illustrating the effects of the increase of damping on the onset of saddle-node bifurcations are given in Figure 8.9(c) for $\zeta = 0.05$ and in Figure 8.9 (d) for $\zeta = 0.075$. The rest of the parameters are held fixed at their previous values. Comparing these figures mutually, as well as with the bifurcation set given in

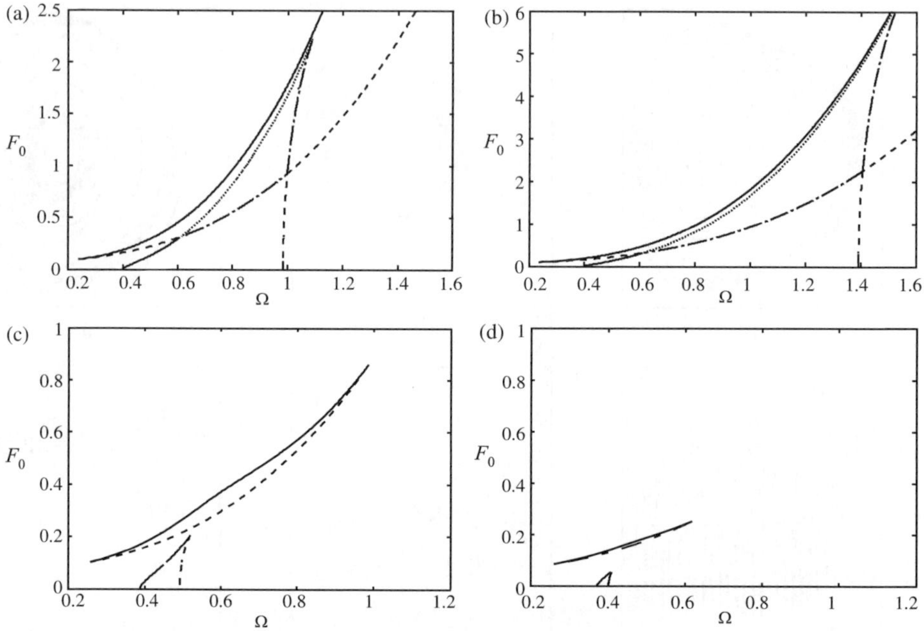

Figure 8.9 An approximate analytical saddle-node bifurcation set in the (Ω, F_0) *plane for* $\gamma = 0.0783$, $F = 0.1$*: (a)* $\zeta = 0.0125$*; (b)* $\zeta = 0.00625$*; (c)* $\zeta = 0.05$*; (d)* $\zeta = 0.075$*. Legend as in Figure 8.3.*

Figure 8.3, it can be concluded that the basic bifurcation structure is crucially dependent on this change in the damping ratio. First, the pairs of bifurcation curves do not intersect each other, i.e., one does not cross over into the frequency region where the others occur. Consequently, there is no region corresponding to five steady states, which implies that making the damping ratio higher reduces the maximum number of the steady states. Secondly, the regions between the branches of each pair becomes narrower. Finally, Figure 8.9(d) shows that if the damping ratio is increased appropriately, there is only a small range of the values of the constant force for which saddle-node bifurcations can occur. For the majority of the values of the constant force from the range investigated, there is no saddle-node bifurcation, i.e., the system does not experience jump phenomena, but does exhibit linear-like behaviour, with FRCs that are single-valued functions. This implies that a suitable choice of damping can be used as a mechanism to eliminate the occurrence of jumps in the system.

8.3.3 Some secondary resonance responses

In the preceding section, the harmonic oscillations in the primary resonance have been discussed, whose frequency is equal to that of the external force. However, as shown in Chapter 5, a nonlinear system can have a wide variety of periodic oscillations in addition to those that have the same frequency as the external force, which are related to *secondary resonance*. Thus, in the asymmetric oscillator, oscillations that have a period twice that of the external force can occur [3]. It is possible for the steady-state response to consist of a term having the same frequency as the excitation and a term whose frequency is exactly one-half of the frequency of excitation. For this reason, this is known as the 1/2 (one-half) *subharmonic resonance* response. By a procedure analogous to that presented in Section 8.3.2, the periodic solutions can be determined and their stability investigated by considering the variational equations of the Hill type [3]. For fixed parameter values, two types of the 1/2 resonance response can be found, which have a phase shift of 180° with respect to each other.

Since the equation of motion (8.2.1) has a cubic term, subharmonic oscillations of order 1/3 can also occur. Following the same procedure as outlined above, three kinds of 1/3-harmonic oscillations can be obtained with a mutual phase shift of 120°. The harmonic, 1/2-harmonic oscillations and 1/3-harmonic responses of the oscillator (8.2.1) with $\zeta = 0.025$, $\gamma = 1$, $F_0 = 0.005$ and $F = 0.14$ are investigated in detail in [3], including defining the fixed points, their stability, trajectories of the stable solutions and the basin of attraction.

The 1/2 subharmonic resonance and its stability limit is closely related to the appearance of *chaotic motion* in the asymmetric oscillator, which is discussed in detail in Section 8.5.

8.4 Regular response of the single-well Helmholtz–Duffing oscillator

In this section the single-well Helmholtz–Duffing (SWHD) oscillator governed by Equation (8.2.11) is investigated. The frequency–amplitude relationship and the approximate solution of the primary resonance response are obtained. In addition to this, the behaviour of the secondary resonance responses is also briefly discussed.

8.4.1 Primary resonance response via perturbation method

To analyse the primary resonance response, the procedure developed by Benedettini and Rega [5] is used. It involves the fourth-order multiple scale perturbation method, in which the higher-order terms are retained in the asymptotic expansions. These are essential for the correct description of the system behaviour when there is a large-amplitude response. First, it is assumed that the solution of Equation (8.2.11) has the form

$$z = \varepsilon \bar{z} \tag{8.4.1}$$

where ε is a small parameter, i.e., $\varepsilon \ll 1$. In this way, the nonlinearity in Equation (8.2.11) generates a term of $O(\varepsilon^3)$. To ensure that this nonlinearity, the damping and the excitation appear at the same stage in the perturbation scheme, the following ordering is needed

$$\zeta = \varepsilon^2\bar{\zeta}, \quad F = \varepsilon^3\bar{F} \tag{8.4.2}$$

Then, Equation (8.2.11) becomes

$$\ddot{z} + \varepsilon^2 2\zeta\dot{z} + z + \varepsilon c_2 z^2 + \varepsilon^2 c_3 z^3 = \varepsilon^2 F\cos\Omega t \tag{8.4.3}$$

where the over-bar has been omitted for simplicity.

Following the underlying idea of the method of multiple scales, multiple independent variables (scales) are introduced

$$T_n = \varepsilon^n t, \quad n = 0, 1, 2, 3, 4. \tag{8.4.4}$$

An approximate solution to Equation (8.4.3) is sought in the following form of an expansion in powers of ε

$$z(t, \varepsilon) = \sum_{n=0}^{4} \varepsilon^n\, z_n(T_0, T_1, T_2, T_3, T_4) \tag{8.4.5}$$

Further, the close proximity of the excitation frequency and the natural frequency is expressed by setting the following relationship between them

$$\Omega = 1 + \varepsilon^2\sigma_1 + \varepsilon^4\sigma_2 \tag{8.4.6}$$

where the quantitative measures of this proximity are the *detuning parameters* σ_1 and σ_2, which need to be determined. Since the excitation in Equation (8.4.3) is of $O(\varepsilon^2)$, the lowest term in the expansion (8.4.6) has been assumed to be of $O(\varepsilon^2)$ for consistency.

Substituting Equations (8.4.4)–(8.4.6) into Equation (8.4.3) and collecting the terms with the same power of ε, the following system of the linear differential equations is obtained

$$\begin{aligned}
D_{00}z_0 + z_0 &= 0,\\
D_{00}z_1 + z_1 &= -2D_{01}z_0 - c_2 z_0^2,\\
D_{00}z_2 + z_2 &= -2D_{01}z_1 - 2D_{02}z_0 - D_{11}z_0 - 2c_2 z_0 z_1 - c_3 z_0^3\\
&\quad -2\zeta D_0 z_0 + F\cos(T_0 + \sigma_1 T_2 + \sigma_2 T_4),\\
D_{00}z_3 + z_3 &= -2D_{01}z_2 - 2D_{02}z_1 - D_{11}z_1 - 2D_{03}z_0 - 2D_{12}z_0\\
&\quad -c_2\left(z_1^2 + 2z_0 z_1\right) - 3c_3 z_0^2 z_1 - 2\zeta(D_0 z_1 + D_1 z_0),\\
D_{00}z_4 + z_4 &= -2D_{01}z_3 - 2D_{02}z_2 - D_{11}z_2 - 2D_{03}z_1 - 2D_{12}z_1 - 2D_{04}z_0\\
&\quad -2D_{13}z_0 - D_{22}z_0 - 2c_2(z_0 z_3 + z_1 z_2)\\
&\quad -3c_3\left(z_0 z_1^2 + z_0^2 z_2\right) - 2\zeta(D_0 z_2 + D_1 z_1 + D_2 z_0)
\end{aligned} \tag{8.4.7a–e}$$

where the partial differential operator are defined with respect to time variables given by Equation (8.4.4) as $D_i = \partial/\partial T_i$, $D_{ik} = \partial^2/(\partial T_i \partial T_k)$.

Since Equation (8.4.7a) has the form of the equation of motion of a *harmonic oscillator*, its solution can be written down in the form

$$\begin{aligned} z_0 &= A(T_1, T_2, T_3, T_4)e^{jT_0} + A^*(T_1, T_2, T_3, T_4)e^{-jT_0} \\ &\equiv A(T_1, T_2, T_3, T_4)e^{jT_0} + \text{c.c.} \end{aligned} \tag{8.4.8}$$

where *c.c.* stands for a complex conjugate. Substituting this solution into Equation (8.4.7b) yields

$$D_{00}z_1 + z_1 = -2jD_1Ae^{jT_0} - c_2(A^2e^{2jT_0} + AA^*) + \text{c.c.} \tag{8.4.9}$$

The requirement to eliminate the *secular terms* means that $D_1A = 0$, which implies $A \equiv A(T_2, T_3, T_4)$. Hence, the solution of Equation (8.4.9) is

$$z_1 = \frac{c_2}{3}A^2e^{2jT_0} - c_2AA^* + \text{c.c.} \tag{8.4.10}$$

If the following polar form

$$A(T_2, T_3, T_4) = \frac{1}{2}a(T_2, T_3, T_4)e^{j\varphi(T_2, T_3, T_4)} \tag{8.4.11}$$

is introduced, with a and φ being the amplitude and phase of oscillations, the solution (8.4.10) gets the form

$$z_1 = \frac{c_2}{6}a^2\cos 2(T_0 + \varphi) - \frac{c_2}{2}a^2 \tag{8.4.12}$$

Substituting the solutions for z_0 and z_1 into Equation (8.4.7c) leads to

$$D_{00}z_2 + z_2 = ST_2 \cdot e^{jT_0} - \left(\frac{2c_2^2}{3} + c_3\right)A^3e^{3jT_0} + \text{c.c.} \tag{8.4.13}$$

where ST_2 stands for the following secular terms

$$ST_2 = -2jD_2A + 4c_2^2A^2A^* - \frac{2c_2^2}{3}A^3A^* - 3c_3A^2A^* - 2\zeta jA + \frac{F}{2}e^{j(\sigma_1 T_2 + \sigma_2 T_4)} \tag{8.4.14}$$

They can be eliminated if they are set to zero. This, together with Equation (8.4.11) results in

$$-jD_2a + aD_2\varphi + \frac{10c_2^2}{24}a^3 - \frac{3}{8}c_3a^3 - \zeta ja + \frac{F}{2}e^{j\phi} = 0 \tag{8.4.15}$$

where the following new variable has been introduced to obtain an autonomous system

$$\varphi = \sigma_1 T_2 + \sigma_2 T_4 - \phi \tag{8.4.16}$$

Separating the real and imaginary parts in Equation (8.4.15) gives

$$\begin{aligned} D_2 a &= -\zeta a + \frac{F}{2}\sin\phi, \\ a D_2 \phi &= a\sigma_1 + c_4 a^3 + \frac{F}{2}\cos\phi \end{aligned} \tag{8.4.17a, b}$$

where the constant c_4 has been introduced for brevity. The expression for this constant is given in Table 8.1. This table also contains the expression for other constants that are introduced subsequently.

Table 8.1 Expressions for the coefficients c_4–c_{14}

Coefficient	Expression
c_4	$\frac{5}{12}c_2^2 - \frac{3}{8}c_3$
c_5	$\frac{335}{1728}c_2^4 - \frac{143}{192}c_2^2 c_3 - \frac{3}{256}c_3^2$
c_6	$\frac{19}{72}c_2^2 - \frac{3}{16}c_3$
c_7	$\frac{485}{1728}c_2^4 - \frac{173}{192}c_2^2 c_3 + \frac{15}{156}c_3^2$
c_8	$\frac{1}{48}c_2^2 + \frac{1}{32}c_3$
c_9	$\frac{59}{432}c_2^3 - \frac{31}{96}c_2 c_3$
c_{10}	$\frac{1}{432}c_2^3 + \frac{1}{96}c_2 c_3$
c_{11}	$-\frac{19}{72}c_2^3 + \frac{5}{8}c_2 c_3$
c_{12}	$\frac{29}{36}c_2^2 - \frac{9}{8}c_3$
c_{13}	$-\frac{11}{72}c_2^2 + \frac{3}{16}c_3$
c_{14}	$\frac{1}{36}c_2^2 + \frac{3}{8}c_3$

For a steady-state solution, the left-hand sides of Equations (8.4.17a,b) are equal to zero, which, after the elimination of ϕ, yields the expression for the first detuning parameter

$$\sigma_1 = -c_4 a^2 \mp \frac{1}{2}\sqrt{\frac{F^2}{a^2} - (2\zeta)^2} \tag{8.4.18}$$

The solution for z_2 follows from Equation (8.4.13) with the secular terms eliminated, and is equal to

$$z_2 = \left(\frac{c_2^2}{12} + \frac{c_3}{8}\right) A^3 e^{3jT_0} + \text{c.c.} \tag{8.4.19}$$

i.e.,

$$z_2 = c_8 a^3 \cos 3(T_0 + \varphi) \tag{8.4.20}$$

By proceeding in the same manner, it follows that

$$D_{00} z_3 + z_3 = -2jD_3 A e^{jT_0} + NST_3 + \text{c.c.} \tag{8.4.21}$$

where NST_3 stands for nonsecular terms. The requirement of having no secular terms is fulfilled if $D_3 A = 0$, which means that A is independent of T_3. Therefore, $A \equiv A(T_2, T_4)$. The nonsecular terms NST_3 are not listed for brevity, but their content is such that the solution for Equation (8.4.21) is

$$\begin{aligned} z_3 = {} & c_9 a^4 \cos 2(T_0 + \varphi) + c_{10} a^4 \cos 4(T_0 + \varphi) + c_{11} a^4 \\ & + \frac{2\zeta c_2 a^2}{9} \sin 2(T_0 + \varphi) + \frac{2c_2 F a}{9} \cos(2T_0 + 2\varphi + \phi) \end{aligned} \tag{8.4.22}$$

By analysing the right-hand side of Equation (8.4.7e), it can be concluded that the terms $D_{01}z_3, D_{02}z_2, D_{11}z_2, D_{03}z_1, D_{12}z_1, D_{13}z_0, D_0 z_2$ and $D_1 z_1$ are nonsecular terms. One of the remaining secular terms is $D_{22}z_0 = -a\,\varphi'^2 \cos(T_0 + \varphi)$. During the simplification $D_2\phi = \sigma_1 - D_2\varphi$ is used, which stems from Equation (8.4.16). Furthermore, in the steady state $D_2\phi = 0$, which leads to $\varphi'^2 = \sigma_1^2$, so that $D_{22}z_0 = -a\,\sigma_1^2 \cos(T_0 + \varphi)$. As a result, Equation (8.4.7e) can be written as

$$D_{00} z_4 + z_4 = ST_{4c} \cos(T_0 + \varphi) + ST_{4s} \sin(T_0 + \varphi) + NST_4 \tag{8.4.23}$$

where

$$\begin{aligned} ST_{4c} = {} & 2aD_4\varphi - 2\zeta D_2 a - \left(2c_2 c_{11} + c_2 c_9 + \frac{c_2^2 c_8}{6} + \frac{3c_3 c_8}{4}\right) a^5 \\ & - \frac{2c_2^2 F a^2}{9} \cos\phi - \frac{13 c_2^2 c_3 a^4}{24} + a\sigma_1^2, \\ ST_{4s} = {} & 2D_4 a + 2\,\zeta a D_2\varphi + \frac{2c_2^2 F a^2}{9} \sin\phi - \frac{2\zeta c_2^2 a^3}{9} \end{aligned} \tag{8.4.24a,b}$$

and, as before, $NST4$ stands for nonsecular terms.

By imposing the requirement of no secular terms, i.e., by equating Equations (8.4.24a,b) to zero, and transforming the resulting equations gives

$$\begin{aligned} D_4 a &= 2c_6\zeta a^3 + \frac{\zeta F}{2}\cos\phi - \frac{c_2^2 F a^2}{9}\sin\phi, \\ aD_4\phi &= a\sigma_2 + \frac{1}{2}a\sigma_1^2 + c_5 a^5 + \zeta^2 a - \frac{\zeta F}{2}\sin\phi - \frac{c_2^2 F a^2}{9}\cos\phi \end{aligned} \tag{8.4.25a,b}$$

For the steady-state solutions, the right-hand sides of Equations (8.4.25a,b) are equal to zero, and such equations give the second detuning parameter

$$\sigma_2 = -\frac{\sigma_1^2}{2} - c_5 a^4 - \zeta^2 \pm \sqrt{\frac{\zeta^2 F^2}{4a^2} + \frac{c_2^4 F^2 a^2}{81} - 4c_6\zeta^2 a^4} \tag{8.4.26}$$

By substituting Equations (8.4.18) and (8.4.26) into Equation (8.4.6), the frequency-amplitude equation is obtained

$$\Omega = 1 - c_4 a^2 \mp \frac{1}{2}(1 + c_4 a^2)\sqrt{\frac{F^2}{a^2} - 4\zeta^2 - c_7 a^4 - \frac{\zeta^2}{2} - \frac{F^2}{8a^2} \pm \sqrt{\frac{\zeta^2 F^2}{4a^2} + \frac{c_2^4 F^2 a^2}{81} - 4c_6\zeta^2 a^4}} \tag{8.4.27}$$

This equation gives the relationship between the steady-state amplitude. It is explicit, unlike the ones considered in Section 8.3.2, which were calculated numerically from a system of coupled implicit algebraic equations. The dependence of the frequency on the amplitude given by Equation (8.4.27) is of the fourth power, different from that of the second power given in Section 7.4, as well as in [11].

In the time domain, the steady-state response is obtained substituting Equations (8.4.8), (8.4.12), (8.4.20) and (8.4.22) into Equation (8.4.5), to give

$$\begin{aligned} z(t) &= -\frac{c_2}{2}a^2 + a\cos(\Omega t - \phi) + \frac{c_2}{6}a^2\cos 2(\Omega t - \phi) + c_8 a^3\cos 3(\Omega t - \phi) \\ &\quad + a^4(c_9\cos 2(\Omega t - \phi) + c_{10}\cos 4(\Omega t - \phi) + c_{11}) \\ &\quad + \frac{2\zeta c_2 a^2}{9}\sin 2(\Omega t - \phi) + \frac{2c_2 F a}{9}\cos(2\Omega t - \phi) \end{aligned} \tag{8.4.28}$$

where the phase shift is found from Equation (8.4.17a,b) for $D_2 a = 0$ and $D_2\phi = 0$, which gives

$$\tan\phi = -\frac{\zeta}{\sigma_1 + c_4 a^2} \tag{8.4.29}$$

It can be seen that the response contains even and odd higher harmonics up to the fourth harmonic as well as two DC terms, indicating that the oscillatory motion is not centred at $z = 0$. This is in qualitative agreement with the asymmetry of phase trajectories discussed in Section 8.2. and shown in Figure 8.1(h).

8.4.2 Frequency-response curves

8.4.2.1 Undamped motion

The undamped forced steady-state oscillations are, according to Equation (8.4.17a,b), either inphase or out-of-phase with the excitation as Equations (8.4.17a,b) gives $\phi = 0$ or $\phi = \pi$ for $\zeta = 0$. The former case corresponds to the left branch of the FRC, with the upper signs in Equation (8.4.27)

$$\Omega_{\mathrm{L}} = 1 - c_4 a^2 - \frac{1}{2}\left(1 + c_4 a^2\right)\frac{F}{a} - c_7 a^4 - \frac{F^2}{8a^2} + \frac{c_2^2 F a}{9} \tag{8.4.30}$$

The latter case corresponds to the right branch of the FRC and the lower signs in Equation (8.4.27)

$$\Omega_{\mathrm{R}} = 1 - c_4 a^2 + \frac{1}{2}\left(1 + c_4 a^2\right)\frac{F}{a} - c_7 a^4 - \frac{F^2}{8a^2} - \frac{c_2^2 F a}{9} \tag{8.4.31}$$

Figure 8.10 shows both the left and the right branch of the FRC for the parameter values $c_2 = 0.79896$, $c_3 = 0.26396$ taken from [5], corresponding to a suspended cable.

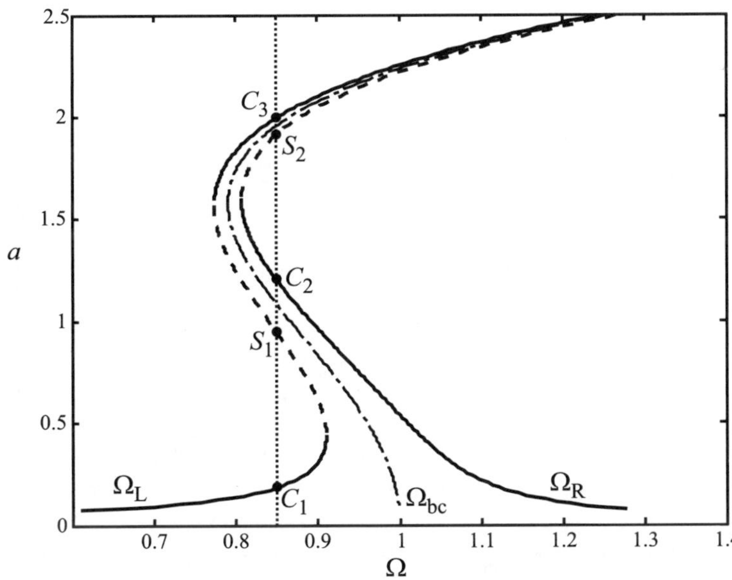

Figure 8.10 Branches of a frequency–response curve Equations (8.4.30) and (8.4.31) and a backbone curve Equation (8.4.32), labelled by a dashed-dotted line, for $c_2 = 0.79896$, $c_3 = 0.26396$, $F = 0.05$ and $\zeta = 0$.

It can be seen that the frequency response branches do not meet, but approach the *backbone curve* for higher amplitudes, the expression for which is

$$\Omega_{\text{bc}} = 1 - c_4 a^2 - c_7 a^4 \tag{8.4.32}$$

This expression has been obtained from Equation (8.4.30) or Equation (8.4.31) for $F = 0$. This figure also illustrates the multivaluedness of the solution, as discussed in Section 8.3.2. It can be seen that there is a frequency region in which there are five *coexisting attractors*, labelled by C_1, C_2, C_3, S_1 and S_2. The state plane a–ϕ containing them and the corresponding domain of attraction is considered in Section 8.4.3.

8.4.2.2 Damped motion

The frequency–response curves defined by Equations (8.4.27) are plotted in Figure 8.11 for damped motion with a constant magnitude of the excitation and different values of damping. In the damped case, the response is bounded. As the damping increases, the frequency response unbends to the left, becoming an oscillator with purely softening characteristics and eventually turning into a single-valued curve. Such a transformation due to the influence of damping is also discussed in Section 8.3.2.4 by means of bifurcation sets. It should be noted that, unlike in the undamped case, when the frequency shift is constant, for the damped motion, this shift changes in accordance with Equation (8.4.29).

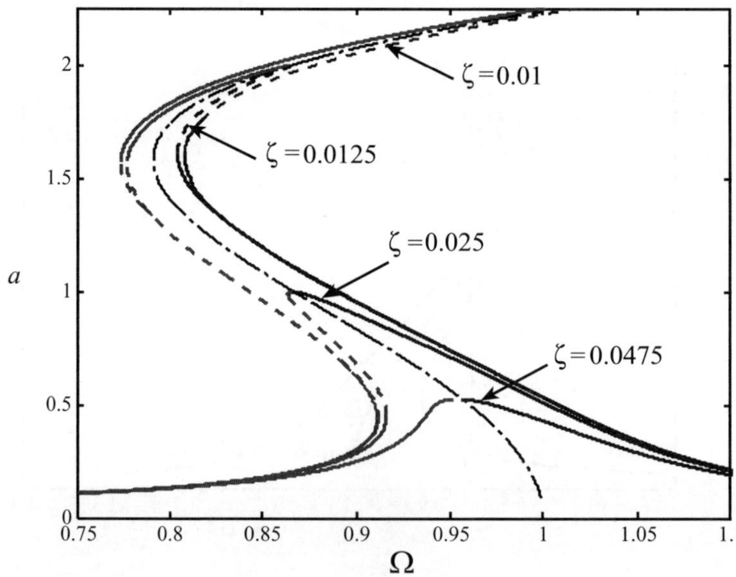

Figure 8.11 Frequency–response curve Equation (8.4.27) for $c_2 = 0.79896$, $c_3 = 0.26396$, $F = 0.05$ *and different values of the damping parameter.*

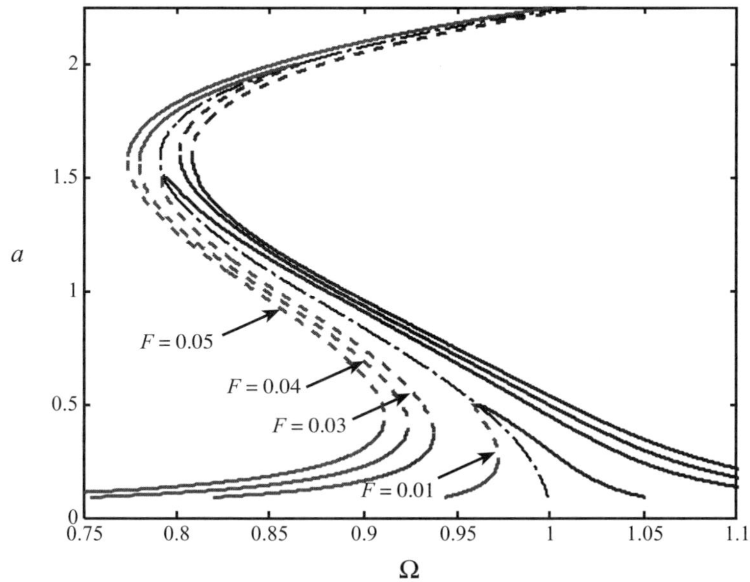

Figure 8.12 Frequency–response curve Equation (8.4.27) for $c_2 = 0.79896$, $c_3 = 0.26396$, $\zeta = 0.025$ *and different values of the damping parameter.*

Figure 8.12 shows how the shape of the FRCs change as the magnitude of the harmonic excitation varies. It can be seen that it influences the maximum number of steady states, similar to that in Figure 8.2.

Although the stability of these steady-state solutions has not been investigated separately, the unstable parts of the FRCs plotted in Figures 8.10 and 8.11 are shown as dashed lines. Note that, as mentioned in Section 8.3.2.1, the stability limit exists at the location of the vertical tangencies of a FRC. Shapes of the FRCs can be related to those shown in Figure 8.5 with a different number and character of the steady-state solution for a different frequency range: one stable solution, three solutions – two stable and one unstable, and five solutions – three stable and two unstable.

8.4.3 Analysis of the steady-state response: coexisting attractors

One of the typical characteristics of nonlinear systems is the coexistence of solutions, which has already been considered in Section 8.3.2.3 for a PC oscillator, where the corresponding basin of attraction was obtained by directly integrating the equation of motion (8.2.1). It is shown here how the influence of the initial conditions on the resulting response can be examined on the basis of the analytical approach shown in Section 8.4.1. For this purpose, the first-order differential equation defining the amplitude a and phase ϕ are used, written in terms of the variations on the *timescales* T_k

$$\dot{a} = \sum_{k=1}^{4} \varepsilon^k D_k a, \quad \dot{\phi} = \sum_{k=1}^{4} \varepsilon^k D_k \phi \qquad (8.4.33\text{a, b})$$

For transient solutions, Equations (8.4.17a,b) need to be taken into account as well as those corresponding to $D_4 a$ and $D_4\phi$ derived for nonzero $D_2 a$ and $D_2\phi$. This leads to

$$\begin{aligned}\dot{a} &= \frac{\zeta F}{4}\cos\phi + \frac{F}{2}\left(1-\frac{\sigma_1}{2}+\frac{c_{12}a^2}{2}\right)\sin\phi - 2\zeta a\left(\frac{1}{2}-c_{13}a^2\right),\\ \dot{\phi} &= \sigma - \frac{\zeta F}{4a}\sin\phi + \frac{F}{2a}\left(1-\frac{\sigma_1}{2}-\frac{c_{14}a^2}{2}\right)\cos\phi + \frac{\zeta^2}{2} + a^2\left(c_4+c_7a^2\right)\end{aligned} \tag{8.4.34a, b}$$

with the coefficients c_{12}-c_{14} defined in Table 8.1.

8.4.3.1 Undamped motion

In the case of undamped motion, Equations (8.4.34a,b) simplify to

$$\begin{aligned}\dot{a} &= \frac{F}{2}\left(1-\frac{\sigma_1}{2}+\frac{c_{12}a^2}{2}\right)\sin\phi,\\ \dot{\phi} &= \sigma + \frac{F}{2a}\left(1-\frac{\sigma_1}{2}-\frac{c_{14}a^2}{2}\right)\cos\phi + a^2\left(c_4+c_7a^2\right)\end{aligned} \tag{8.4.35a,b}$$

For the steady-state solutions, the left-hand side of these equation are equal to zero, yielding the displacements of three solutions for $\phi = 0$ and two solutions for $\phi = \pi$. If $\sigma \approx \sigma_1 = -0.15$, they correspond to C_1, S_1, C_3 and C_2, S_2, respectively, labelled in Figure 8.10. These points are shown also in the state plane a–ϕ in Figure 8.13. In addition, trajectories plotted by solving Equations (8.4.35a,b) numerically for different initial conditions are also given. The strong influence of the initial condition on the motion is evident. Points C_1, C_2 and C_3 are centres and points S_1 and S_2 are saddles. Two *separatrices* pass through each *saddle point*, dividing the state plane into regions with the same shape of trajectories. The arrows indicate the direction of motion of the representative point from the initial position during time.

8.4.3.2 Damped motion

The trajectories of the damped motion are plotted in the state plane a–ϕ in Figure 8.14, obtained by integrating Equations (8.4.34a,b) numerically for different initial conditions. Compared with the state plane for the undamped motion given in Figure 8.13, this state plane is significantly different. The centres turned into foci, towards which trajectories are attracted. Two saddles with separatrices still exist. One can also notice the branches of the separatrices that are attracted to the foci themselves. Note that the coordinates of the fixed points, foci and saddles are slightly different now from those corresponding to the fixed points of the undamped motion (centres and saddles) shown in Figure 8.13.

To demonstrate that there are instances in which a small change in the initial conditions produces a large change in the response of the system, Equations (8.4.34a,b) are solved for two sets of the initial conditions $IC = (a(0);\ \phi(0))$ that differ slightly from each other. They are chosen to be $IC_1 = (2.63;\ 3\pi/2)$ and $IC_2 = (2.635;\ 3\pi/2)$.

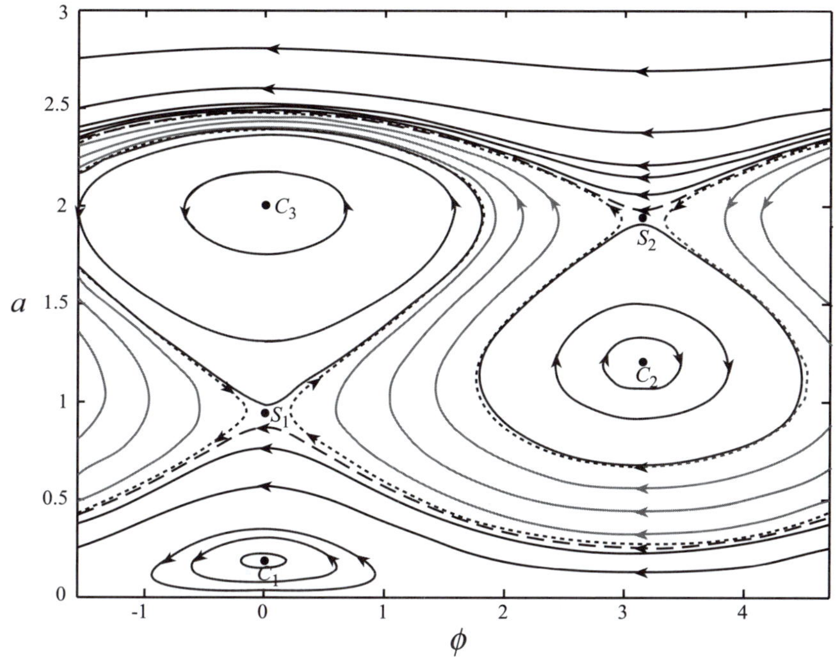

Figure 8.13 State plane for the undamped response for the parameters value as in Figure 8.10 and $\sigma = -0.15$.

The corresponding trajectories are labelled in the state plane in Figure 8.14 by a dotted and dashed line, respectively. The corresponding time histories $a(t)$ and $\phi(t)$ are plotted in Figure 8.15, giving the evidence of the profoundly different responses obtained for slightly changed initial amplitudes.

8.4.4 Some secondary resonance responses

The studies of the behaviour of a nonlinear SWHD oscillator corresponding to the elastic cable subjected to harmonic forcing in the neighbourhood of the secondary resonances can also be performed by using a similar perturbation approach described in Section 8.4.1. They were reported in references [16] and [17], which also contains many useful references of the background work related to secondary resonance behaviour of the systems with quadratic and cubic nonlinearities.

The analysis conducted in [16] is related to 1/2 and 1/3 subharmonic resonance conditions. The authors studied the same problem as in [5], but introduced different nondimensionalisation and applied a second-order perturbation approach with various ordering for the damping and excitation parameters to find approximate steady-state solutions. In the case of both 1/2 and 1/3 subharmonic resonances, critical values of the excitation occur below which only trivial stable solutions exist and a finite-amplitude subharmonic is never excited. Above such thresholds, regions exist

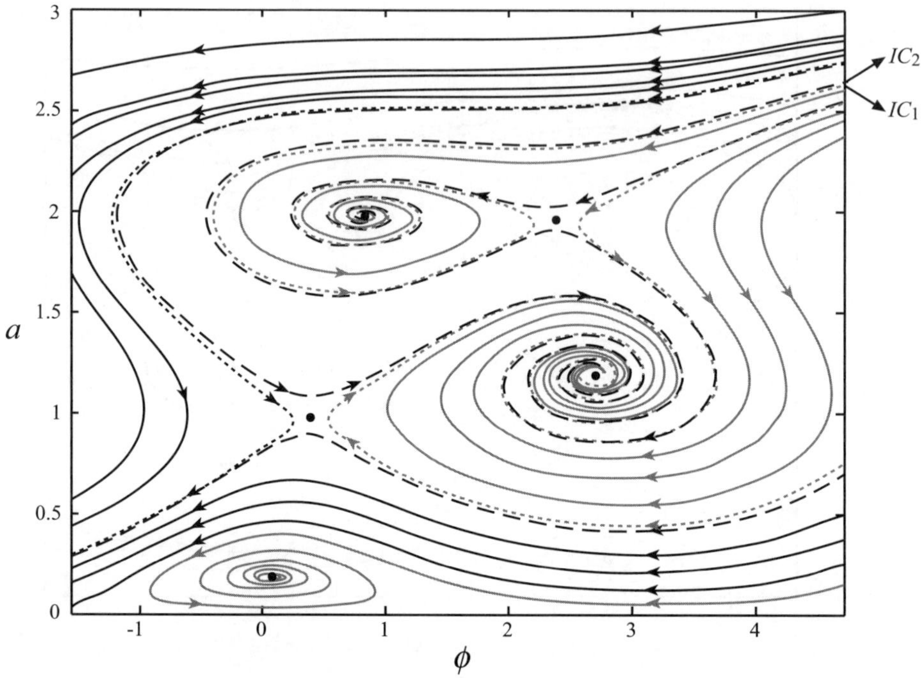

Figure 8.14 State plane for the undamped response for the parameters values $c_2 = 0.79896$, $c_3 = 0.26396$, $F = 0.05$, $\zeta = 0.01$ and $\sigma = -0.15$.

for nontrivial stable solutions or the coexistence of both trivial and nontrivial stable solutions. In the latter case, the initial conditions define which kind of oscillation actually develops.

Studies of oscillations in the neighbourhood of the *superharmonic resonance* of order two and three were reported in [17] with the same nondimensionalisation as in [16] and via second-order perturbation analyses. The investigations performed showed that the second-order resonance has a significantly stronger effect than the third-order resonance and the amplitude of the relevant superharmonic component is higher than that at the excitation frequency. In the case of the prevailing cubic nonlinearity, i.e., for taut cables, the response is of a hardening type; sagged cables experience softening behaviour, with higher amplitudes of the superharmonic response. As in the case of the primary resonance discussed above, these phenomena are associated with the significant role played by the quadratic nonlinearity due to the asymmetry.

8.5 Chaotic response of the pure cubic oscillator

Chaotic behaviour of the PC oscillator modelled by Equation (8.2.1) was first reported by Ueda [18–21]. Ueda's report on the appearance of *strange attractors* raised many

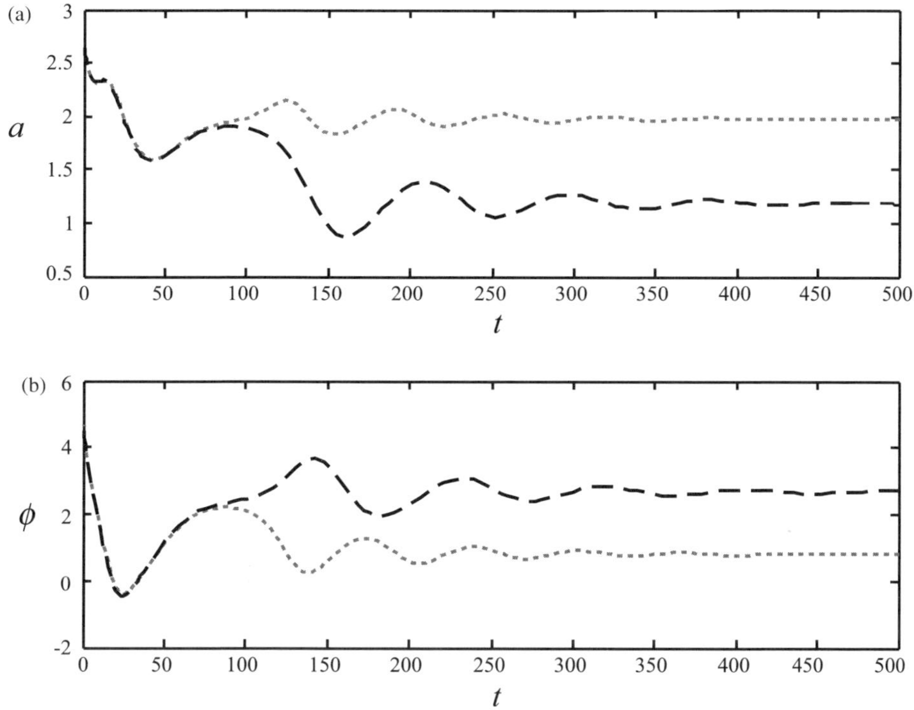

Figure 8.15 Effect of the slightly different initial conditions $IC = (a(0); \phi(0))$ *on the response of the system for* $IC_1 = (2.63;\ 3\pi/2)$*, labelled by a dotted line and* $IC_2 = (2.635;\ 3\pi/2)$*, labelled by a dashed line: (a)* $a(t)$*; (b)* $\phi(t)$*.*

questions, such as: how can *chaos* be related to primary and secondary resonances, what are the routes of its development from the *regular motion* and how can these routes be described analytically in terms of approximate models? Szemplinska-Stupnicka and her coworkers [22–25] showed analytically and confirmed numerically that chaos found by Ueda is related to the stability limit of 1/2 subharmonic resonance in the PC oscillator and that chaotic motion is preceded by a sequence of *period-doubling bifurcations*. This is discussed in this section.

8.5.1 A cascade of period-doubling bifurcations as a route to chaos: analytical considerations

A period-doubling bifurcation is examined by using the results given in Section 8.3 with the primary resonance response being defined by Equations (8.3.11a-c) and with the perturbation (8.3.13), so that the linearised variational equation is written as

$$\ddot{u} + 2\zeta\dot{u} + (\hat{\sigma}_0 + \hat{\sigma}_1 \cos(\Omega t + \theta) + \hat{\sigma}_2 \cos 2(\Omega t + \theta))u = 0 \tag{8.5.1}$$

where

$$\hat{\sigma}_0 = 3\gamma A_0^2 + \frac{3\gamma}{2}A_1^2, \quad \hat{\sigma}_1 = 6\gamma A_0 A_1, \quad \hat{\sigma}_2 = \frac{3}{2}\gamma A_1^2 \tag{8.5.2}$$

As commented earlier in Section 8.3, due to the parametric forcing, multiple resonance can occur. The first unstable region occurs at the frequency $\Omega \approx 2\sqrt{\hat{\sigma}_0}$ [11]. The approximate theory of Hill's equation allows the assumption that the solution at the stability boundary is given by

$$u(t) = \hat{B}\cos\left(\frac{\Omega}{2}t + \hat{\varphi}\right) \tag{8.5.3}$$

The form of Equation (8.5.3) shows that bifurcation from T-periodic solution (8.3.10) to a $2T$-periodic solution can appear, where $T = 2\pi/\Omega$. In order to find the unstable regions on the resonance curves for the first harmonic A_1, the solution (8.5.3) is inserted into Equation (8.5.1). Then, the harmonic balance method is applied, leading to

$$\begin{aligned} -\frac{\Omega^2}{4} + \hat{\sigma}_0 + \frac{\hat{\sigma}_1}{2}\cos(\theta - \hat{\varphi}) &= 0, \\ \zeta\Omega + \frac{\hat{\sigma}_1}{2}\sin(\theta - \hat{\varphi}) &= 0 \end{aligned} \tag{8.5.4a, b}$$

A nontrivial solution exists if the determinant

$$\hat{\Delta}_1 = \left(\hat{\sigma}_0 - \frac{\Omega^2}{4}\right)^2 + \zeta^2\Omega^2 - \frac{\hat{\sigma}_1^2}{4} \tag{8.5.5}$$

vanishes, i.e., $\hat{\Delta}_1 = 0$. In the case $\hat{\Delta}_1 < 0$, there are two real eigenvalues, one is negative and one is positive, which corresponds to instability. Therefore, instability occurs for $\hat{\Delta}_1 < 0$. This criterion is shown in Figure 8.16 on the frequency–response

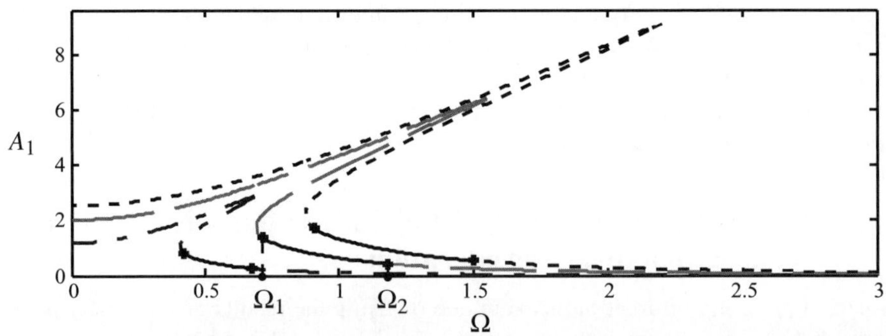

Figure 8.16 Frequency–response curve for the first harmonic A_1 for $\zeta = 0.025$ and: $F_0 = 0.02$, $F = 0.1$ (dashed-dotted line); $F_0 = 0.1$, $F = 0.5$ (dashed line); $F_0 = 0.2$, $F = 1$ (dotted line).

curves for the first harmonic, where Ω_1 and Ω_2 denote critical points for which $\hat{\Delta}_1 = 0$ is satisfied. The unstable regions are calculated by using $\hat{\Delta}_1 < 0$ and denoted in Figure 8.16 as thick lines. They lie on the lower branches of A_1 between the frequencies Ω_1 and Ω_2.

Now, the period-doubling bifurcation at the critical points Ω_1 and Ω_2 is investigated by examining the existence and stability of the steady-state solution (8.5.3) in the local neighbourhood of these two points. To do this, the complete form of the variational equation is considered

$$\begin{aligned} &\ddot{u} + 2\zeta\dot{u} + (\hat{\sigma}_0 + \hat{\sigma}_1\cos(\Omega t + \theta) + \hat{\sigma}_2\cos 2(\Omega t + \theta))u \\ &\quad + 3\gamma(A_0 + A_1\cos(\Omega T + \theta))u^2 + \gamma u^3 = 0 \end{aligned} \tag{8.5.6}$$

Its solution is found by the method of averaging [11] by assuming it in the form (8.5.3), but with $\hat{B} = \hat{B}(t)$ and $\hat{\varphi} = \hat{\varphi}(t)$. Thus, differentiating it once with respect to time gives

$$\frac{du}{dt} = -\hat{B}\frac{\Omega}{2}\sin\left(\frac{\Omega}{2}t + \hat{\varphi}\right) \tag{8.5.7}$$

with the following constraint

$$\frac{d\hat{B}}{dt}\cos\left(\frac{\Omega}{2}t + \hat{\varphi}\right) - \hat{B}\frac{d\hat{\varphi}}{dt}\sin\left(\frac{\Omega}{2}t + \hat{\varphi}\right) = 0 \tag{8.5.8}$$

By differentiating Equation (8.5.7) once, substituting it into Equation (8.5.6), solving the resulting equation together with Equation (8.5.8), the following system of the averaged equations can be obtained:

$$\begin{aligned} \frac{d\hat{B}}{dt} &= -\frac{\hat{B}}{\Omega}\left(\zeta\Omega + \frac{\hat{\sigma}_1}{2}\sin(\theta - 2\hat{\varphi})\right) \equiv W_1(\hat{B}, \hat{\varphi}), \\ \frac{d\hat{\varphi}}{dt} &= \frac{1}{\Omega}\left(\hat{\sigma}_0 - \frac{\Omega^2}{4} + \frac{\hat{\sigma}_1}{2}\cos(\theta - 2\hat{\varphi}) + \frac{3\gamma}{4}\hat{B}^2\right) \equiv W_2(\hat{B}, \hat{\varphi}) \end{aligned} \tag{8.5.9a,b}$$

Steady-state motion $d\hat{B}/dt = 0$, $d\hat{\varphi}/dt = 0$ occurs when

$$\begin{aligned} &\zeta\Omega + \frac{\hat{\sigma}_1}{2}\sin(\theta - 2\hat{\varphi}) = 0, \\ &\hat{\sigma}_0 - \frac{\Omega^2}{4} + \frac{\hat{\sigma}_1}{2}\cos(\theta - 2\hat{\varphi}) + \frac{3\gamma}{4}\hat{B}^2 = 0 \end{aligned} \tag{8.5.10a,b}$$

Eliminating the angle $(\theta - 2\hat{\varphi})$ from Equations (8.5.10a,b), gives

$$\hat{\Delta}\left(\Omega^2, \hat{B}^2\right) = \left(\hat{\sigma}_0 - \frac{\Omega^2}{4} + \frac{3}{4}\gamma\hat{B}^2\right)^2 + \zeta^2\Omega^2 - \frac{\hat{\sigma}_1^2}{4} = 0 \tag{8.5.11}$$

The bifurcation parameters in the neighbourhood of two critical frequencies are defined as

$$\mu_1 = \Omega^2 - \Omega_1^2 \tag{8.5.12}$$

and

$$\mu_2 = \Omega_2^2 - \Omega^2 \tag{8.5.13}$$

The expression (8.5.11) can be expanded into a power series, yielding

$$\hat{\Delta}\left(\Omega^2, \hat{B}^2\right) \approx \hat{\Delta}\left(\Omega_{1,2}^2, 0\right) + \frac{\partial \hat{\Delta}}{\partial \Omega^2}\left(\Omega^2 - \Omega_{1,2}^2\right) + \frac{\partial \hat{\Delta}}{\partial \hat{B}^2}\hat{B}^2 = 0 \tag{8.5.14}$$

On the basis of Equation (8.5.5), one concludes $\hat{\Delta}\left(\Omega_{1,2}^2, 0\right) = 0$. Then, the amplitudes close to Ω_1 and Ω_2 are found, respectively, to be

$$\hat{B}^2 = \frac{1}{3\gamma}\left(1 - \frac{2\zeta^2}{\hat{\sigma}_0 - \frac{\Omega_1^2}{4}}\right)\mu_1, \quad \hat{B}^2 = -\frac{1}{3\gamma}\left(1 - \frac{2\zeta^2}{\hat{\sigma}_0 - \frac{\Omega_2^2}{4}}\right)\mu_2 \tag{8.5.15}$$

To answer the question whether the bifurcation solutions (8.5.15) are stable or not, the steady-state values of the amplitude is perturbed $\hat{B}^* = \hat{B} + \delta\hat{B}$, as well as the phase $\hat{\varphi}^* = \hat{\varphi} + \delta\hat{\varphi}$. Then, Equations (8.5.9a,b) yield the following variational equations:

$$\begin{aligned} \frac{d\delta\hat{B}}{dt} &= \frac{\partial W_1}{\partial \hat{B}}\delta\hat{B} + \frac{\partial W_1}{\partial \hat{\varphi}}\delta\hat{\varphi}, \\ \frac{d\delta\tilde{\varphi}}{dt} &= \frac{\partial W_2}{\partial \hat{B}}\delta\hat{B} + \frac{\partial W_2}{\partial \hat{\varphi}}\delta\hat{\varphi} \end{aligned} \tag{8.5.16a,b}$$

which give

$$\begin{aligned} \frac{d\delta\hat{B}}{dt} &= \frac{\hat{B}\hat{\sigma}_1}{\Omega}\cos(\theta - 2\hat{\varphi})\delta\hat{\varphi}, \\ \frac{d\delta\hat{\varphi}}{dt} &= \frac{3\gamma}{2\Omega}\hat{B} + \frac{\hat{\sigma}_1}{\Omega}\sin(\theta - 2\hat{\varphi})\delta\hat{\varphi} \end{aligned} \tag{8.5.17a,b}$$

A characteristic equation, after being combined with Equations (8.5.10a,b) is

$$\kappa^2 + 2\zeta\kappa + \frac{3\gamma}{\Omega_{1,2}^2}\hat{B}^2\left(\hat{\sigma}_0 - \frac{\Omega_{1,2}^2}{4}\right) = 0 \tag{8.5.18}$$

where κ is the characteristic exponent.

According to the Routh–Hurwitz criterion, its roots κ have negative real parts if

$$\frac{3\gamma}{\Omega_{1,2}^2}\hat{B}^2\left(\hat{\sigma}_0 - \frac{\Omega_{1,2}^2}{4}\right) > 0 \tag{8.5.19}$$

On analysing Equation (8.5.5) it can be concluded that both in the vicinity of μ_1 and μ_2 this condition is satisfied. Thus, both period-doubling bifurcations – the one close to

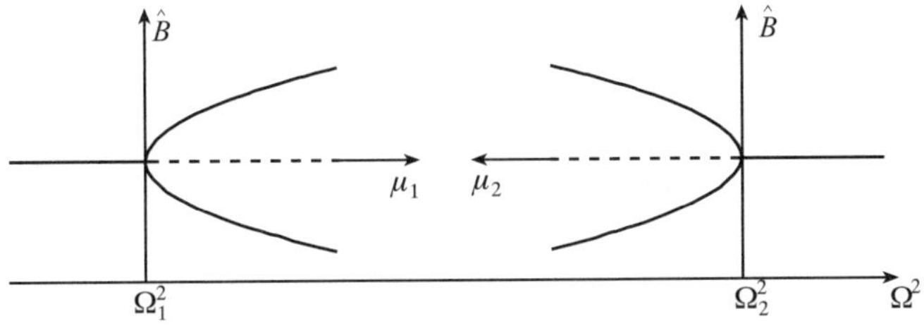

Figure 8.17 Period-doubling bifurcation diagrams.

μ_1 while increasing the frequency and the one close to μ_2 while decreasing it – are stable. The corresponding schematic *bifurcation diagram* in shown in Figure 8.17, where the stable branches are denoted by a solid line and the unstable branches by a dashed line.

8.5.1.1 Further period-doubling bifurcations

At frequencies that are far from Ω_1 and Ω_2 the assumption that A_0 and A_1 are constant is not appropriate and the $2T$-periodic solution in the whole range $\Omega_1 < \Omega < \Omega_2$ is sought in the form

$$y^{(1)}(t) = A_0 + A_1 \cos(\Omega t) + A_{1/2} \cos\left(\frac{\Omega t}{2} + \varphi\right) \tag{8.5.20}$$

where A_0, A_1, $A_{1/2}$ and φ are unknown and need to be determined. By substituting Equation (8.5.20) into the equation of motion (8.2.1) and applying the harmonic balance method, the coupled system of algebraic equations defining the unknown variables can be derived. For the perturbation $u^{(1)}(t)$ of the solution (8.5.20), the following variational equation can be derived

$$\ddot{u}^{(1)} + 2\zeta\dot{u}^{(1)} + 3\gamma\left(\begin{array}{c}\sigma_0^{(1)} + \sigma_1^{(1)}\cos\dfrac{\Omega}{2}t + \sigma_2^{(1)}\sin\dfrac{\Omega}{2}t + \sigma_3^{(1)}\cos\left(\dfrac{3\Omega}{2}t + \varphi\right)\\ + \sigma_4^{(1)}\cos\Omega t + \sigma_5^{(1)}\sin\Omega t + \sigma_6^{(1)}\cos 2\Omega t\end{array}\right)u^{(1)} = 0 \tag{8.5.21}$$

where $\sigma_0^{(1)} - \sigma_6^{(1)}$ stand for certain functions of A_0, A_1, $A_{1/2}$ and φ. What is seen from this equation is the existence of the terms with the frequency $\Omega/2$ and $3\,\Omega/2$. It follows that in the first unstable region associated with these terms, the solution with the $\Omega/4$ and $3\,\Omega/4$-harmonic components occurs. This implies a build-up of the harmonic components with the frequencies $\Omega/4$ and $3\Omega/4$, which should be added to the solution (8.5.20) when seeking the steady-state solution. Following analogous procedures, the possibility of a build-up of the components the denominators of which have the general form $2^n T$ can be detected, where $n = 1, 2, \ldots$, which suggests a possible cascade of period-doubling bifurcations [22,25].

8.5.2 A cascade of period-doubling bifurcations: numerical simulations

The analysis in the previous section shows that there is a possibility of period-doubling bifurcation and gives the system parameters at which it may occur. To verify the results of the approximate theory, numerical simulations were carried out for the equation of motion given by Equation (8.2.1) and some selected values of the system parameters to get some qualitative measures, such as *phase portraits* and *Poincaré maps*. A very fine frequency resolution of the order of 10^{-6} was required but the reported values are approximated to order 10^{-2}.

The bifurcation diagram corresponding to one of the frequency–response curves plotted in Figure 8.16 is shown in Figure 8.18 for decreasing frequency from $\Omega = 1.20$.

It can be seen that by decreasing frequency the first period-doubling bifurcation is observed at $\Omega = 1.19$, followed by higher period-doublings $4T$ at $\Omega = 1.07$ and $8T$ at $\Omega = 1.05$. This confirms the appearance of a cascade of period-doubling bifurcations, developing into chaotic motion at $\Omega = 1.04$.

Responses characterised in the phase plane projections and Poincaré maps are shown in Figure 8.19. The sampling time for the Poincaré maps was $T = 2\pi/\Omega$, so that the number of points n marked indicates the period of the response nT. It can be seen that there is good agreement between the value of the first period-doubling

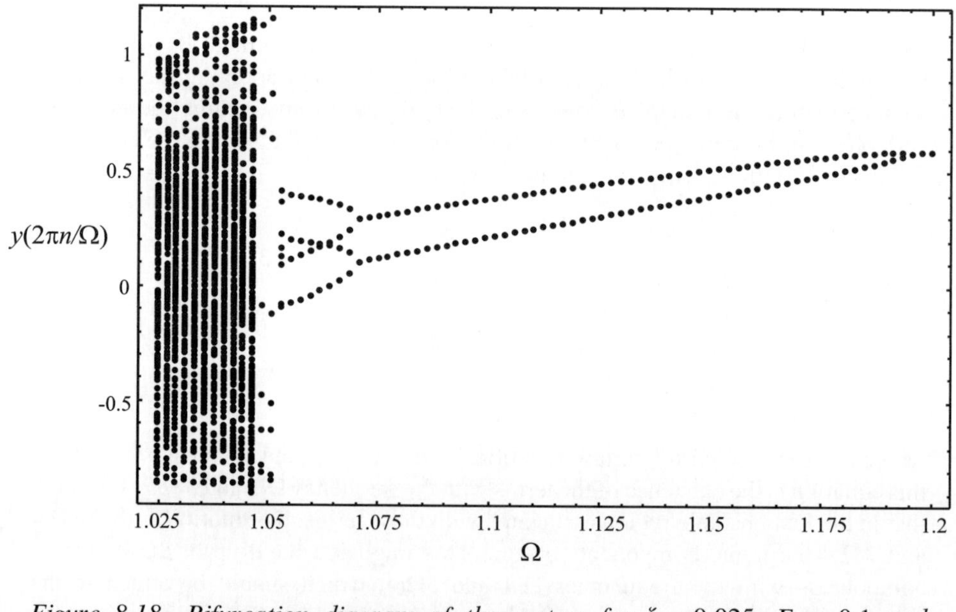

Figure 8.18 Bifurcation diagram of the system for $\zeta = 0.025$, $F_0 = 0.1$ and $F = 0.5$ and for decreasing frequency Ω.

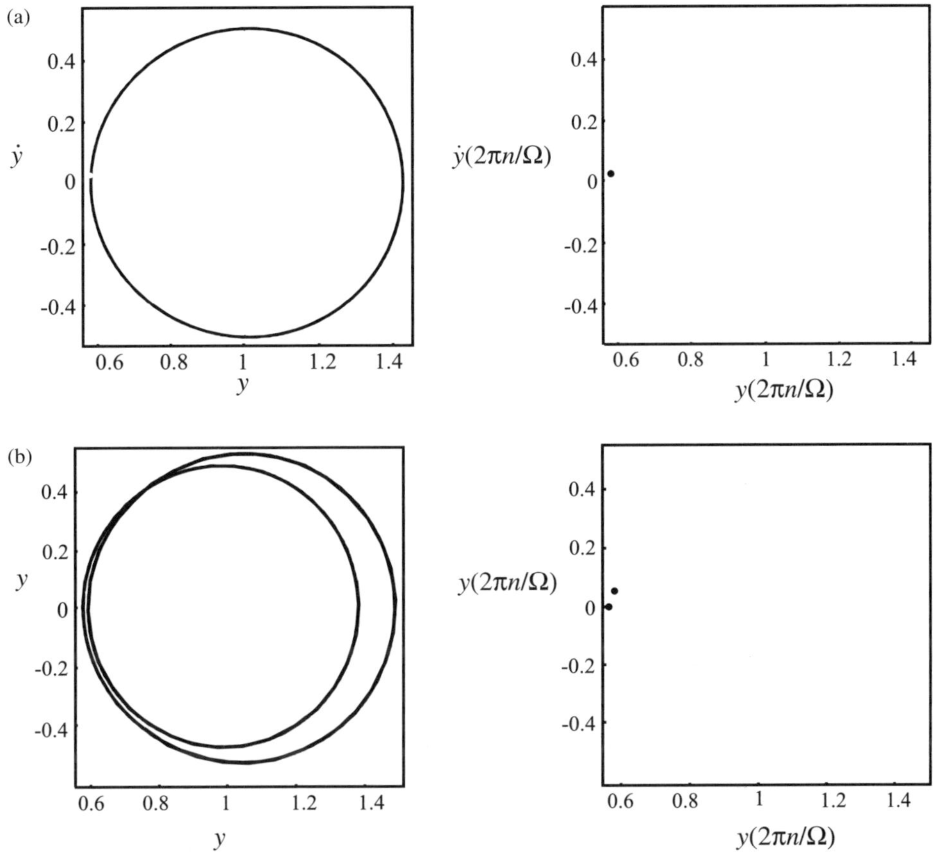

Figure 8.19 Phase projections and Poincaré maps for $\zeta = 0.025$, $F_0 = 0.1$ *and* $F = 0.5$*: (a) period-one motion* $\Omega = 1.2$*; (b) period-two motion* $\Omega = 1.19$*. Phase projections and Poincaré maps for* $\zeta = 0.025$, $F_0 = 0.1$ *and* $F = 0.5$*: (c) period-four motion* $\Omega = 1.07$*; (d) period-eight motion* $\Omega = 1.05$*; (e) chaotic motion* $\Omega = 1.04$.

bifurcation obtained in this approach and that one found previously using Equations (8.3.11a-c) and (8.5.5) with $\Omega_2 = 1.19$ (see Figure 8.16). Figure 8.19(e) shows the phase-plane projections and the Poincaré map corresponding to chaotic motion. The former contains the orbit that tends to fill out a section of the phase plane in a rather complicated manner. If the numerical simulation had been continued, the plane would be more densely filled by orbits. The latter shows a two-dimensional cross-section of a strange attractor. Had the simulation been continued, more and more points would be added, filling out the strange attractor in an orderly manner.

Chaotic motion can also be verified by examining some quantitative measure, such as the *Lyapunov exponents* (λ) [26,27]. They give the rate of divergence ($\lambda > 0$)

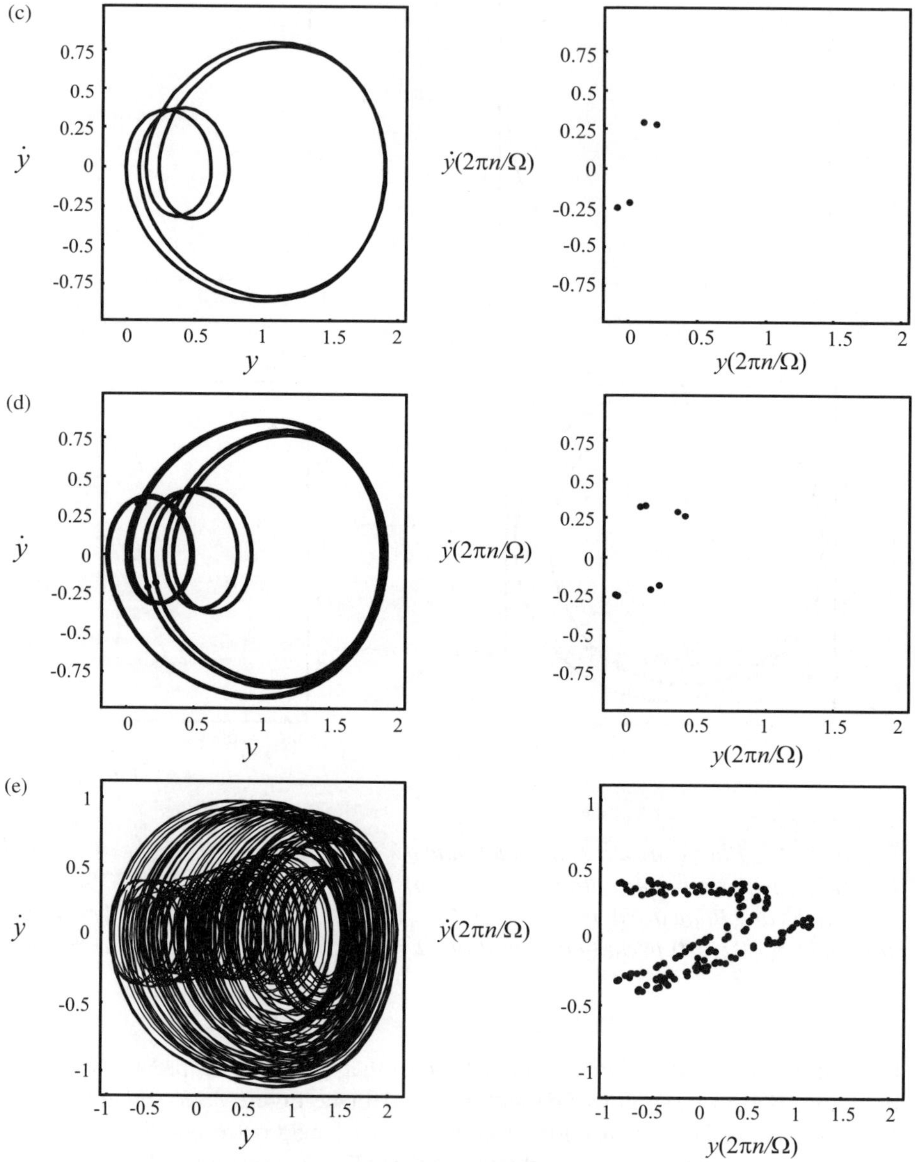

Figure 8.19 (Continued)

or convergence ($\lambda < 0$) of nearby trajectories in phase space. One positive Lyapunov exponent results in an exponential separation of trajectories, and, if properly computed, represents one of the strongest indicators of chaotic motion.

In general, for a system the equation of motion of which can be written down as a set of N autonomous first-order ordinary differential equation, exactly N Lyapunov exponents can be found, with $\lambda_1 \geq \lambda_2 \geq \cdots \geq \lambda_N$. Since the equation of motion of the system under consideration (8.2.1) can be expressed as

$$\begin{aligned}\dot{y}_1 &= y_2,\\ \dot{y}_2 &= -2\zeta y_2 - \gamma y_1^3 + F_0 + F\cos y_3,\\ \dot{y}_3 &= \Omega\end{aligned} \tag{8.5.22}$$

where $y \equiv y_1$, the system is 3-dimensional, and has three Lyapunov exponents. The behaviour of such systems is chaotic, if $\lambda_1 > 0,\ \lambda_2 = 0,\ \lambda_3 < 0$ and $\lambda_3 < -\lambda_1$ [27]. For the parameters of the system (8.5.22) with the values corresponding to those yielding Figures 8.18 and 8.19(e) it was obtained that when the frequency Ω decreases and passes $\Omega = 1.04$, the Lyapunov exponents are $\lambda_1 = 0.043,\ \lambda_2 = 0,\ \lambda_3 = -0.093$, which confirms the appearance of chaos.

8.6 Chaotic response of the single-well Helmholtz–Duffing oscillator

In this section the single-well Helmholtz–Duffing oscillator associated with cable dynamics [7,28] is considered in order to identify when and how chaotic responses are related to system parameters. The system under consideration corresponds to a suspended cable modelled by Equation (8.2.11) with $c_2 = 35.953$ and $c_3 = 534.53$. It should be noted that this oscillator has the same physical parameters as the one considered in Section 8.4.2, but these coefficients are different due to the different ways in which the nondimesionalisations have been performed [5,7]. (Also, see Chapter 2.)

Figure 8.20 shows a global picture of the system response in the parameter space defined by the excitation frequency and the excitation magnitude, originally reported in [7] (note that P used therein corresponds to F used herein). This plot was obtained by performing numerical integrations for zero initial conditions. It illustrates a rich and complex system response in terms of the regions of different periodic and chaotic motion. There are several distinguishable regions. Starting from higher frequencies, it can be seen that there are two regions located in the neighbourhood of the 1/2 and 1/3 subharmonic resonances, i.e., $\Omega = 2$ and $\Omega = 3$. The third region links the zone in the neighbourhood of the superharmonic resonances $\Omega = 1/3$ and $\Omega = 1/2$, ranging from the primary resonance zone $\Omega = 1$ approximately, to a zero frequency. Unlike in the first two regions, where some fairly well-defined zones exist, in the third region, many transition zones are noticeable as well as zones with chaotic response. The appearance of chaos in these three regions is associated with different values of the forcing

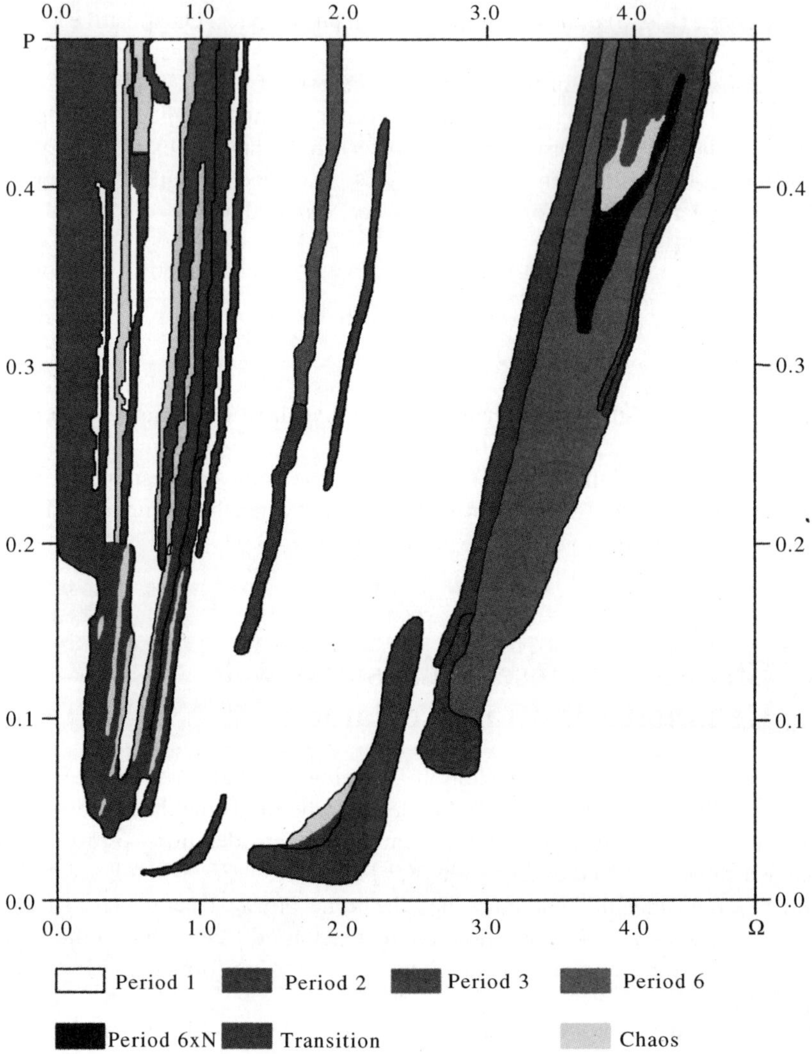

Figure 8.20 Periodic and chaotic response in the excitation frequency-magnitude of the excitation force plane. Note that $P \equiv F$ in Equation (8.2.11) and $\zeta = 0.05$. Reprinted from [7], Copyright 1991, with permission from Elsevier. See Plate 4 for a coloured version of this figure.

magnitude. They are very low in the 1/2 subharmonic region, slightly above the magnitudes yielding a period-2 response. Quite the opposite, the forcing magnitudes leading to chaos in the 1/3 subharmonic region are considerably higher. In the superharmonic range, there are almost parallel stripes of chaotic regions, distributed along all the range of the forcing magnitudes considered.

8.6.1 Routes to chaos

In three identified regions different *routes to chaos* and shapes of strange attractors are observed. They are identified through qualitative dynamic measures, phase plane portraits, Poincaré maps and power spectra (Figure 8.21), which furnish reliable quantitative indicators for chaos, too. The interested reader is referred to [7] for further global quantitative measures for these case, such as Lyapunov exponents and dimension of attractors.

In the 1/2 subharmonic region, chaos is well established as shown and confirmed by all of the considered measures (Figure 8.21(a)): a strange attractor occurs in the Poincaré map and a continuous broadband *frequency spectrum* with clear spikes at the dominating frequencies. It was found that in this range, the transition from stable period-1 motions occurs via a sudden jump, while the one from stable period-2 motions is smooth. This smooth transition was observed as consisting of period-doubling bifurcations, but it also exhibits solutions with different periodicity.

In the 1/3 subharmonic region, transition to chaos from the left is realised by means of a period-6 response and responses originating from it, having a period that is a multiple of it as transition from the left is smoother, characterised by a sequence of

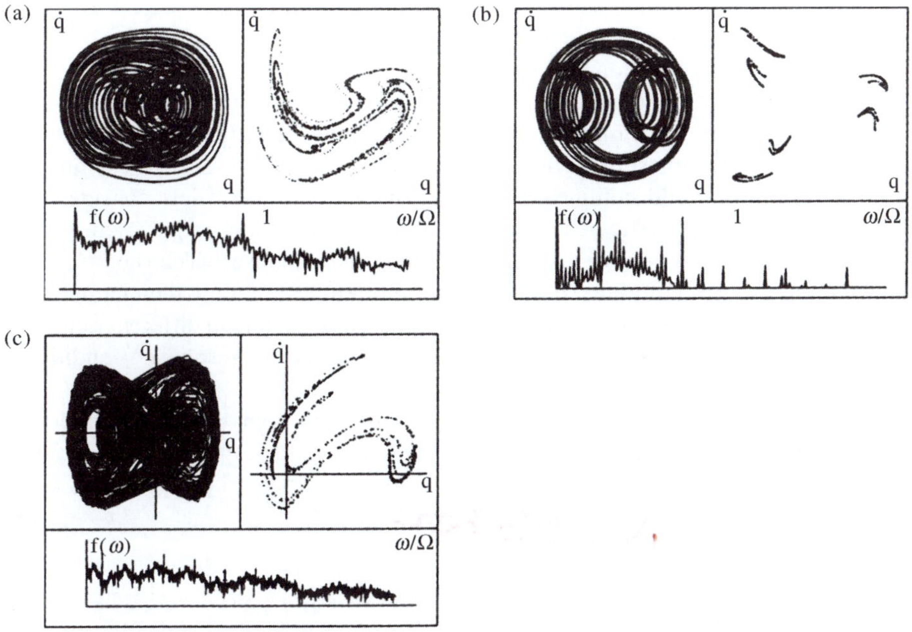

Figure 8.21 Chaotic attractors for $P \equiv F = 0.04$ in: (a) 1/2 subharmonic range ($\Omega = 1.74$); (b) 1/3 subharmonic range ($\Omega = 3.82$); (c) superharmonic range ($\Omega = 0.55$). Note that $q \equiv y$ in Equation (8.2.11). Reprinted from [28], Copyright 1995, with permission from Springer Science + Business Media.

the solutions with the doubled period, which sometimes contains period multiples other than period-6 response. Figure 8.21(b) shows the strange attractor that exhibits six distinct groups, originating from the period-6 solution around which the motion fluctuates.

In the superharmonic range at low forcing magnitudes, transition to chaos occurs with a *period-doubling cascade* from a period-1 solution when decreasing the frequency, while it is of a sudden type when increasing the frequency. At high forcing magnitudes, different kinds of transitions to chaotic response were observed. One strange attractor corresponding to the stronger chaotic response is shown in Figure 8.21(c), and it is similar to the one found in the 1/2 subharmonic range, but it has a thinner shape.

8.7 Summary

In this chapter single-well asymmetric nonlinear harmonically excited oscillators have been considered. One oscillator has pure cubic nonlinearity and a constant force acting on it, which models the dynamics of a quasi-zero stiffness vibration isolator. The system has a single-well potential, and the corresponding equation of motion can be expressed as an equation with quadratic and cubic nonlinearity. An equation of motion of this type is called the Helmholtz–Duffing equation. The second oscillator considered is related to cable dynamics, which can also be described by the Helmholtz–Duffing equation.

The analyses of the dynamics of these two systems have included their regular motion and *chaotic oscillations*. The results presented highlight the richness and variety of their dynamics. In particular, owing mostly to the presence of asymmetry, i.e., both even and odd nonlinearities, the dynamics of the models considered is characterised by: hysteretic behaviour related to the amplitude-frequency response, sudden jumps of the system steady-state amplitude, the response that can be hardening or softening, but also of a mixed nonlinearity type (first softening, then hardening), multiple coexisting regular attractors, strange chaotic attractors in different ranges of forcing parameter values associated with meaningful external resonance conditions, and different routes to chaos.

The techniques used for these investigations, both analytical and numerical, illustrate the variety of approaches that can be used for these purposes.

References

[1] P.W. Anderson, More is different. *Science, New Series*, 177, 393–396, 1972.

[2] I. Kovacic, M.J. Brennan, T.P. Waters, A study of a non-linear vibration isolator with quasi-zero stiffness characteristic. *Journal of Sound and Vibration*, 315, 700–711, 2008.

[3] C. Hayashi, *Nonlinear Oscillations in Physical Systems*, McGraw-Hill, New York, 1964.

[4] G. Rega, F. Vestroni, F. Beneditini, Parametric analysis of large amplitude free vibrations of a suspended cable. *International Journal of Solids and Structures*, 20, 95–105, 1984.

[5] F. Benedettini, G. Rega, Non-linear dynamics of an elastic cable under planar excitation. *International Journal of Non-Linear Mechanics*, 22, 497–509, 1987.

[6] F. Benedettini, G. Rega, Numerical Simulations of chaotic dynamics in a model of an elastic cable. *Nonlinear Dynamics*, 1, 23–38, 1990.

[7] G. Rega, F. Benedettini, A. Salvatori, Periodic and chaotic motions of an unsymmetrical oscillator in nonlinear structural systems. *Chaos, Solitons and Fractals*, 1, 39–54, 1991.

[8] D.W. Jordan, P. Smith, *Nonlinear Ordinary Differential Equations*, Oxford University Press, New York, 1999.

[9] G.A. Korn, T.M. Korn, *Mathematical Handbook for Scientists and Engineers*, McGraw-Hill, New York, 1961.

[10] A.H. Nayfeh, B. Balachandran, *Applied Nonlinear Dynamics*, John Wiley & Sons, Inc., New York, 1994.

[11] A.H. Nayfeh, D.T. Mook, *Nonlinear Oscillations*, John Wiley, New York, 1979.

[12] W. Szemplinska-Stupnicka, Higher harmonic oscillations in heteronomous non-linear systems with one degree of freedom. *International Journal of Non-Linear Mechanics*, 3, 17–30, 1968.

[13] J.M.T. Thompson, H.B. Stewart, *Nonlinear Dynamics and Chaos*, John Wiley & Sons, Chichester, 1986.

[14] I. Kovacic, M.J. Brennan, B. Lineton, Effect of a static force on the dynamic behaviour of a harmonically excited quasi-zero stiffness system. *Journal of Sound and Vibration*, 325, 870–883, 2009.

[15] I. Kovacic, M.J. Brennan, B. Lineton, On the resonance response of an asymmetric Duffing oscillator. *International Journal of Non-Linear Mechanics*, 43, 857–867, 2008.

[16] G. Rega, F. Benedettini, Planar non-linear oscillations of elastic cables under subharmonic resonance conditions. *Journal of Sound and Vibration*, 132, 367–381, 1989.

[17] F. Benedettini, G. Rega, Planar non-linear oscillations of elastic cables under superharmonic resonance conditions. *Journal of Sound and Vibration*, 132, 353–366, 1989.

[18] Y. Ueda, Explosion of strange attractors exhibited by Duffing's equation. *Annals of the New York Academy of Sciences*, 357, 422–433, 1980.

[19] Y. Ueda, Randomly transitional phenomena in the systems governed by Duffing's equation. *Journal of Statistical Physics*, 20, 181–196, 1979.

[20] Y. Ueda, Steady motions exhibited by Duffing's equation: a picture book of regular and chaotic motions, P.J. Holmesed. *New Approaches to Nonlinear Problems in Dynamics*. SIAM, Philadelphia, pp. 311–322, 1980.

[21] Y. Ueda, Chaotically transitional phenomena in the forced negative-resistance oscillator. *Institute of Electrical and Electronic Engineers Transaction on Circuits and Systems*, 28, 217–223, 1981.

[22] W. Szemplinska-Stupnicka, J. Bajkowski, The 1/2 subharmonic resonance and its transition to chaotic motion in a non-linear oscillator. *International Journal of Non-Linear Mechanics*, 21, 401–419, 1986.

[23] W. Szemplinska-Stupnicka, Secondary resonance and approximate models of rout to chaotic motion in non-linear oscillators. *Journal of Sound and Vibration*, 113, 155–172, 1987.

[24] W. Szemplinska-Stupnicka, P. Niezgodzki, The approximate approach to chaos phenomena in oscillators having single equilibrium position. *Journal of Sound and Vibration*, 141, 181–192, 1990.

[25] W. Szemplinska-Stupnicka, A discussion of an analytical method for controlling chaos in Duffing's oscillator. *Journal of Sound and Vibration*, 178, 276–284, 1994.

[26] A. Wolf, J.B. Swift, H.L. Swinney, J.A. Vastano, Determining Lyapunov exponents from a time-series. *Physica D*, 16, 285–317, 1985.

[27] J.J. Thomsen, *Vibrations and Stability, Advanced Theory, Analysis, and Tools*. 2nd edn, Springer Verlag, Berlin Heildelberg, 2003.

[28] G. Rega, Bifurcation and chaotic in the Helmholtz–Duffing oscillator. J. Awrejcewicz ed. *Bifurcation and Chaos: Theory and Application*, Springer-Verlag, Berlin Heidelberg, pp. 191–215, 1995.

Appendix

Translation of sections from Duffing's original book

Keith Worden[1] and Heather Worden

[1]*Department of Mechanical Engineering, University of Sheffield, Mappin St, Sheffield, S1 3JD, United Kingdom*

This appendix contains various sections of Duffing's book entitled "Erzungene Schwingungen bei veränderlicher Eigenfrequencz und ihre technische Bedeutung." This book is written in Duffing's native tongue, German, and has been cited many times since 1918, the year it was published. To the editors' knowledge this book has never been translated into English. To make the sections of the book presented in this appendix accessible to a wider audience, they have been translated into English by Keith and Heather Worden. Extensive footnotes have been provided by the translators to ensure that the appendix can be read as a stand-alone document. The figures have been redrawn by Ivana Kovačić to match the original figures as closely as possible.

The sections have been chosen to give a flavour of the book and, in the editors' opinion, they reflect the most important and the most interesting aspects of the work. The sections of the book that are not presented here contain many details of the way in which Duffing solved the various forms of the nonlinear equation that bears his name.

The Duffing Equation: Nonlinear Oscillators and their Behaviour, First Edition. Edited by Ivana Kovacic and Michael J. Brennan.

The translated sections are as follows:

Section	Page numbers in the Appendix	Page numbers in Duffing's book
Book cover	325	cover
Forward	326	iii–iv
Contents	328–329	v–vi
Introductory Overview	330–333	1–7
II. Pseudoharmonic Oscillation; Section 12 (part)	334–336	39, 40, 42, 43
III. Experiments	337–347	76–87
VI. Technical significance of pseudoharmonic oscillations	348–354	99–108

Forced Oscillations with Variable Natural Frequency and their Technical Significance

by

Georg Duffing
Engineer

with 23 Illustrations

Braunschweig
Printed and published by Freidrich Vieweg and Son
1918

Foreword

The impetus for the present study first presented itself to me through the observation of machines.

The engineer who is pursuing detailed investigations during operation, of the behaviour of the machines he has built, is necessarily led to give his full attention to the periodic motions which frequently and unexpectedly happen. Out of this necessity, for a series of years, there have arisen investigations of the oscillations of machines, whose results agreed with reality, provided that the condition was met that the natural frequency of the dynamical system was independent of the excitation.

One of the most striking points of this theory was the concept of resonance, which strictly speaking only has meaning under the aforementioned condition. If the natural frequency of the oscillating system is not constant, then special phenomena occur, whose detailed investigation should be initiated through the present work in the interests of engineering.

The simplest system in which this condition of a constant natural frequency is not fulfilled, would presumably be the symmetrical circular pendulum, whose movement has been studied by excellent mathematicians for a long time. The theory of free pendulum oscillations was brought by Weierstrass into a form which is probably unsurpassable in its simplicity. After I had the opportunity to become familiar with this material in a presentation by Mr Professor H.A. Schwarz in Berlin University on "Applications of Elliptic Functions", I dared hope that it would not be too difficult a step from this point to achieve the laws of general forced oscillation - at least for moderate excitations. It was a deception, the problem offers great difficulties in general, so that the solution was not possible through purely mathematical methods. I was forced to use approximations and to check my solutions through experiment, since even the investigation of the convergence of infinite series is not yet resolved. For the research I was favoured with strong support from Mr Professor Eugen Meyer, because he made it possible for me carry out these works in the secure laboratory of the Koenigliche Technische Hochschule in Berlin. He is thanked here for his help through advice and practical work.

I think the application of elliptic functions, which has proved useful for control calculations, is not current with many colleagues in this subject. The convergence methods are represented in such detail therefore, that even readers who do not want to spend so much time on the study of the subject, can reach a full understanding of my results.

In this case, the developments of the second part up to number 10 can be ignored, and also number 14 should be omitted. The control calculations made by using elliptic functions can then be replaced by approximations of high-order, which although of a more elementary nature, would be no less arduous.

It is very much hoped that the work also awakens interest in mathematical circles for this topic, which apparently, to be completely resolved, requires extraordinary tools and more time than is at the disposal of an engineer for this purpose. The solution in the present work will suffice for the understanding of many technical procedures.

The method of successive approximation, which has proved itself here as a usable tool, should also render good service in many other technical problems.

I express my particular thanks to Professor F. Emde and Professor K.W. Wagner for their valuable support with proof-reading and advice.

Berlin-Schoeneberg, September 1917

Georg Duffing

Contents

Introductory Overview

The processes that take place in real technological constructions are generally far too complicated for their precise description to be possible by simple means. Therefore, even oscillating systems have been idealised in order to get serviceable pictures; that is, one has simplified the differential equations via the theory of small oscillations following the procedure of the classical engineers, so that the solution presents itself in a sufficiently clear form for technical purposes. The basis of this method was the theory of harmonic oscillation with its extremely simple results. This theory is based mainly on the assumption that the restoring force which draws the oscillating body back to the resting point is proportional to its deflection from the resting point.

There are certain indications, however, that something about the theory so created, in no way agrees with reality, and that another explanation is necessary; these are: the behaviour of a number of synchronous three-phase machines in parallel operation, also the phenomena in an electromagnetic oscillator circuit in the presence of iron under the influence of an applied external voltage, and also the passage of rapidly running machines through their critical states.

The synchronous three-phase electrical machines were driven by gas machines. The frequency of the drive force and the so-called natural frequency were sufficiently separated that only moderate oscillations occurred if the drive engines were in a steady state. If this condition was however disturbed by some even slightly stronger ignitions, then even after the combustion had become regular again, the oscillations became still larger and larger, so that the machines finally became out of step. According to the results of the theory, the oscillations should have regained their normal size in the course of time after the resumption of the normal combustion, as a result of the damping.

The circumstance at first remained unexplained, one had to work with this fact and take care to eliminate every disturbance of the burning. It should not be claimed that the deviation from the law of harmonic oscillation was the only cause of the error. At any rate however this deviation harbours a problem, as will be more precisely explained in Section 30, page 104.

An electromagnetic oscillator circuit, which is formed by the placing in series of a condenser and a choke coil, produces the phenomenon of harmonic oscillation precisely, if the choking coil contains no iron. However, if the coil has an iron core, the results are qualitatively totally changed. Primarily, the current strength is not an unambiguous function of the applied voltage and frequency, as will also be shown by our later calculations.

This case has been dealt with in detail by O. Martienssen[1]. The experiments described by him produce the following, at first sight remarkable, behaviour. If you keep the frequency of the applied voltage constant over the oscillation circuit, and let the effective value of the voltage steadily increase from zero, then the dependence on the voltage of the current strength is represented by a certain curve. On the other hand, if you allow the voltage to decline again from the highest value attained, you get a

[1] Physik. Zeitschr. 1910, S. 48: Über neue Resonanzerscheinungen in Wechselstromkreisen.

different curve, which only coincides in particular places with the aforementioned curve, so that in fact in a particular area two different current strengths are associated with the same voltage. This behaviour is simply caused by the variability of the natural frequency of the system, because in the case of the iron-free oscillation circuit with constant natural frequency $\omega = 1/\sqrt{CL}$ all processes are explained precisely by harmonic oscillation.

It is Martienssen's contribution that he properly recognised and described these consequences of the variable natural frequency for the first time.

A power or work machine is subject to periodic elastic deformations in the steady state. These deformations are formed more strongly with certain frequencies of the drive impulse than they are with others, which is generally described as the resonance or the critical operating condition of the machine.

According to the current reigning theory of these phenomena, the machine ought not to show appreciably different behaviour slightly above or slightly below the critical speed. In most cases however, the operation is much calmer above resonance than at the same distance below resonance. The cause of this deviation from theory is once again the dependence of the natural frequency on the amplitude.

The variability of the natural frequency is a result of the equations of motion of the system; consequently the phenomena must allow themselves to be found in a simple mechanical system, whose construction and equations of motion result in variable natural frequency.

In this connection the symmetrical circular pendulum presents itself as the simplest system, whose movement has been the subject of study of mathematicians for years, although only in so far as we are considering free oscillation, with forced oscillations, one restricted oneself to infinitely small amplitudes.

Since in the case of a pendulum all infinitely small oscillations possess the same frequency, $\sqrt{g/L}/(2\pi)$, independent of the amplitude, following the theory of harmonic oscillation, a periodic state of motion ought not to be possible where there is an excitation of the same frequency, the amplitudes ought to grow with time unrestrictedly. In fact in this case the amplitudes grow initially; but linked with this growth is a decrease in the natural frequency, the pendulum moves away from the resonance so to speak, and in this way makes a periodic equation of motion possible again at the excitation frequency $\sqrt{g/L}/(2\pi)$. If one, kinematically speaking, reduced the drive frequency while retaining the force amplitude, and in this way tried to attain resonance, then the pendulum would move away from resonance through further enlargement of the amplitude at resonance.

However, if one starts with or exceeds the driving frequency over the value of $\sqrt{g/L}/(2\pi)$, then an amplitude appears which agrees very closely with that following from the theory of harmonic oscillation and which always declines with further raising of the forcing frequency.

From these purely qualitative reflections, the difference between the circumstances below and above resonance in the pendulum already follows. Arithmetical investigation has yielded the following properties of "pseudoharmonic oscillation" for all systems with the same structure (variable frequency), as they are characterised here:

1. When there is agreement of the forcing frequency with that of small oscillations (resonance), a periodic condition of movement with finite amplitude presents itself.

2. When there is harmonic oscillation with the above prerequisite there occurs a phase displacement of a quarter-period between excitation and response. When there is pseudoharmonic oscillation this is no longer the case, except when you have the provision of it being simple harmonic excitation and only the first harmonic of the response is taken into consideration.

3. The response in the case of stationary oscillation is not an unambiguous function of the frequency and size of the excitation.

If the frequency of the excitation is smaller than the natural frequency of small oscillations, and one increases the size of the excitation from the small oscillation up, then the response grows continuously with the excitation, until at a certain excitation the response jumps to a higher value[2]. In the case of an excitation below this critical value, for certain particular frequencies the response amplitude is not unambiguously determined by the excitation. In general, several periodic conditions of movement are associated with a particular excitation, among which, some show stable, and some unstable behaviour.

For an excitation which is larger than that critical value, the amplitude is determined unambiguously.

Arithmetical pursuit of oscillation processes for systems of this type (the majority of all real systems that are capable of oscillation) meets great difficulties in its most general form. The present work therefore limits itself to the treatment of periodic processes (of stationary or permanent oscillations) that are important for engineering.

The results on free oscillation are not new and are yielded by simple application of elliptic functions. Forced oscillations can only be represented up to now by successive approximations, but with any desired level of precision, as shown.

Concerning the arrangement of the material, for the sake of completeness the full differential equation of harmonic oscillation has been dealt with in Chapter I, and in fact in a form that does not seem to be known in wider circles.

Chapter II is devoted to pseudoharmonic oscillation.

The convergence of the approximation processes used has not been investigated; since apparently very much more extensive works are necessary for that, its investigation has been postponed for a later opportunity. Examination of the arithmetic results through primitive experimental arrangements as in Chapter III has so far not yielded contradictions, a precise corroboration through experiments with precision apparatus is desirable and will hopefully be not long awaited.

The conspicuous phenomenon, that no phase shift of a quarter period has occurred with the onset of resonance, has given cause for the investigation of the influence of damping, at least for the symmetrical case in Chapter IV.

[2] *Translators' note*: Duffing is clearly speaking of a 'softening' cubic nonlinearity where the coefficient of the cubic nonlinear term is negative. This is consistent with his later notation.

The consideration of stability in Chapter V should only serve to bring to attention the difficulties of this question, which can only be answered completely when the general solution is available.

Regarding the older work on the subject, we can say the following: The investigation of free pseudoharmonic oscillation can be regarded as resolved by the Weierstrass theory of elliptic functions.

A description of other works on free oscillations is in the "Repertorium der Physik" of R.H. Weber and R. Gans, Vol. 1, p. 205, Leipzig 1915, where the main features of the work of J. Horn[3] are also reproduced. For our purposes, of other works, the work of F. Braun[4] is worthy of note. Braun describes free oscillations using Jacobian elliptic functions, a method that Biermanns[5] has recently taken up.

There is little on forced oscillations: the work of Rayleigh[6], which refers to the differential equation

$$\ddot{x} + \alpha x - \beta x^2 = E \cos pt + F \cos(qt - \varepsilon)$$

gives an incorrect result, which is meaningless for $p = \sqrt{\alpha}$ or $q = \sqrt{\alpha}$. The reason is that the result (2)[7] taken as the first approximation is not an approximation in the case of $p = \sqrt{\alpha}$ or $q = \sqrt{\alpha}$.

C. Schaeffer dealt with the same equation in his "Introduction to Theoretical Physics"[8]; the solution is unusable in the same way, since also in that reference, all the approximation functions become meaningless when there is agreement of the forcing frequency with the natural frequency for infinitely small oscillations [p. 145, equations 124 to 128]. Equally little can the equation 121, p. 144 hold for free oscillations, since the approximation function when integrated turns out not to be periodic.

The present work makes no claim to be complete. It is the attempt to open up unknown ground by dealing with individual questions, which in the end should find their answer in the general solution of the differential equation. Whether the difficulties which oppose general solution are so great that this goal cannot be attained in the foreseeable future, will not be discovered until there are further advances. Joint working of different sides and in experimental as well as arithmetic areas is required for that, and the desire that these might begin as early as possible was the reason for presenting the work, even though incomplete, on the scientific technique.

[3] Zeitschr. f. Math. u. Phys. **47**, 400 (1902) and **49**, 246 (1903).

[4] Pogg. Ann. d. Phys. u. Chem. **151**, 51 u. 250 (1874).

[5] Arch. f. Elektrotechn. Vol. III, Issue 12 p. 345 (1915): Der Schwingunskreis mit eisenhaltiger Inductivität.

[6] Theory of Sound, 2. edn, Vol. 2, p. 79. Macmillan and Co., London, 1984.

[7] *Translators' note*: This refers to the equation in the main text $\ddot{x} + \alpha x = 0$.

[8] Leipzig, Veit & Co., 1914.

Editors' Note. In the following section, which corresponds to part of Section 12 of Duffing's book, he describes for the first time a method to obtain the steady-state amplitude of the harmonically excited oscillator with cubic nonlinearity.

II. Pseudoharmonic Oscillation

12. We will use the same approach with equation (24) for non-harmonic oscillation[9]

$$\ddot{x} + \alpha x - \gamma x^3 = k \sin \omega t \tag{24a}$$

in which γ should not be zero. In this case, the process of successive approximation continues and is infinite.

Assuming that $x_0 = A \sin \omega t$ and noting that

$$\sin^3 \omega t = \frac{3}{4} \sin \omega t - \frac{1}{4} \sin 3\omega t$$

we obtain the first approximation (having the same form as equation (21))[10]

$$x_1 = -\frac{1}{\omega^2}\left(k - \alpha A + \frac{3}{4}\gamma A^3\right) \sin \omega t + \frac{\gamma}{36\omega^2} \sin 3\omega t \tag{34}$$

Now, we now assume that

$$-\frac{1}{\omega^2}\left(k - \alpha A + \frac{3}{4}\gamma A^3\right) = A \tag{34a}$$

It is possible to find a real value of A satisfying this condition, which will terminate the process yielding the solution of equation (24) in the form of a Fourier series.

Taking into account

$$\sin^2 \omega t \sin 3\omega t = -\frac{1}{4} \sin \omega t + \frac{1}{2} \sin 3\omega t - \frac{1}{4} \sin 5\omega t,$$

$$\sin \omega t \sin^2 3\omega t = \frac{1}{2} \sin \omega t + \frac{1}{4} \sin 5\omega t - \frac{1}{4} \sin 7\omega t,$$

$$\sin^3 3\omega t = \frac{3}{4} \sin 3\omega t - \frac{1}{4} \sin 9\omega t$$

[9] *Translators' note*: 'The same approach' refers to the method of successive approximation used in the previous section to find the response of the linear harmonically excited oscillator $\ddot{x} + \alpha x = k \sin \omega t$. By writing it in the form $\ddot{x} = -\alpha x + k \sin \omega t$, the process of succesive approximation starts with $\ddot{x}_1 = -\alpha x_0 + k \sin \omega t$, where it is assumed that $x_0 = A \sin \omega t$. Integrating this expression, the periodic solution for x_1 is found to be $x_1 = -\frac{1}{\omega^2}(k - \alpha A) \sin \omega t$. The integration constants are taken to be zero to ensure periodic response. Setting $x_1 = x_0$, gives $A = -\frac{k}{\omega^2 - \alpha}$, which is the well-known result and the process stops.

[10] *Translators' note*: Duffing is referring to the form of the solution with the first and third harmonics, given earlier in his text for the same equation, but obtained by using a different method.

we find the first two terms of the second approximation, which are

$$x_2 = -\frac{\sin\omega t}{\omega^2}\left(k-\alpha A+\frac{3}{4}\gamma A^3-\frac{\gamma^2}{48\omega^2}A^5+\frac{\gamma^3}{864\omega^4}A^7\right)$$
$$-\frac{\sin 3\omega t}{9\omega^2}\cdot\left(-\frac{1}{4}\gamma A^3-\frac{\alpha\gamma}{36\omega^2}A^3+\frac{\gamma^2}{24\omega^2}A^5+\frac{\gamma^4}{62208\omega^6}A^9\right)+\ldots$$

Using equation (34a), we have

$$x_2 = \left(A+\frac{\gamma^2}{48\omega^4}A^5-\frac{\gamma^3}{864\omega^6}A^7\right)\sin\omega t$$
$$+\left(\frac{1}{36\omega^2}\gamma A^3+\frac{\alpha\gamma}{324\omega^4}A^3-\frac{\gamma^2}{216\omega^4}A^5-\frac{\gamma^4}{559872\omega^8}A^9\right)\sin 3\omega t+\ldots \quad (35)$$

In the majority of cases, the difference between x_2 and x_1, and even more, between x_3 and x_2 is negligible, so that the approximate periodic solution of equation (24a) is

$$x = A\sin\omega t+\frac{1}{36\omega^2}\gamma A^3\sin 3\omega t \quad (36)$$

where the real values of A satisfy the following cubic equation

$$\frac{3}{4}\gamma A^3+(\omega^2-\alpha)A+k=0 \quad (36a)$$

This equation can be written in the form

$$\left(1-\frac{\omega^2}{\alpha}\right)A-\frac{k}{\alpha}=\frac{3}{4}\frac{\gamma}{\alpha}A^3 \quad (37)$$

Then, we notice that in the case when $\omega=\sqrt{\alpha}$ (for small values of k and A), this result agrees with equation (33) on page 35, which has been derived in a different way[11].

⋮

To investigate the roots of the cubic equation (37) in A, we will express it as follows

$$\zeta=\left(1-\frac{\omega^2}{\alpha}\right)A-\frac{k}{\alpha} \quad \text{and} \quad \zeta=\frac{3}{4}\frac{\gamma}{\alpha}A^3$$

and plot each side of the equation as a function of A, so that the roots are the abcissa values of the intersections of the straight line

$$\zeta=\left(1-\frac{\omega^2}{\alpha}\right)A-\frac{k}{\alpha}$$

[11] *Translators' note*: The following calculations of the approximate solution for some particular values of the system parameters are not given here.

with the cubic parabola

$$\zeta = \frac{3}{4}\frac{\gamma}{\alpha}A^3.$$

This is how Figure 4 is obtained, from which it can be seen that: when $\omega = \sqrt{\alpha}$ and $\omega > \sqrt{\alpha}$ every k yields a unique value of A; for $\omega < \sqrt{\alpha}$, equation (37) can have one or three real roots depending on the value of k. This implies unusual behaviour of the system. If the system under the excitation $k \sin \omega t$ has periodic motion that corresponds roughly to the harmonic oscillation $A_1 \sin \omega t$ (convergence of the process provided), and one increases quite gradually (strictly speaking: infinitely slowly) the frequency of the excitation with steady k, the amplitude increases at first slowly, until for a certain value of ω the periodic motion stops and appears after longer time as roughly harmonic oscillation of the opposite phase and the bigger amplitude A_3[12].

The same behaviour appears if one increases gradually the magnitude of excitation k with an unchanged ω.

This result is confirmed by the experiment that will be discussed later on.

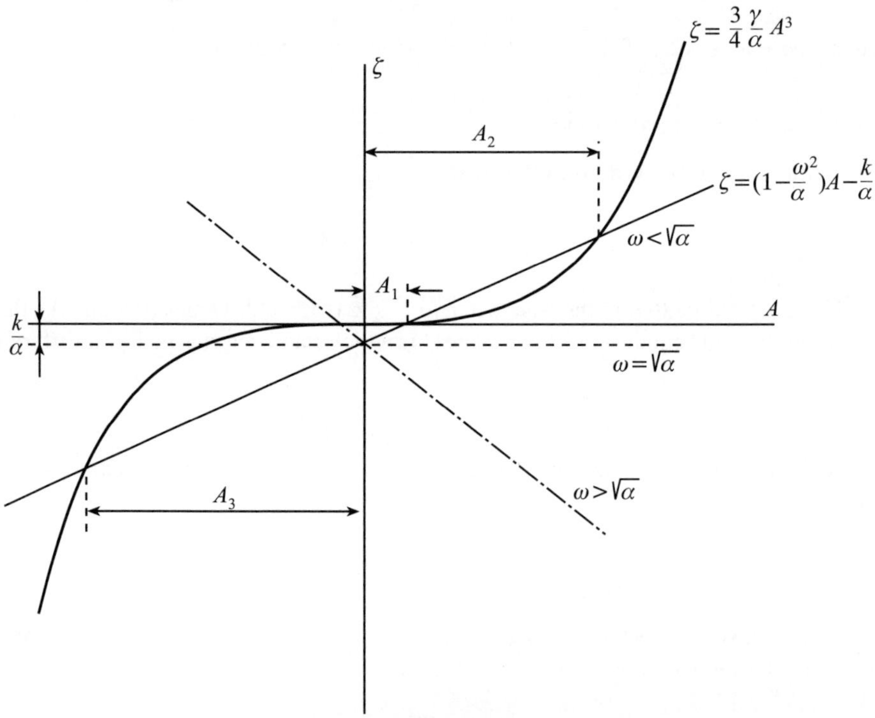

Figure 4

[12] *Translators' note*: Here, Duffing is describing the jump-up phenomenon, which is now well known in a softening system when the frequency is increased quasi-statically.

Editors' Note. The following section describes the experiments that Duffing conducted to verify his analytical results. He used a pendulum as described throughout his book. The pendulum was driven by a harmonic force that was provided by a heavy pendulum as shown in Figure 13 in this section. The pendulum could be made to act as a symmetric device that has a restoring force consisting of a linear and a cubic term as functions of displacement, or an asymmetric pendulum by the addition of a spring so that the restoring force has linear, quadratic and cubic terms as a function of displacement.

III. Experiments

21. The purpose of the experiments was:

1. To verify the results of the calculation by measuring the responses of an oscillating system whose constitution corresponds to the differential equation (1)[13];
2. to establish whether in the case of corresponding values of k and ω the three states of motion resulting from the calculation can really be represented[14];
3. whether the behaviour described above of the system at point 5 (Figure 11, p. 72) really applies[15];
4. to check whether other motions, in addition to the periodic motions described above, might be created.

A circular pendulum which oscillates freely about an equilibrium under the influence of gravity and a constant tangential force, or is subjected to a variable imposed

[13] *Translators' note*: Given in the main body of the text $\ddot{x}+\alpha x = R(t)$.

[14] *Translators' note*: This is referring to the general nonlinear equation of motion $\ddot{x}+f(x) = k\cos\omega t$, where $f(x) = \alpha x-\beta x^2-\gamma x^3$. The 'three states of motion' correspond to the three possible response amplitudes in the frequency range where more than one response is possible.

[15] *Translators' note*: This is the first reference to Figure 11, which is included here out of its place in the book; some explanation is needed.

This footnote is critical for the understanding of Figures 15 and 19 (and the associated discussion) in the following. Figure 11 is a graphical representation of which response amplitudes are possible at a given forcing frequency ω and amplitude k. Duffing establishes in equation (71) of Section 19 of the book that the allowed response amplitudes A are specified by the solutions of,

$$\left(1-\frac{\omega^2}{\alpha}\right)A-\frac{k}{\alpha} = A^3\left(\frac{3}{4}\frac{\gamma}{\alpha}+\frac{\frac{\beta^2}{\alpha^2}}{1-\frac{3}{2}\frac{\gamma}{\alpha}A^2}\right)$$

It is seen that when $\beta = 0$, this equation corresponds to equation (37) given in Section 12.

Now, he observes that in a graph of ζ against A, the allowed amplitudes will be the points of intersection of the curve C,

$$\zeta_1 = A^3\left(\frac{3}{4}\frac{\gamma}{\alpha}+\frac{\frac{\beta^2}{\alpha^2}}{1-\frac{3}{2}\frac{\gamma}{\alpha}A^2}\right)$$

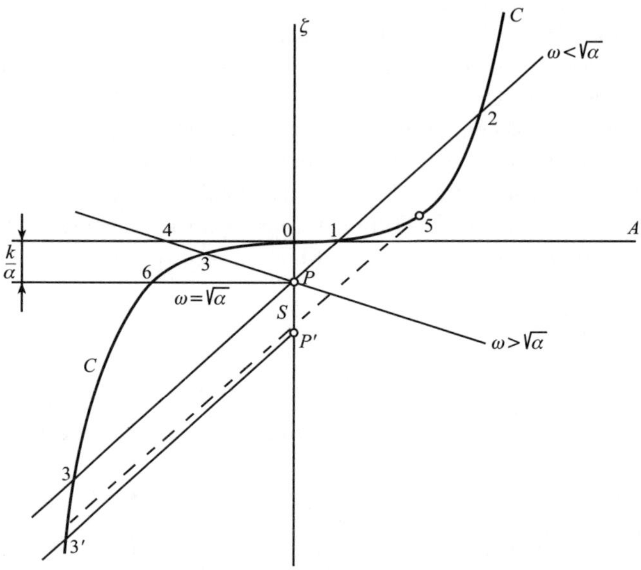

Figure 11

tangential force $k \sin \omega t$ is probably regarded as the simplest system which corresponds to the differential equation (1) p. 23[16]. As we have already seen in the derivation of equation (70)[17], this system does not precisely correspond to our

---

and the straight line,

$$\zeta_2 = \left(1 - \frac{\omega^2}{\alpha}\right)A - \frac{k}{\alpha}$$

This is again a lovely construction done in a similar way to that in Figure 4. The curve is fixed completely by the coefficients of the nonlinearity and only the straight line changes with the forcing conditions. On this graph, certain intersection point numbers are used to indicate possible solutions. As the system under discussion is an undamped softening system, if the forcing frequency is less than the natural frequency of the underlying linear system, three possible response amplitudes can be obtained and these are labelled 1, 2 and 3 in Figure 11 (with 1 the unstable solution). As the forcing amplitude increases in this case, solutions 1 and 2 can coalesce, and the point of coalescence is labelled 5. If the forcing amplitude is taken further, only one stable solution remains, and this is labelled 3′ as it is the continuation of the previous solution 3 which still exists. The 'behaviour' Duffing refers to above is presumably the disappearance of solution 5 and the (potentially large) jump to solution 3′. If the forcing frequency is above the linear natural frequency, a single response amplitude is possible and this is also labelled 3 (On the line joining *P* to point 4 in Figure 11). In modern terms, much of the behaviour summarised in Duffing's Figure 11 would be discussed in the context of jumps in the frequency-response curve. Also, one should remember that Duffing did not have the language of bifurcation theory to work with.

[16] *Translators' note*: Refers to main text, $\ddot{x} + \alpha x - \beta x^2 - \gamma x^3 = k \cos \omega t$. It is actually labelled there as equation (1a), with equation (1) representing the unforced case.

[17] *Translators' note*: Again refers to an equation elsewhere, the important point is that polynomial terms of all orders appear in the restoring force.

condition in the case of small oscillations, until the terms with x^4, x^5 etc. are disregarded.

The exact equation of motion is:

$$L\ddot{\psi} + g \sin \psi = c + k \sin \omega t, \quad \text{where} \quad c = g \sin \psi_0 \tag{73}$$

[cf. Equation (70)].

The integration of this differential equation has not yet been achieved, not even in the simplest case:

$$L\ddot{\psi} + g \sin \psi = c$$

Only the special case:

$$L\ddot{\psi} + g \sin \psi = 0$$

has been worked through completely (cf. Weierstrass (Estset), Gesamte Werke (Complete works), Vol. 6, p. 247).

The solution in the simplest form, represented by elliptic functions, is:

$$\sin \psi = \sqrt{\frac{2H}{L}} \frac{\sigma(u)\sigma_2(u)}{\sigma_3(u)^2}$$

where

$$u = \sqrt{\frac{g}{L}}t, \quad H = \frac{v_0^2}{2g}$$

and v_0 means the velocity at the lowest point (Figure 12). (Chapter III.)

If one wishes to solve equation (73) by successive approximation in a particular case, then one gives it the form:

$$\ddot{x} = -g \sin \frac{x}{L} + c + k \sin \omega t$$

and uses the expression (64), p. 59 as the first approximation[18], after A, B and C are determined numerically. The following approximation values are then determined via the expansion orders:

$$\ddot{x}_2 = -g \sin \frac{x_1}{L} + c + k \sin \omega t$$

$$\ddot{x}_3 = -g \sin \frac{x_2}{L} + c + k \sin \omega t$$

etc.

[18] *Translators' note*: From elsewhere in the text,

$$x_1 = A \sin \omega t + \frac{1}{4\omega^2}\left(\frac{\beta}{2}A^2 + \frac{3}{2}\gamma A^2 B\right)\cos 2\omega t + \frac{\gamma}{36\omega^2}A^3 \cos \omega t + C$$

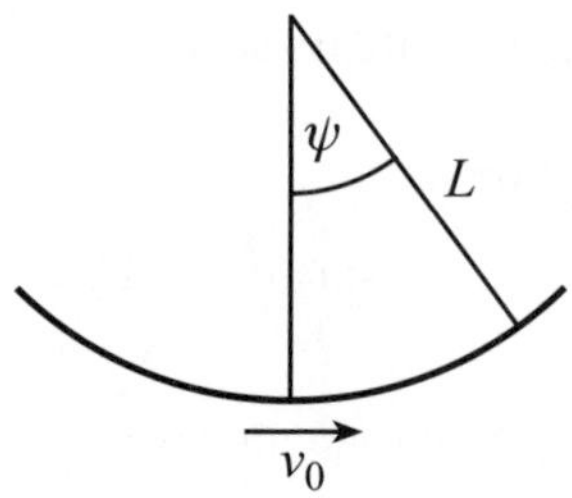

Figure 12

The integrations are carried out graphically without difficulty.

In a particular case I investigated, the differences showed themselves as so insignificant that it is justified to consider the circular pendulum as equivalent to a system that corresponds to our conditions, for the present purpose.

Experimental setup

22. A small pendulum as in Figure 13a, whose ideal length could be altered by displacement of the upper weight *G*, was laid on the edge *S* and received its impetus via silk threads which were passed over the small drums *T*.

One thread was coupled to a spring F_2 (Figure 13b), which was connected to ground at *A*. The other silk thread coupled via an identical spring F_1 to the very heavy

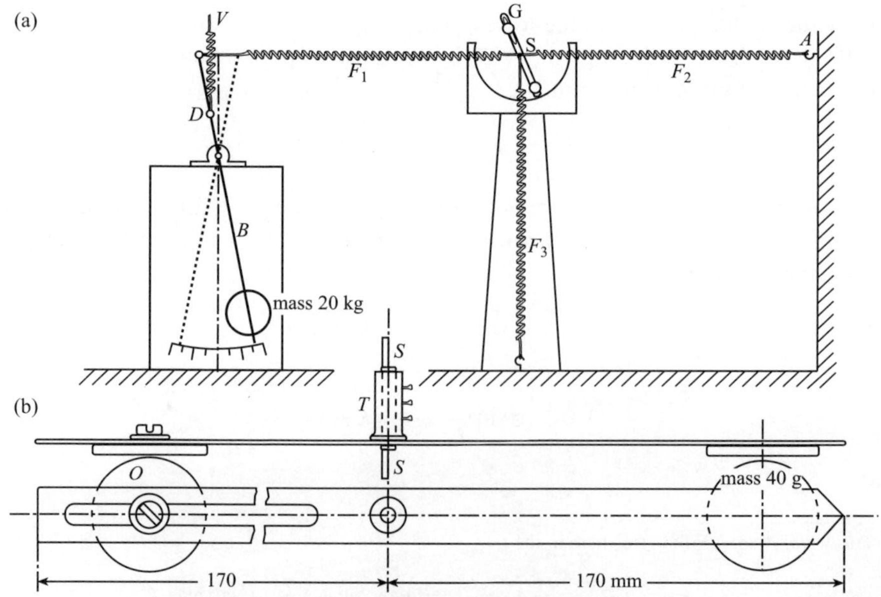

Figure 13

drive pendulum B, which was on ball-bearings. A third thread was loaded with a vertical spring F_3, so that any equilibrium of the pendulum could be produced.

The large pendulum was kept in constant excitation during the experiment by means of small impulses, which were performed by hand during the procedure through the midpoint onto the lever D.

In order to be able to slightly alter the period of oscillation of the large pendulum, it was loaded with a vertical spring V which was connected, under adjustable constant tension, at point D.

The displacements of both pendulae were read off on as large scales as possible.

23. The equations of motion of the experimental system

These can be easily written using Figure 14.

The equilibrium of the small pendulum (under the angle ψ_0 to the perpendicular) is expressed by

$$0 = gS\sin\psi_0 + c(T_{20} - T_{10} - T_{30})$$

Here gS means the static moment of the pendulum body (with drum and weights), c means the radius of the drive drum, T_{10}, T_{20}, T_{30} are the spring tensions. If one indicates the moment of inertia of the pendulum mass with respect to the axis of rotation by J, then the deceleration moment of the angle ψ in the state of motion is:

$$J\frac{d^2\psi}{dt^2} = gS\sin\psi_0 + c(T_2 - T_1 - T_3)$$

If f_1, f_2, f_3 are the spring constants, one has:

$$\begin{aligned} T_1 - T_{10} &= f_1(\xi - c(\psi - \psi_0)), \\ T_2 - T_{20} &= f_2 c(\psi - \psi_0), \\ T_3 - T_{30} &= -f_3 c(\psi - \psi_0) \end{aligned}$$

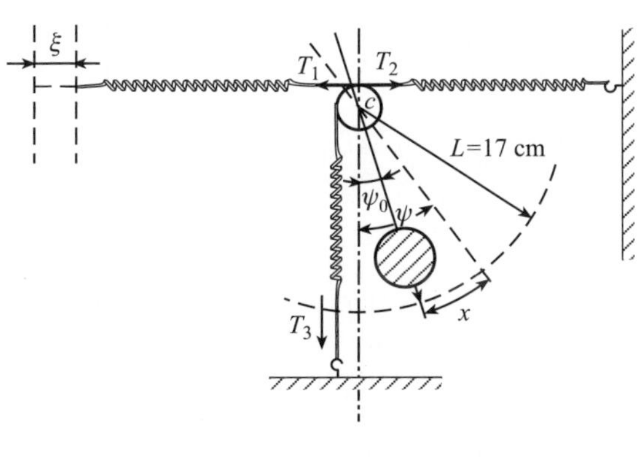

Figure 14

From this result is produced:

$$\frac{d^2\psi}{dt^2} = \frac{gS}{J}(\sin\psi - \sin\psi_0) + \frac{c^2}{J}(f_1 + f_2 + f_3)(\psi - \psi_0) = \frac{cf_1}{J}\xi$$

as the equation of motion of the pendulum, if the law is also taken into consideration, in which the forced motion ξ occurs.

In the case of relatively large mass of the pendulum B one can ignore the effect of S on B and one is entitled to the view that ξ changes over time independently of the movement of the small trial pendulum.

The law of this forced movement has the form $\xi = a \sin \omega t$ to a very good approximation, where a is a constant which is fixed by the amplitude of the pendulum B.

Through the introduction of

$$x = L(\psi - \psi_0)$$

and

$$\frac{x}{L}\cos\psi_0 - \frac{x^2}{2L^2}\sin\psi_0 - \frac{x^3}{6L^3}\cos\psi_0$$

in place of $\sin\psi - \sin\psi_0$, one gets the equation of motion in the form

$$\frac{d^2x}{dt^2} + \left(\frac{gS}{J}\cos\psi_0 + \frac{c^2}{J}(f_1 + f_2 + f_3)\right)x,$$

$$-\frac{1}{2L}\frac{gS}{J}\sin\psi_0 x^2 - \frac{1}{6L^2}\frac{gS}{J}\cos\psi_0 x^3 = \frac{cf_1 L}{J}a \sin\omega t$$

or

$$\ddot{x} + \alpha x - \beta x^2 - \gamma x^3 = k \sin\omega t$$

in which,

$$\left.\begin{aligned} \alpha &= \frac{gS}{J}\cos\psi_0 + \frac{c^2}{J}(f_1 + f_2 + f_3), \\ \beta &= \frac{1}{2L}\frac{gS}{J}\sin\psi_0, \\ \gamma &= \frac{1}{6L^2}\frac{gS}{J}\cos\psi_0, \\ k &= \frac{cf_1 L}{J}a \end{aligned}\right\} \qquad (74)$$

The coefficient α can be easily determined experimentally by letting the system perform very small free oscillations and observing the period of oscillation.

If one releases all the threads from the drum then $\psi_0 = 0$ and forms the communicated α, the value gS/J results. If one replaces the thread connections and observes small free oscillations ($k = 0$), the value $(c^2/J)(f_1 + f_2 + f_3)$ is produced. The values β, γ are easy to calculate.

The value of k results from the static condition $x = s$ which corresponds to the maximal value of ξ. One has

$$\alpha s - \beta s^2 - \gamma s^3 = k$$

The value $x = s$ is observed from the scale of the small pendulum and k is calculated from that. In most cases it is sufficient to assume $k = \alpha s$. Then the value k/α of equation (69)[19] is nothing other than the static force, which corresponds to the maximal value of ξ.

A measurement of spring tensions, weights, moments of inertia was not necessary. It is only necessary to determine: the frequencies of the trial pendulum at very small deflections, both inside the system and isolated; the frequencies of the drive pendulum at the deflections used; the static deflections of the trial pendulum corresponding to the respective maximal ξ and the dynamic deflections in the oscillation state. The amplitudes of the experimental pendulum are given in millimetres and measured at a circular scale of radius 170 mm.

Only two examples are presented from the series of experiments.

I. Symmetrical Pendulum

$$\psi = 0, \ f_3 = 0 \quad \text{(spring } F_3 \text{ removed).}$$

24. The number of oscillations per minute of the small pendulum in association with the system where the pendulum B was held still was set up at

$$60 \frac{\sqrt{\alpha}}{2\pi} = 33.15 \ \text{min}^{-1}$$

The time taken for 100 oscillations was measured with the stopwatch.

The number of oscillations per minute $60(\omega/2\pi)$ of the drive pendulum B was measured in the same way during oscillation of the whole system.

$60 \frac{\omega}{2\pi}$	Static deflection	Dynamic deflection	$1 - \frac{\omega^2}{\alpha}$	Point in diagram
33.31 min^{-1}	1.2 mm	45 mm	−0.010	1
33.28 min^{-1}	2.4 mm	65 mm	−0.008	2
33.11 min^{-1}	4.8 mm	92 mm	+0.002	3

[19] *Translators' note*: This equation number from elsewhere covers both,

$$\zeta = \frac{3}{4}\frac{\gamma}{\alpha} A^3 + AB\left(2\frac{\beta}{\alpha} + 3\frac{\gamma}{\alpha}B\right)$$

and

$$\zeta = \left(1 - \frac{\omega^2}{\alpha}\right)A - \frac{k}{\alpha}$$

The experimental results were checked as in equation (37)[20]. The value $(c^2/J)(f_1+f_2+f_3)$ showed itself to be negligibly small, so that

$$\frac{3}{4}\frac{\gamma}{\alpha}=\frac{1}{8L^2},\qquad \zeta=\frac{1}{8}\frac{A^3}{L^2}$$

could be set.

In Figure 15a the abscissae A are at their natural size, the ordinates ζ have been increased to five times their size for the sake of clarity. As was expected, the responses come out smaller than according to the calculation in consequence of the unavoidable resistances to movement (air resistance, friction at the edges). With hindsight, agreement between experiment and calculation up to approximately 10 percent of the calculation result can be described as good. The second part of equation (36)[21] is negligible compared to the first for $A = 10$ cm, as a simple calculation shows.

Incidentally it should be noted that this case allows a comparison with the exact natural frequencies of (73) p. 77, because here the constant part of equation (73) is $c = 0$. The results of the precise calculation can hardly be distinguished graphically from the results of equation (37), so for example, results for $x = 10$ cm,

$$\psi=\frac{10}{17}=0.58824$$

which corresponds to an angle of 33° 42′ 12″. The exact period of the pendulum is for this:

$$T=2\pi\sqrt{\frac{L}{g}}\times 1.02205,\qquad \omega=\sqrt{\frac{g}{L}}\times\frac{1}{1.02205}$$

$$\frac{\omega^2}{\alpha}=0.95731,\qquad 1-\frac{\omega^2}{\alpha}=0.04269$$

a result that graphically agrees exactly with our Figure 15a, if one introduces the point with the coordinates A and $\zeta=(1-\omega^2/\alpha)A$ for $A = 10$ cm.

II. Assymetrical Pendulum

$$\psi_0=37°4',\qquad L\psi_0=110\,\text{mm}$$

---

[20] *Translators' note*: From Section 12

$$\left(1-\frac{\omega^2}{\alpha}\right)A-\frac{k}{\alpha}=\frac{3}{4}\frac{\gamma}{\alpha}A^2$$

[21] *Translators' note*: From Section 12

$$x=A\sin\omega t+\frac{1}{36}\frac{\gamma}{\omega^2}A^3\sin 3\omega t$$

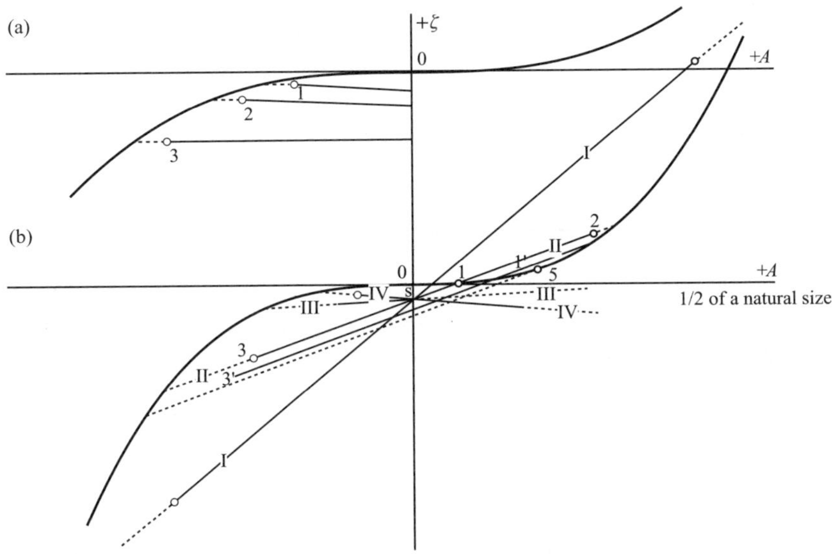

Figure 15

25. The angle ψ_0 was made as large as possible, so as to get a strongly bent curve ζ in the coordinate system ζ. The silk threads of F_3 could not have withstood an increase over the applied mass, also the knife-edges were not strong enough for a greater load.

Before the experiment were determined:

$\frac{gS}{J} = 17.95\, s^{-1}$	by measurement of the period of the unconnected pendulum (small oscillations)
$\alpha = 14.67\, s^{-2}$	by measurement of the oscillations per minute of the connected pendulum when $\psi_0 = 37°4'$, $60\frac{\sqrt{\alpha}}{2\pi} = 36.57\ \text{min}^{-1}$

with

$$\frac{gS}{J}\cos\psi_0 = 17.95 \times 0.7979 = 14.32\ \text{s}^{-2}$$

and the results following equation (74) are:

$$\frac{c^2}{J}(f_1 + f_2 + f_3) = \alpha - \frac{gS}{J}\cos\psi_0 = 0.35\ \text{s}^{-2}$$

It further follows that:

$$\beta = \frac{1}{2}\frac{17,95}{17} \times 0.6027 = 0.3182\ \text{cm}^{-1}\ \text{s}^{-2}$$

$$\gamma = \frac{1}{6}\frac{17,95}{17^2} \times 0.7979 = 0.0083\ \text{cm}^{-2}\ \text{s}^{-2}$$

The experiments were carried out with constant k, corresponding to the static condition $s = 1.1$ mm. The following were measured:

$60\frac{\omega}{2\pi}$	$1-\frac{\omega^2}{\alpha}$	Straight line	Dynamic amplitude		
33.40 min^{-1}	0.166	I ... I	$\overline{107}$ mm	7 mm	−91 mm
35.21 min^{-1}	0.073	II ... II	$\overline{69}$ mm	16 mm	−61 mm
36.36 min^{-1}	0.011	III ... III	–	–	−29 mm
36.86 min^{-1}	−0.016	IV ... IV	–	–	−23 mm

The excitations which correspond to the numbers with lines above them allowed themselves to be held only for a short time, while all other excitations could be held any length of time. Therefore the first ones correspond to states of motion which are either unstable or are very sensitive to small disturbances. The numbers entered are the arithmetical average of the positive and negative amplitudes, so they ought to agree with the coefficients A of (64), p. 59 (*Translators' note*: see footnote 17), when $\frac{\gamma}{3c\omega^2}A^3 \sin 3\omega t$ is ignored.

For a control we characterise A following (71), p.70[22], from the curve:

$$\zeta = A^3\left(\frac{3}{4}\frac{\gamma}{\alpha} + \frac{\frac{\beta^2}{\alpha^2}}{1-\frac{3}{2}\frac{\gamma}{\alpha}A^2}\right)$$

The results are given in Figure 15b.

The repeat of the control calculation for individual points using more precise methods, which I also carried out following the given procedure for equation (73), p. 77, should be unnecessary, since the corrections which had to be applied to the result following equation (71), p. 70 (*Translators' note*: footnote 21) came out as completely insignificant.

As the graph Figure 15b shows, experiment agrees much less well with calculation than in the symmetrical case. The cause of this behaviour is the deformation of the weak knife-edge because of the load through spring F_3, by which the resistances to movement were considerably increased in comparison to the symmetrical case. That this disagreement is particularly striking for negative A is explainable because here the reactions at the edges come out larger than for positive A.

The states of motion which correspond to the points 1, 2 and 3 were produced repeatedly by holding the small pendulum and letting go at the corresponding

[22] *Translators' note*: From elsewhere in the text, this equation number encompasses,

$$\left(1-\frac{\omega^2}{\alpha}\right)-\frac{k}{\alpha} = A^3\left(\frac{3}{4}\frac{\gamma}{\alpha} + \frac{\frac{\beta^2}{\alpha^2}}{1-\frac{3}{2}\frac{\gamma}{\alpha}A^2}\right)$$

and,

$$B = \frac{A^2}{2}\frac{\frac{\beta}{\alpha}}{1-\frac{3}{2}\frac{\gamma}{\alpha}A^2}$$

moment. The states 1 and 3 allowed themselves to be held at length, state 2 however did not.

Further to this, the forced motion practised by the large pendulum rose gradually, up to an amount which corresponded to a static excitation $s' = 2s$. Through this the stable state 1 should be carried over to the state $1'$. This proved impossible, because shortly, before $1'$ had been reached, the movement of the period $2\pi/\omega$ stopped, and the stable state $3'$ set in after the passage of a longer complicated transition movement. This behaviour is explainable by the strong bending of the real curve in comparison to the calculated one. With a very exactly set up pendulum with as small as possible resistances to movement, one will be able to get nearer to point 5[23].

Considering the rather primitive condition of the experimental apparatus, which in view of the time constraints, could not have been improved in a short time, one can be satisfied with the agreement between calculation and experiment. The imprecision lies on the side of experiment, and an improvement and refinement of this is to be striven for above all else in the future.

The suspected behaviour of the system at point 5, about which the calculation up to now does not give a resolution, was confirmed indisputably by experiment, and this was its main purpose. For every size of the excitation k there exists such a 'critical point'.

The existence of further periodic movements could not be established.

[23] *Translators' note*: Although the point numbers in these paragraphs do in fact refer to Figure 15, the understanding of the figures is in terms of the previously mentioned Figure 11. The important point for the discussion is that Duffing is trying to establish experimentally if two solutions coalesce at a critical 'point 5' and the system jumps to a different stable solution at $3'$.

Editors' Note. The following section describes the engineering problems that motivated Duffing in his studies. He observed similar behaviour in three engineering problems and had the foresight to model them by the simple nonlinear equation that bears his name. Moreover, he observed that the harmonically forced pendulum also behaved qualitatively in the same dynamical manner that the engineering systems did. The first problem involves an electrical circuit containing an iron core inductor described by Martienssen and Biermanns; the other two problems involve the dynamics of three-phase electrical generators.

VI. Technical significance of pseudoharmonic oscillation

28. One could raise the objection that the preceding deductions would have more interest for the representative of theoretical mechanics and physics, than for use in engineering. The following discussion is for those who should still believe in the possibility of isolating engineering from the aforementioned disciplines.

When one finds oneself far enough from resonance - which mostly the engineer tries to do - then the formulae of simple harmonic oscillation are sufficient.

However when one wants to get out of a danger, it is good to know which way one has to go. We have recognised that the system behaves differently with respect to an excitation under resonance to one that is over resonance. Already the gain of this one realisation gives specifically technical value to the observation.

In engineering, oscillating systems occur where the coefficients β, γ vanish relative to the coefficient α. Torsion waves with *centric* masses, and in the electrotechnical field, oscillator circuits without iron, will come very close to this case. If the oscillation circuit contains iron, then it shows not only quantitatively, but also qualitatively quite different phenomena; as has been demonstrated in the investigations by Martienssen and Biermanns.

Biermanns expressed the connection between the magnetisation current i and the accompanying flux ψ through the iron-containing oscillator circuit through the formula:

$$i = \frac{1}{L}\psi + \frac{1}{\lambda}\psi^3$$

and so achieved the differential equation

$$\frac{d^2\psi}{dt^2} = -\frac{1}{KL}\psi - \frac{1}{\lambda}\psi^3$$

in which K, L, λ are constants.

The differential equation has the same form as our equation (3)[24], and is integrated by Biermanns via the Jacobian elliptic functions. At the end of the work it is

[24] *Translators' note*: Duffing sometimes has multiple copies of equation numbers, this one refers to the equation on page 23 of the original manuscript,

$$\frac{d^2x}{dt^2} = \alpha x - \gamma x^3$$

mentioned by Biermanns that also the more general case

$$i = \frac{1}{L}\psi + \frac{1}{\lambda_1}\psi^3 + \frac{1}{\lambda_2}\psi^5$$

is accessible by dealing with elliptic functions, that however the given method leads to quite extraordinarily extensive calculations. The methods of approximation given in the present work stand out by not having excessively increased calculations even with the adoption of higher terms in the expression for the restoring force.

The great value of the experimental investigation of the iron-containing oscillator circuit through Martienssen was indicated in the ensuing discussion at the lecture by Petersen (Electrotechn. Zeitschrift. 1915, issue 28).

In the same discussion K.W. Wagner mentions the instability of the operating state corresponding to our point 5 (Figure 11), and he assumes a constant applied clamping voltage. The proof that Wagner gives for this (Elektrotechn. Zeitschr. 1916, p. 149) is extraordinarily clear and beautifully simple; however since average values of current and voltage are considered, it is not immediately transferrable to the case of our experimental pendulum. The proof stands on a purely kinematic basis, since a periodic state of motion is compared with neighbouring states of motion, which in the same way are required to have a proviso of being periodic. It can therefore be no more valid as a sufficient answer to the question about the stability of the saturated iron oscillator circuit as our explanations 27, although the conclusions of Wagner are completely in harmony with the phenomena in the pendulum experiment.

29. An important case, that lies in the border area between electrical technology and engineering (as far as such a description is still used today), is represented by the synchronous 3-phase machine working at a constant busbar voltage (compare the representation of Pichelmayer, 'Wechselstromerzeuger', Goschen 1911, p. 21).

With such a machine, assuming unsaturated iron, one has the case of the circular pendulum if there is constant magnetic excitation: in reality, because of the iron saturation, this case is very nearly realised and can be judged close to resonance according to our equation (1) (*Translators' note*: footnote 15).

If one considers an ideal synchronous three-phase generator (zero Ohmic resistance of the armature windings, constant total reactance $= x$, full drum machine), then the electrical power:

$$L_e = eJ\sqrt{3}\cos\phi = \sqrt{3}\frac{eE\sin\psi}{x}$$

follows from the voltage triangle *OAB* (Figure 18) in a recognised way.

When there is constant excitation of the magnetic field, E is constant and the electrical power is therefore proportional to $\sin\psi_0$, where ψ_0 means the angle (measured in phase degrees), around which the magnetic field is in advance of the rotating field of the *anchor*.

The kinetic energy of the rotor would be

$$T = \frac{1}{2}\Theta w^2$$

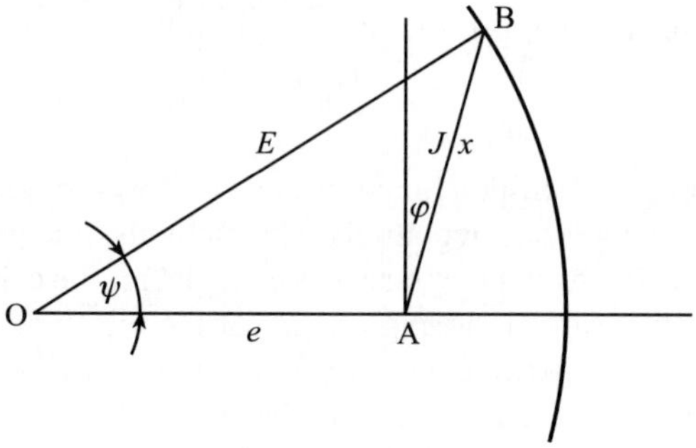

Figure 18

where Θ means the moment of inertia and w the angular velocity. A necessary condition is that the measurement system is the absolute electromagnetic cgs system.

The equation of motion of the rotor results from the power theorem

$$\frac{dT}{dt} = \Sigma L$$

which is a generalisation of the purely mechanical statement

$$\frac{dT}{dt} = \Sigma(Xv_x + Yv_y + Zv_z)$$

in which X, Y, Z mean the force components and v_x, v_y, v_z the velocity components belonging to them.

Here $\Sigma L = L_m - L_e$ is the propulsion power less the electrical power summoned up by the generator.

If D_m indicates the size of the drive torque at time t, $L_m = D_m.w$.

One gets the equation of motion[25]

$$\Theta w \frac{dw}{dt} = D_m w - \sqrt{3}\frac{Ee}{x}\sin\psi$$

[25] Following Pichelmayer, Wechselstromerzeuger, Göschen 1911, p. 67, the electrical power in the case of 3-phase machines with single poles is dependent on our angle of the form:

$$L_e = a\sin\psi + b\sin\psi\cos\psi$$

In the case of this form of L_e the equation of motion, which now no longer agrees with the equation of motion of the circular pendulum, is reduced to the form,

$$\frac{d^2x}{dt^2} = \alpha x - \beta x^2 - \gamma x^3 = k\sin\omega t$$

if small sizes of higher order are ignored. (*Translators' note*: we have changed the text a little here to avoid having a footnote within a footnote.). Following the numerical determination of the coefficient α, β, γ, the formulae (71), p. 70 gives information about the behaviour of the oscillating system.

The torque D_m is composed of the constant torque of the steady state D_0 and a variable part D, it is

$$D_m = D_0 + D$$

where

$$D_0 w_0 = \sqrt{3}\frac{Ee}{x}\sin\psi_0$$

Therefore the equation of motion is:

$$\Theta w \frac{dw}{dt} = \frac{w}{w_0}\sqrt{3}\frac{Ee}{x}\sin\psi_0 + Dw - \sqrt{3}\frac{Ee}{x}\sin\psi$$

Further

$$w = w_0 + \frac{1}{p}\frac{d\psi}{dt}$$

where p means the number of pole pairs.

Now w/w_0 is immediately approximated by 1. Under this condition one gets the equation of motion in the form:

$$\frac{\Theta w_0}{p}\frac{d^2\psi}{dt^2} + \sqrt{3}\frac{Ee}{x}(\sin\psi - \sin\psi_0) = Dw_0$$

or

$$g\frac{\Theta w_0}{p}\frac{x}{\sqrt{3}Ee}\frac{d\psi}{dt^2} + g(\sin\psi - \sin\psi_0) = Dw_0 g\frac{x}{\sqrt{3}Ee}$$

If one introduces the ideal pendulum length

$$L = g\frac{\Theta w_0}{p}\frac{x}{\sqrt{3}Ee}$$

and takes into consideration only one part D of the expression $D_1 \sin\omega t$, where w depends on ω in a particular way that is determined by the type of drive machine, our equation of motion is:

$$L\frac{d^2\psi}{dt^2} + g(\sin\psi - \sin\psi_0) = k\sin\omega t$$

where

$$k = D_1 w_0 g\frac{x}{\sqrt{3}Ee} = g\frac{D_1}{D_0}\sin\psi_0$$

The ideal pendulum length L depends on the machine and operation constants, k is dependent on the drive conditions.

One can therefore illustrate the behaviour of the machine with an asymmetrical pendulum.

We deal with this equation of motion in accordance with Sections (17) and (18) through the approximation (72), p.70. Then with $\alpha = (g/L)\cos\psi_0$ we get the approximate static deflection

$$s = \frac{k}{\alpha} = L\frac{D_1}{D_0}\tan\psi_0 \quad \text{or} \quad \frac{s}{L} = \frac{D_1}{D_0}\tan\psi_0$$

30. If we take as an example the 3-phase generator, which has a phase advance $\psi_0 = 40^\circ$ with full, almost induction-free, load, and which without changing the excitation even on an empty run (strictly speaking: with very small load) should produce no perturbations in parallel operation. For these two states for example:

$$\alpha_1 = \frac{g}{L}\cos 40^\circ = \frac{g}{L} \times 0.7660, \quad \alpha_2 = \frac{g}{L}\cos 0^\circ = \frac{g}{L},$$

$$\frac{\alpha_1}{\alpha_2} = 0.7660$$

In reality the drive of such a machine is not of the type where one could be satisfied with the single part $k \sin \omega t$ on the right side of our differential equation, instead there is still validity in the influence of parts with $\sin 2\omega t$, $\cos 2\omega t$ ... etc.

From the point of view of the linear differential equation, which is how the problem has been regarded up to today, the law of superposition is valid, that is one determines the forced oscillation for each individual part of the trigonometric series representing the excitation and adds the results. If the sum of these individual results does not overstep a certain size, whose fixing is the work of electrical engineering, one expects undisturbed operation following this theory.

It follows from our observations on pseudoharmonic oscillation, in which as we have seen the movement is not unambiguously determined by the excitation, that despite the above such installations can fail.

Strictly speaking, the consideration of several parts on the right side of the differential equation allows a new broader investigation (cf. Nr. 31). If we look at the influence of an individual part for itself in the following, the result is only valid under the proviso that the other parts are really suppressed. One cannot derive an overall result through addition, probably however the process could suffice to show the problem.

In order to have a particular case in view, we will take an operation which, apart from the fundamental oscillation with 100 cycles per minute, also contains a harmonic with 200 cycles per minute.

Now the two frequencies $\sqrt{\alpha_1}$, $\sqrt{\alpha_2}$, whose relationship is $\sqrt{\alpha_1/\alpha_2} = 0.8752$, should be arranged between the two drive frequencies:

$$\omega_1 = \frac{100}{60} 2\pi = 10.472 \text{ s}^{-1}$$

(corresponding to 100 cycles per minute),

$$\omega_2 = \frac{200}{60} 2\pi = 20.944 \text{ s}^{-1}$$

(corresponding to 200 cycles per minute),

through which then

$$L = \frac{g}{\alpha}\cos\psi_0$$

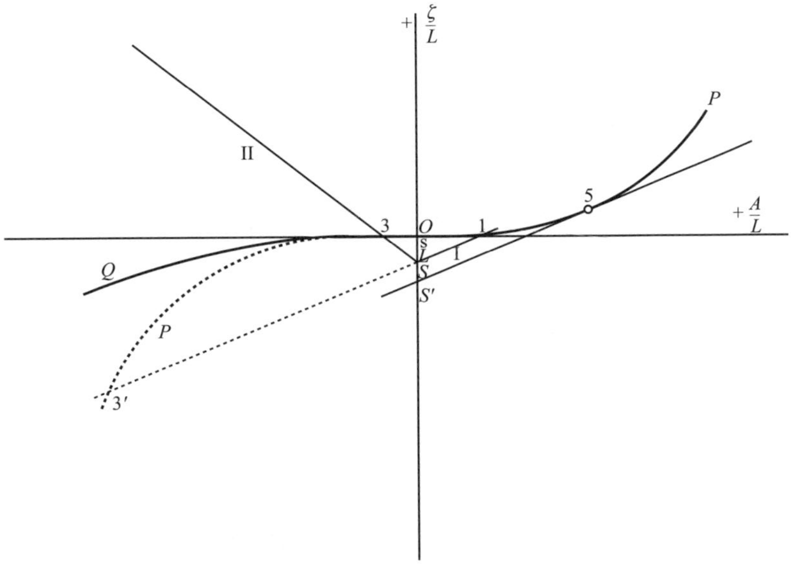

Figure 19

and also with that

$$\Theta = \frac{pL}{gw_0}\frac{\sqrt{3}Ee}{x}$$

is determined. In this action we are reliant on trial and error.

If one assumes the following relationships:

$$\omega_1 = 10.472\,\text{sec}^{-1}, \quad \sqrt{\alpha_1} = 14.93\,\text{s}^{-1}, \quad \sqrt{\alpha_2} = 17.06\,\text{s}^{-1},$$

$$\omega_2 = 20.944\,\text{s}^{-1}$$

then

$$1 - \frac{\omega_1^2}{\alpha_1} = 0.508, \quad 1 - \frac{\omega_2^2}{\alpha_2} = 0.508$$

It would be a coincidence if one was in reality to arrive at these numbers. The calculation of the machine constants (particularly of the reactance x) is encumbered with a certain uncertainty which is further increased by the fluctuations in the composition of the material and in the machine operation. The amount of the error is not exaggerated by using ± 8 percent in the calculation of the values of $\sqrt{\alpha}$.

So perhaps we should reckon with

$$\alpha_1 = 222.83 \times 0.85\,\text{s}^{-2}, \quad \alpha_2 = 290.91 \times 0.85\,\text{s}^{-2}$$

$$1-\frac{\omega_1^2}{\alpha_1}=0.421, \quad 1-\frac{\omega_2^2}{\alpha_2}=-0.775$$

and these conditions can be presented in a diagram (Figure 19).

We draw the curve *POQ*, whose equation is[26].

$$\frac{\zeta}{L}=\frac{A^3}{L^3}\left(\frac{1}{8}+\frac{1}{4}\frac{\tan^2\psi_0}{1-\frac{1}{4}}\frac{A^2}{L^2}\right)$$

corresponding to $\psi_0=0$ (in neutral).

$$\frac{\zeta}{L}=\frac{1}{8}\frac{A^3}{L^3}$$

Through the point *S*, that corresponds to the value $s/L=0.1$, we put the straight lines *I* and *II*, corresponding to the values:

$$1-\frac{\omega_1^2}{\alpha_1}=0.421, \quad 1-\frac{\omega_2^2}{\alpha_2}=0.775$$

One sees that the intersection points 1 and 3 lie on the almost straight line parts of the curves *P* and *Q*, so that the pendulum swing can be calculated practically without error following the formulae of harmonic oscillation.

But one also sees in addition that point 1 cannot move closer to point 5, without running the risk of point 1 changing into point 3′. Since the curves *P* and *Q* are sensitive to resistances to movement, as experiment has shown, in reality they are more strongly bent than in the drawing, and the risk of changing into phase 3′ exists long before reaching point 5.

So in order to be more careful with the condition that is represented by *I*, one will move away from resonance without consideration of the fact that condition *II* must then approach resonance.

However it should be stressed that the behaviour described of the drive under resonance only exists when there is positive γ [equation (71), p. 70].

With negative γ the behaviour of the system reverses, corresponding to the form of the curves *POP* and *OQ*. The value $s/L=0.1$ produces $D_1/D_0-0.1$, $\cot\psi_0=0.119$.

The amplitude of the relevant harmonics should not amount to more than 12 percent of the average torque, which factor should be attended to, particularly in operation of piston machines. The use of a light flywheel, if it is at all possible, therefore demands great care during calculation and dimensionalisation.

---

[26] *Translators' note*: The following discussion is again based on Figure 11. Duffing establishes a curve *POQ* and two straight lines *I* and *II*; the intersection points of the curve and the lines show which response amplitudes are possible. The problem he is concerned about is that the stable response labelled 1, can reach the critical point 5 and thereafter jump to the solution 3′.

Glossary

This contains a list of some definitions and terms used in this book. The aim of providing such a list is to help the reader who is not a specialist in the field to understand the content without having to search the literature. The terms in bold below are written in italics in the main text when they appear for the first time in each chapter. They are also written in italics in the Index. The definitions are based on some of those given in references [1–6] listed at the end of the Glossary.

Anharmonic oscillator: An oscillator that has a nonlinear relationship between force and displacement.

Attractor: A point or collection of points in the phase space (*See definition*) where all the initial states tend to approach the steady-state. There are several types of attractors, such as a point attractor, a periodic or quasiperiodic attractor and a chaotic attractor (*See definition*). The point attractor corresponds to a stable equilibrium or fixed point (*See definition*), the periodic attractor relates to asymptotically stable periodic motion (*See definition*) or to a limit cycle (*See definition*) and the quasiperiodic attractor to asymptotically stable quasiperiodic motion (*See definition*).

Autonomous dynamical system: A continuous dynamical system described by a set of ordinary differential equations in which time does not appear explicitly. If time appears explicitly, the dynamical system is said to be **nonautonomous**.

Backbone curve: The curve depicting the dependence of the nonlinear 'natural' frequency with amplitude during free oscillation.

Basin of attraction: The set of initial conditions leading to a given attractor.

Bifurcation: A qualitative change in the topology of the phase portrait (*See definition*), which occurs during the quasistatic variation of a control/bifurcation parameter through its critical value. This can be graphically illustrated in a form of a **bifurcation diagram**, which shows the possible fixed points or periodic orbits of a system as a function of the control/bifurcation parameter near the bifurcation point. A **bifurcation set** displays a collection of the different local bifurcations that take place in a given range of the chosen control parameter.

Break-loose frequency: The frequency of oscillation at which the force across a Coulomb damper is just sufficient to initiate relative sliding across the interface.

The Duffing Equation: Nonlinear Oscillators and their Behaviour, First Edition. Edited by Ivana Kovacic and Michael J. Brennan.

Chaos; chaotic motion: An expression for a specific type of irregular motion of a deterministic system, which fundamentally differs from **regular motion**, which includes equilibrium solutions, periodic and quasiperiodic solutions. Yoshisuke Ueda used the expression 'randomly transitional phenomena' to describe this type of motion. James A. Yorke and his coauthor Tien-Yien Li coined the term 'chaos' in a paper published in 1975 entitled 'Period Three Implies Chaos'.

Chaotic attractor; chaotic solution; chaotic trajectory; chaotic oscillations: An attractor with a positive Lyapunov exponent, i.e., with extreme sensitivity to initial conditions. The terms 'strange' and 'chaotic' attractors are sometimes used synonymously, although rigorously they are not.

Chaotic saddle: The union of infinitely many saddle orbits arising in the Smale horseshoe map (*See definition*) and possessing the main features of chaos such as, the extreme sensitivity to initial conditions.

Coexisting attractors: Two different attractors which are attained by a system depending on the motion initial conditions for the same value of the control parameters.

Conservative system: Autonomous mechanical system with no energy dissipation. The behaviour of this system can be completely described by a single scalar function, called a Lagrangian, which is defined as the kinetic energy of the system minus its potential energy.

Control of chaos: The elimination of chaos (*See definition*) from the system dynamics or, alternatively, the exploitation of chaotic motion by means of appropriate control techniques and methods.

Coulomb damping: A simple model of the nonlinear damping force provided by dry friction acting at the interface of two solids. This force, assumed to be of constant magnitude, acts in a direction opposite to the relative sliding velocity. Thus, the force is discontinuous at zero sliding velocity.

Crises: Global bifurcations in which a chaotic attractor suddenly disappears or changes in size when it touches the boundary of its basin of attraction (*See definition*) or of one of its sub-basins.

Cross-well chaos; cross-well chaotic motion; cross-well chaotic response; cross-well chaotic attractor: A chaotic attractor scattered over neighbouring wells in a multiwell (typically, two-well) potential system.

Detuning parameter: A small parameter giving the detuning of a system variable (typically, the excitation frequency) with respect to a perfect dynamical condition (typically, a resonance).

Dynamical integrity; integrity of dynamical systems: The occurrence of topological properties of a given region in phase space (*See definition*) and their persistence with a varying control parameter (a common interpretation is the persistence of the nonfractal character of a basin of attraction).

Equivalent viscous damping: A concept involving the approximation of a nonviscous damping force by an equivalent linear viscous damping force. The equivalence is based on equal energy dissipation per cycle during harmonic motion across the damper.

Fixed point; equilibrium point; equilibrium: The point $\hat{\mathbf{x}}$ is a fixed or equilibrium point of the dynamical system given by $d\mathbf{x}/dt = f(\mathbf{x})$, if $f(\hat{\mathbf{x}}) = 0$. Equilibria can be classified by evaluating the Jacobian matrix (*See definition*) at each of the equilibrium points of the system, and then finding the resulting eigenvalues. An equilibrium point is **hyperbolic** if none of the eigenvalues have zero real part. If all eigenvalues have negative real part, the equilibrium is a **stable node**. If at least one has a positive real part, the equilibrium is an **unstable node**. If there are multiple positive eigenvalues, the fixed point is an **unstable inflected node**, but if there are multiple negative eigenvalues, the *fixed point* is a **stable inflected node**. If at least one eigenvalue has a negative real part and at least one positive real part, the equilibrium is a **saddle point**. Complex eigenvalues indicate a **centre**, when the real part is equal to zero, i.e., when the eigenvalues are purely imaginary, a **stable spiral** (**focus**), when the real part is negative, or an **unstable spiral** (**focus**), when the real part is positive.

Floquet theory: The theory that is concerned with the solution of linear ordinary differential equations with periodic coefficients.

Flow: This is formed by the trajectories in time of the phase space of a continuous dynamical system.

Forced oscillations; forced vibration: Motion that occurs if a system is continuously driven by an external agency. **Forced harmonic oscillations** occur when the external agency is time-harmonic.

Fractal dimension: A generalised notion of dimension for a set of points to deal with fractals that may not be self-similar. There are various definitions, each one sharing the idea of variation of measure with the change in scale at a vanishing small scale.

Free oscillations; free vibration: This is the natural response of a system to some initial displacement and/or impulse (initial velocity). The response is completely defined by the properties of the system.

Frequency-response diagram (curve); amplitude–frequency-response (curve); response curve: A plot showing how the steady-state amplitude of displacement changes with frequency.

Frequency spectrum; power spectrum: The square of the amplitude of the Fourier transform of a time history. It provides information about the relative importance of the frequencies existing in the response. Periodic motion and quasiperiodic motion have a discrete frequency spectrum, unlike chaotic motion that has a continuous frequency spectrum.

Global bifurcation: A bifurcation in which the characteristics cannot be determined only through a local analysis near a fixed point. To investigate a global bifurcation, a global analysis is required, which, for example, can predict the occurrence of a homoclinic orbit (*See definition*).

Global integrity measure; GIM: The normalised hypervolume (area in the two-dimensional case) of the safe basin (*See definition*).

Hamiltonian system: This is a dynamical system whose equations of motion can be derived from a scalar function called a **Hamiltonian** or Hamiltonian function $H(q_i, p_i, t)$, where $i = 1, \ldots, n$, where n is equal to the number of degree of freedom, q_i are generalised coordinates, p_i are generalised momenta and t is time. The equations of motion are then defined by $\dot{p}_i = -\partial H/\partial q_i$, $\dot{q}_i = \partial H/\partial p_i$. The Hamiltonian of a conservative system (*See definition*) does not depend on time.

Harmonic oscillator: An oscillator that has a linear relationship between force and displacement.

Heteroclinic bifurcation; heteroclinic tangency: The global bifurcation occurring at the tangency of the stable manifold (*See definition*) and the unstable manifold (*See definition*) of two different saddles, which marks the onset of transient chaos.

Heteroclinic orbit; heteroclinic path: Any phase path that joins two different equilibrium points.

Hill's equation: An ordinary differential equation of the form $\frac{d^2x}{dt^2} + f(t)x = 0$, where $f(t)$ is a periodic function.

Homeomorphic: Two objects are homeomorphic if they can be deformed into each other by a continuous, invertible mapping.

Homoclinic bifurcation; homoclinic tangency: The global bifurcation occurring at the tangency of the stable manifold (*See definition*) and unstable manifold (*See definition*) of a given saddle, which marks the onset of transient chaos.

Homoclinic intersection; homoclinic tangling; homoclinic transversal intersection: The bunch of intersecting stable and unstable manifolds of a certain saddle, which is at the heart of chaos.

Homoclinic orbit; homoclinic path; homoclinic solution; homoclinicity: Any phase path that joins an equilibrium point to itself is known as a homoclinic path.

Hopf bifurcation: A local bifurcation in which a limit cycle is either destroyed or created as a control parameter is varied. The former is called a **subcritical Hopf bifurcation**, and the latter a **supercritical Hopf bifurcation**. It is named after mathematician and astronomer E. Hopf (1902–1983), who studied this bifurcation for dynamical systems with more than two dimensions in the 1940s.

Hysteresis; hysteretic behaviour: The phenomenon in which the system response is different for the case when parameter values are increased or decreased. This phenomenon is observed, for example, during frequency sweeps of a hardening or softening nonlinear oscillator.

Integrity factor; IF: The normalized radius of the largest hypersphere (circle in the two-dimensional case) entirely belonging to the safe basin (*See definition*).

Invariant manifold: A manifold (e.g., a curve, a surface, etc.) in phase space that is mapped onto itself under a mapping or a continuous time evolution. The most significant invariant manifolds are the **stable manifolds** and the **unstable manifolds** of saddle solutions. **A stable manifold is** an invariant manifold whose points converge to a saddle-point forward in time. It is important because it usually represents the boundary between the basins of attraction of different attractors. **An unstable manifold** is an invariant manifold whose points converge to a saddle-point backward in time. It converges to whatever attractor in the relevant basin and represents the skeleton of a chaotic attractor.

Jacobi elliptic functions: These are the standard forms of elliptic functions. There are three basic ones sn, cn and dn, each of which depends on two variables $\mathrm{sn} = \mathrm{sn}(u, k^2)$, $\mathrm{cn} = \mathrm{cn}(u, k^2)$ and $\mathrm{dn} = \mathrm{dn}(u, k^2)$, where u is called the argument and k^2 is called the modulus. They arise from the inversion of the elliptic integral of the first kind.

Jacobian; Jacobian matrix: This is the matrix of all first-order partial derivatives of a vector-valued function. It represents the best linear approximation to a differentiable function near a given point. Suppose that G: $\mathbf{R}^n \rightarrow \mathbf{R}^m$ is a function from Euclidean n-space to Euclidean m-space, given by m real-valued component functions, $y_1(x_1, \ldots, x_n), \ldots, y_m(x_1, \ldots, x_n)$. If the partial derivatives of all these functions exist, they can be organised in an m-by-n matrix, the Jacobian matrix of F is as follows:

$$\mathbf{J}_G = \begin{bmatrix} \partial y_1/\partial x_1 & \cdots & \partial y_1/\partial x_n \\ \vdots & \cdots & \vdots \\ \partial y_m/\partial x_1 & \cdots & \partial y_m/\partial x_n \end{bmatrix}$$

Japanese attractor: A strange attractor discovered by Yoshisuke Ueda in the 1960s, and named the 'Japanese Attractor' by French Mathematician, David Ruelle (*See* Figure G.1). It was obtained by solving the equation $\frac{d^2x}{dt^2} + 0.1\dot{x} + x^3 = 12\cos t$.

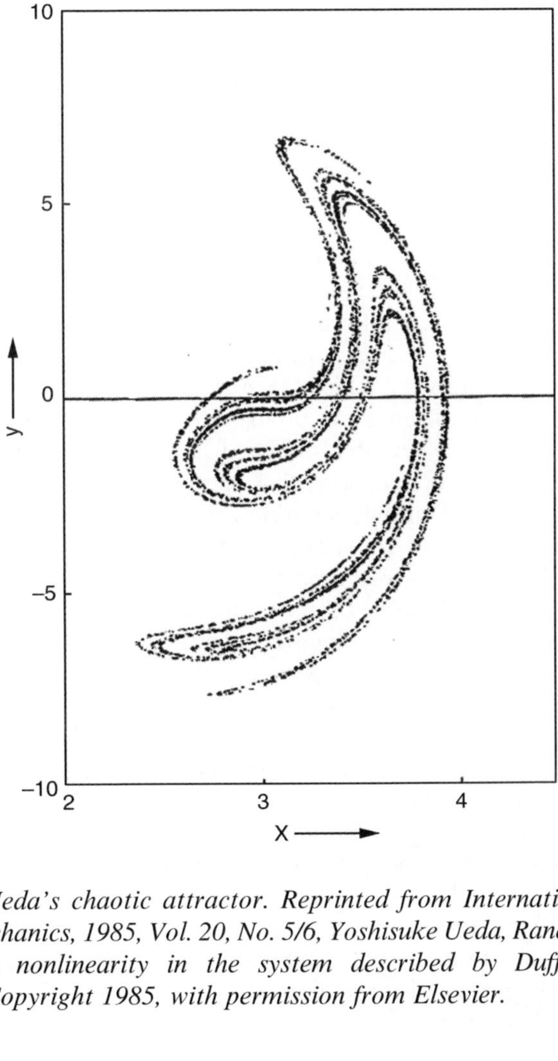

Figure G.1 Ueda's chaotic attractor. Reprinted from International Journal of Non-linear Mechanics, 1985, Vol. 20, No. 5/6, Yoshisuke Ueda, Random phenomena resulting from nonlinearity in the system described by Duffing's equation, pp. 481–491. Copyright 1985, with permission from Elsevier.

Jump phenomenon; jump: A sudden (discontinuous) change of the amplitude of the response when the frequency is varied very slowly and the rest of the system parameters are kept constant. A jump phenomenon is a simple example of a crisis or saddle-node bifurcation (*See definition*).

Jump points; turning points: The points where the frequency response curve has vertical tangents with associated jumps in the response amplitude. At these points a stable (node) and an unstable (saddle) response amplitude coincide, giving rise to a saddle-node bifurcation (*See definition*).

Level curve: This is a curve in the phase plane that is determined by the initial conditions. The level curve is also called an **integral curve** or a curve of constant energy.

Limit cycle: A closed orbit in phase space, representing periodic motion of the system. It is usually occurs in the periodic motion of nonlinear dissipative systems.

Local bifurcation: A bifurcation in which the characteristics are determined by the analysis of a vector field near a fixed point. Typical types of local bifurcations are: saddle-node bifurcation, transcritical bifurcation, pitchfork bifurcation, and Hopf bifurcation (*See definitions*).

Lyapunov (Liapunov) exponent: A quantitative measure of the exponential divergence or attraction in time of trajectories that correspond to slightly different initial conditions. A positive Lyapunov exponent represents one of the strongest indicators of chaotic motion. It is named after Aleksandr Mikhailovich Lyapunov (1857–1918).

Melnikov criterion; Melnikov method: A criterion based on the intersection of stable and unstable manifolds (*See definition*), which is used to define the critical region in the parameter space for the onset of chaos. The main idea behind this method was proposed by Victor Kozmich Melnikov (1935–2003) in 1963.

Orbit: This is a trajectory of a flow (*See definition*). Some typical orbits are homoclinic orbits and heteroclinic orbits (*See definitions*) as well as periodic orbits corresponding to a periodic solution, i.e., periodic motion (*See definition*).

Parametric excitation: An excitation that appears as a result of a time-varying system parameter. The **Mathieu equation** describes, for example, the case when this excitation appears as a harmonic coefficient in the equation of motion, while in the case of **Hill's equation** (*See definition*) this coefficient is periodic.

Period-doubling bifurcation; flip bifurcation: A local bifurcation in which a limit cycle of the system changes into a cycle of twice the period as a control parameter is varied.

Period-doubling cascade: A sequence of period-doubling bifurcations usually leading to chaos as a control parameter is varied.

Periodic motion: This is a type of motion $x(t)$ such that $x(t) = x(t + T)$, where T is the fundamental period of the motion.

Phase space and portrait: The trajectory as a continuous dynamical system traces out its evolution over time in an abstract phase space, whose coordinates describe the dynamical state of the system at any particular time. Ensembles of trajectories fill the phase space to form a phase portrait. In a dissipative system this portrait will show the structure of the attractors and basins; to emphasize this, it is sometimes called the attractor-basin phase portrait.

Pitchfork bifurcation: A bifurcation in which nontrivial fixed points are either destroyed or created while a system parameter is varied. It is produced when the sign of a real eigenvalue is changed.

Poincaré section and map: Mapping of an intersection point of a trajectory with a surface of sections onto the subsequent intersection point. The flow (continuous time evolution) in phase space is reduced to a map of a lower dimensional plane onto itself. In nonautonomous systems that are T-periodic, the natural choice of a Poincaré map is a stroboscopic map mapping the variables at time t onto those at $t + T$. Named after Jules Henri Poincaré (1854–1912).

Potential well: This is the potential energy of a system plotted as a function of the displacement. In general, it can have several local minima or maxima. In the case when it has one local minimum, it is called a **single-well** or **one-well potential**; if it has two local minima, it is called a **two-well, double-well or twin-well potential**.

Quasiperiodic motion: This is an oscillation containing two or more frequencies that are incommensurate, bearing an irrational relationship to each other.

Quasi-zero stiffness: This is a stiffness element whose force-deflection characteristic has a zero gradient at one position. In a vibration isolator this is chosen so that this occurs at the static equilibrium position.

Resonance: This is a peak in the magnitude of the frequency-response curve (*See definition*). The **primary (principal) resonance** occurs when the system is excited at frequencies close to its natural frequency. **Secondary resonances** include, for example, subharmonic and superharmonic resonances (*See definitions*).

Routes to chaos: Ways in which a system can become chaotic as a result of a change in parameter values. Basically, there are four routes to chaos: period-doubling (discovered by Mitchell Jay Feigenbaum in 1975), intermittency (reported by Yves Pomeau and Paul Manneville in the papers published in 1979 and 1980), subcritical instability and a sequence of global bifurcations (Ruelle–Takens–Newhouse scenario named after their work published in 1978).

Saddle-node bifurcation; fold bifurcation: A local bifurcation in which two fixed points, one unstable and one stable, are created or destroyed along an equilibrium path as a control parameter is varied.

Safe basin: The union, in phase space, of all initial conditions guaranteeing some specific, dynamically acceptable, response performance (it usually involves the convergence in time towards one or more attractors).

Secular terms: Terms that grow without bound as time increases. They appear in the procedure used in perturbation methods. If they exist in the solution given in the form of a power series, the assumption that this series is uniformly valid is violated. The adjective describing these terms was derived from the French word 'siècle' for 'century', as the effect of these terms is appreciable only after a long time, which was seen to have the order of a century in an exaggerated sense.

Self-excited oscillation: This is a resonance produced when systems have a pair of complex conjugate eigenvalues and the real part is positive. Systems with negative damping or with an asymmetric stiffness matrix undergo self-excited oscillations.

Separatrix: A curve (usually, a stable invariant manifold) in the phase plane that separates two different basins of attraction; it is a portion of the basin boundary.

Sink: An asymptotically stable fixed point.

Smale horseshoe: A two-dimensional mapping introduced by Stephen Smale while studying the behaviour of the orbits of the van der Pol oscillator in 1963. It describes the stretching and folding actions that produce mixing of a chaotic motion in a very simple way.

Snap-through oscillations: The large amplitude, regular or nonregular, motion ensuing from oscillations occurring around different unstable equilibrium positions.

Stability: An attractor of a dissipative dynamical system is asymptotically stable in the sense of Lyapunov because all local trajectories flow back to the attractor. Fixed points of a Hamiltonian system can be at most neutrally stable in the sense of Lyapunov, with all local trajectories staying close to the point, though not returning to it. Orbital stability relates to a phase-space criterion in which divergence of the unseen time coordinate is deemed as unimportant.

Steady state: This is a time-invariant motion or equilibrium. A typical start of a dissipative system will experience a transient motion before settling asymptotically onto a stable steady-state solution, called an attractor (*See definition*).

Stick-slip motion: The motion across a Coulomb damper involving both sticking and slipping phases during a cycle of oscillation.

Strange attractor: An attractor with noninteger fractal dimension (the term 'strange' attractor is often used as an alternative to 'chaotic' attractor, although not every strange attractor is chaotic).

Stroboscopic map: A map showing successive states, in the phase space, observed at regular time intervals (equal to the time period of the excitation).

Subduction: A global bifurcation in which a chaotic attractor is replaced by a periodic attractor in an unchanged basin as a control parameter is varied.

Subharmonics: Frequencies that are integral fractions of the excitation frequency.

Subharmonic resonance: The resonance or large-amplitude response that occurs when the excitation frequency is n-times larger than the response frequency, where n is an integer.

Superharmonic resonance: The resonance or large-amplitude response that occurs when the excitation frequency is n-times smaller than the response frequency, where n is an integer.

Superharmonics: Frequencies which are integer multiples of the excitation frequency.

Symmetry-breaking bifurcation: A global bifurcation in which a symmetrical attractor is replaced by two coexisting asymmetrical attractors.

Time scales: Multiple timescales are used to construct uniformly valid approximations to the solutions of perturbation problems, both for small and large values of the independent variables. For example, slow scale (**slow time**) and fast scale (**fast time**) can be introduced and subsequently treated as independent variables.

Transcritical bifurcation: A local bifurcation in which the number of fixed points does not change. This type of bifurcation is characterised by the change of the stability of the fixed points.

Transient motion: The evolutionary motion leading to a steady state (*See definition*).

Ueda's equation: The equation $\frac{d^2x}{dt^2} + k\dot{x} + x^3 = B\cos t$, where k and B are constants, named after Yoshisuke Ueda, who started studying it in 1960s.

Wada basin: A basin in two-dimensional phase space, each boundary point of which is also in the boundary of two other basins. The term stems from the Wada properties of the sets with similar properties. It was used for the first time in 1917 in the publication by the Japanese mathematician Kunizo Yoneyama, who credited the discovery to his teacher Takeo Wada.

References

[1] J.M.T. Thompson, H.B. Stewart, *Nonlinear Dynamics and Chaos*. John Wiley and Sons, 2002.

[2] T. Kapitaniak, S.R. Bishop, *The Illustrated Dictionary of Nonlinear Dynamics and Chaos*, John Wiley and Sons, 1999.

[3] J. Guckenheimerand P. Holmes, *Nonlinear Oscillations, Dynamical Systems, and Bifurcations of Vector Fields*, Springer, New York, 1983.

[4] A.H. Nayfeh, B. Balachandran, *Applied Nonlinear Dynamics*, John Wiley & Sons, New York, 1995.

[5] R.H. Rand, Lecture Notes on Nonlinear Vibrations, version 52, (http://audiophile.tam.cornell.edu/randdocs/nlvibe52.pdf), Accessed 29 March 2010.

[6] J.J. Thomsen, *Vibrations and Stability, Advanced Theory, Analysis, and Tools*. 2nd edn, Springer, 2003.

Index

The terms in *italics* can be found in the **Glossary**, which explains their meaning.